W9-ABN-181

LIE GROUPS: HISTORY, FRONTIERS AND APPLICATIONS

VOLUME I

SOPHUS LIE'S 1880 TRANSFORMATION GROUP PAPER

Translated by Michael Ackerman
Comments by Robert Hermann

MATH SCI PRESS
18 Gibbs Street
Brookline, Massachusetts 02146

(In part a translation of
"Theorie der Transformationsgruppen"
by S. Lie, Math. Ann., 16 (1880), 441 - 528.)

ISBN 0-915692-10-4
Library of Congress Catalog Card Number 75 - 17416

MATH SCI PRESS
18 Gibbs Street
Brookline, Massachusetts 02146

PRINTED IN THE UNITED STATES OF AMERICA

PREFACE

By. R. Hermann

With this volume I begin a project I have thought about for twenty years, the adaptation in modern terms of some of the major works of 19-th century differential geometry and Lie group theory. It is certainly most appropriate to begin with Lie, whose ideas dominate this period and who was one of the most brilliant and imaginative mathematicians of all time. In translating Lie in this way I have more in mind than historical piety. Among all the 19-th century masters, Lie's work is in detail certainly the least known today. Many of his ideas and problems have hardly been touched since his death.

Of course, what are now called "Lie groups" (roughly, what Lie called "finite continuous groups") have been extensively studied since World War II. The key influences in this revival were Chevalley's book "The Theory of Lie Groups", and the "Seminaire Sophus Lie" notes from Paris in the 1950's (mostly written by P. Cartier). These works finally made a substantial portion of the ideas of Lie and E. Cartan accessible to a reader well-trained in the methodology of modern mathematics. This accessibility is only relative--to this day, very few people know very well the geometric and algebraic side of Lie group theory. The research specialization which Lie so forcefully deplores in the Forward translated below has now gone so far that many people now work in esoteric (and often tedious) subspecialties or offshoots of Lie group theory who clearly are not familiar in detail with the simple and broad geometric view of the theory utilized and developed by Lie. Here, as in so much of the rest of mathematics, the relentless urge to do "new" research has submerged the beautiful old ideas. However, at least there are now some excellent introductory treatises to certain portions of Lie group and algebra theory--e.g., those by Sagle and Walde, Samelson, F. Warner and, on the physics side, Gilmore and Miller.

The main feature of Lie's work is the relation between groups and differential equations. It will be clear from the material translated in this and succeeding volumes that Lie thought of himself as the successor to Abel and Galois, doing for differential equations what they had done for algebraic equations. (What is now routinely called the "Galois theory of differential equations" is just a minute part of what Lie had in mind.) This flavor has been almost completely lost, and one of my major goals is to revive it and push its development with the powerful tools of modern mathematics.

Apparently, Lie's work went through three cycles. First, he developed material in rough intuitive and geometric form, which was published in Norwegian journals. Then, a more polished and systematic version was written for Mathematische Annallen, the leading journal of the day. (It is an interesting contrast to today's habits that the journal published what we would call Review Articles, and which editors would now disdain.) Finally, he wrote--in collaboration with junior colleagues--extremely long and detailed books. Unfortunately, the books are difficult for the modern reader, and have probably played a major role in Lie's reputation for incomprehensibility. I have found the papers from the intermediate cycle to be the most accessible, and accordingly, will build my comments around their translation, with the important collaboration of Dr. Michael Ackerman. The key document in this volume is the translation of "Theorie der transformationsgruppen I", Math. Ann. VI, 1880, Collected Works, Vol. 6, p. 1.

The main result of this paper is the classification (under local changes of variables) of the finite dimensional Lie algebras which can act on two dimensional manifolds. These results seem to be forgotten today, and this paper would be worth reviving even if that were the only thing it contained. However, there are also glimpses of more general problems, and it is a pleasure for me (in my comments) to indicate the directions in which Lie's ideas can be generalized and utilized. The next papers in this series will deal, in a similar way, with Lie's papers on differential invariants, differential equations admitting known transformation groups, and infinite Lie groups.

Here is the format to be followed in all this work. I will write some preliminary chapters, outlining the mathematical background that I believe it is useful for the reader to have available to understand the paper in the most general and efficient way. (However, I will usually assume that the reader is familiar with the elements of differentiable manifold theory.) Then, Michael Ackerman's translation of Lie's text will be presented, interspersed with my comments (printed in italic). Finally, I present in the closing chapters my own version of some of the material, its ramifications, extensions, and applications.

I want to emphasize applications because that is very much in the spirit of Lie's work. The reader will see for himself how widely Lie thought of mathematics--and the applications were clearly part of his world. There are two areas where Lie's ideas are in direct touch with current work: Elementary Particle Physics and Mathematical Systems Theory. Since the physics applications have been extensively discussed in my own books (they are referred to by abbreviations listed in the Bibliography), I will concentrate here on Systems Theory, which, in fact, forms the ideal setting for Lie's ideas.

The best explanation Lie himself gave of his work is to be found in the Forwards to his books. Accordingly, I present the most interesting parts of them as introductions to these volumes, beginning here with the Forward to "Geometrie der Berührungstransformationen".

Finally, I want to thank Dr. Michael Ackerman for his collaboration and help. I also thank Arthur Krener and Alan Mayer for their help and suggestions. Karin Young has done a fine job typing the manuscript. A word of thanks also to Mike Spivak, Al Gowan and Nick Loscocco.

LIE'S FORWARD TO "BERÜHRUNGSTRANSFORMATIONEN"

As an introduction to the present work it seems to me to be fitting to preface it with some general remarks.

In the course of time the position of analysis with respect to geometry and to the various branches of applied mathematics has passed through many extensive changes. During the past century the mutual stimulation of the separate disciplines was greatly advanced; and in this, first one and then the other discipline was dominant. This is in the nature of things and in itself is not to be regretted. But in our century, mathematics has split up into many very extensive areas, and this division has often led the representatives of one area to misjudge the importance of others, so that, to the detriment of their own discipline, fruitful ideas from the outside have been ignored.

Permit me to elucidate these remarks by recalling briefly some of the phases of the development of mathematics.

For the ancient Greeks, geometry was the almost exclusively dominant discipline of mathematics. As yet there was no abstract analysis, and even astronomy and mechanics were subordinate to geometry. After noteworthy beginnings by Diophantus and the Indians, there first developed in the

Renaissance an algebra, whose lack among the Greeks is strongly perceptible.

Then, through the introduction of the concept of coordinates and through the creation of analytic geometry by Descartes, an epoch-making bond between geometry and analysis was forged, which was soon to lead to the truly fundamental concept of mathematics, the function concept. The concepts of integral and derivative, whose germ already occurs in Archimedes' geometric investigations, developed gradually and indeed, be it noted, through the treatment of geometric, kinematic and mechanical problems in the works of Kepler, Cavalieri, Descartes, Wallis and especially Fermat. Similar considerations led Newton and Leibniz to discern the inner essence and, in particular, the reciprocal relation of these two concepts and thereby to found the infinitesimal calculus. The theory of differential equations and the calculus of variations had the same geometrical source.

Not less characteristic, though less noted, is the fact that the concept of a transformation, which is becoming more and more prominent in modern mathematics and claims recognition alongside the function concept as an independent fundamental concept, originated with the ancient geometers, though, of course, in a very special form--still, its source was geometric intuition. Also the closely related concept of differential invariant was first encountered in geometry, namely, in the theory of curvature.

These brief remarks already show that in earlier centuries geometry and mechanics exercised the most powerful influence on the development of analysis. That, conversely, analysis had the most powerful reciprocal effect on geometry and mechanics is familiar and need not be further discussed here.

This reciprocal action was advanced and really made possible by the fact that the separate disciplines were relatively inextensive, so that at the end of the preceding century Euler, Lagrange, Laplace and around the turn of the century Gauss and then Cauchy could comprehend all branches of mathematics.

In our century this has gradually changed. The extent of the separate disciplines grew to an extraordinary degree; indeed, new independent sciences were formed, like mathematical physics at the hands of Laplace, Ampère, Fourier, Fresnel, Green, Cauchy, Poisson and Lejeune-Dirichlet.

Geometry, which had been led into new roads in the preceding century by Euler and Monge, also received a mighty impetus, first from the establishment of projective geometry as an independent discipline by Carnot, Brianchon and especially Poncelet and its further development by Möbius, Plücker, Chasles and Steiner and, secondly, from the enrichment of infinitesimal geometry by new, path-breaking ideas through Gauss' theory of curvature and Minding's related results. Both directions contributed to the development of Cayley's invariant theory.

Still mightier was the rise of analysis in the twenties and thirties of this century. Already the investigations of Lagrange and even more of Gauss and Cauchy provided many new and important results — they were distinguished by a perfect form and in particular a classical rigor, which in the preceding century had been lacking to so many mathematicians. But above all, it was Abel who, by his comprehensive problems, his great, deep discoveries and his logical acuity, inaugurated the new epoch. And Abel's lucid exposition of his new theories contributed in an essential way to the possibility of mathematicians' grasping Cauchy's theory of the imaginary and especially Galois' investigations, as deep as they are difficult.

It is true that Abel and his successor Galois hardly dealt with geometry in their brief lives. It is all the more remarkable that it was Abel's ideas which enriched geometry with broad new points of view, just as Galois' ideas have already begun to exercise a similar influence.

The pioneering discoveries only indicated here have given analysis, geometry and mathematical physics so great a content and such an extent that indeed it has become impossible for an individual to comprehend all of mathematics, for even Jacobi and Riemann hardly succeeded in doing this completely.

Riemann, who was influenced by Gauss and Jacobi, even though he is sooner to be considered the successor of Cauchy and Abel, knew how to apply the tools of geometry to analysis in a magnificent way. Even though his astounding mathematical instinct sometimes provided him immediately with results which his time did not allow him to establish definitively by purely logical considerations, still these splendid results are the best testimony to the fruitfullness of his methods.

I should like to consider Weierstrass, Riemann's contemporary, to be also a successor of Abel, not only because of the direction of his investigations, but more so because of his purely analytical method, in which he exerted himself to avoid geometric intuition as a tool. As outstanding as Weierstrass' accomplishments are for the foundations and for the highest domains of analysis, it still seems to me that his one-sided emphasis in analysis has not had an entirely favorable influence on some of his students.

I believe that in this I stand on the side of Klein, who has understood so well how to take fruitful stimuli for analysis from geometric intuition.

In any case, it is regrettable that the great develop-ment of analysis in Germany in the last decades has not been accompanied by corresponding progress in geometry. Prominent German geometers like Möbius, Plücker, von Standt and Grassmann were not appreciated at their true worth.

As indicated above, the splitting up of mathematics has often exercised an unfavorable effect on the representatives of the separate disciplines. Indeed, while some geometers go so far as to consider it meritorious to renounc the tools of analysis completely (more properly said, to as great an extent as possible) in treating geometrical problem on the other hand one finds here and there among the analyst the view that analysis not only can be developed independently of geometry, but that it must, since in their opinion, proofs of analytic propositions by geometric considerations are not unconditionally reliable.

In my scientific endeavors I have always proceeded from the view that on the contrary it is desirable that analysis and geometry should now, as earlier they did, mutually support and enrich each other with new ideas. This view was the theme of my Antrittsvorlesung at the University of Leipzig in 1886.

For more than twenty-five years I have been trying to gain acceptance for this view of mine by means of my own work. It may have been especially characteristic of my work that, after the model of Monge, I apply geometric concepts to analysis, especially those introduced by Poncelet and Plücker, while on the other hand I have extended the ideas of Lagrange, Abel and Galois on the treatment of algebraic equations to geometry, and in particular to the theory of differential equations. ...

(The rest of this Forward refers to the details of the volume for which it was written, and does not concern the subject matter of this Volume.)

TABLE OF CONTENTS

Page

Chapter A

A MODERN DIFFERENTIAL-GEOMETRIC NOTATION FOR LIE'S WORK

1. DIFFERENTIABLE MANIFOLDS AND VECTOR FIELDS

The basic setting will be the modern theory of differential geometry on manifolds. As introductions, I recommend O'Neill [1], Spivak [1], Bishop and Goldberg [1], Loomis and Sternberg [1], Warner [1], or Misner, Thorne and Wheeler [1]. I shall follow most closely the notations of my own books, particularly DGCV, GPS and the Interdisciplinary Mathematics series.

By a <u>manifold</u> I will mean a C^∞ differentiable, finite dimensional manifold. (It is to be understood that they are, as topological spaces, <u>paracompact</u>, i.e., are the countable union of compact subsets.) Maps, functions, transformations, etc. (all more-or-less synonymous) will be C^∞, unless mentioned otherwise.

Such manifolds will usually be denoted by letters like X, Y, Z. Points of X will be denoted by

$$x \quad .$$

F(X) denotes the set of C^∞, real-valued functions on X. It is a commutative associative algebra (under pointwise addition and multiplication). A vector field on X is a derivation $f \to A(f)$ of F(X). They form an F(X)-module, denoted by V(X).

Remark. I have chosen the labelling of these objects, e.g., "X" for a typical manifold, "A" for a vector field, to agree with the notations used in Lie's papers. They are different from the typical notations used in my own books.

A tangent vector to a point $x \in X$ is a directional derivative, i.e., an R-linear map

$$v: F(X) \to R$$

such that:

$$v(f_1 f_2) = v(f_1) f_2(x) + f_1(x) v(f_1)$$

$$\text{for } f_1, f_2 \in F(X) \quad .$$

X_x denotes the vector space of tangent vectors to X at x. The tangent bundle to X is the space of pairs

$$(x, v) \quad ,$$

with $x \in X$, $b \in X_x$. It is denoted by

$$T(X) \quad .$$

The projection map

$$T(X) \to X$$

is the map

$$(x, v) \to x \quad .$$

$T(X)$ can be made into a manifold, so that the projection-map takes $T(X)$ into a vector bundle over X.

A vector field $A \in V(x)$ defines a cross-section map

$$x \to A(x) \in X_x$$

of $T(X)$. To define it, set:

$$A(x)(f) = A(f)(x)$$

$$\text{for } f \in F(X), \quad x \in X \quad .$$

In this way, $V(X)$ becomes identified with $\Gamma(T(X))$, the space of cross-sections of the vector bundle $T(X)$.

2. COORDINATE SYSTEMS

A diffeomorphism

$$\phi: X \to Y$$

is a C^∞ map between two manifolds, such that the inverse map

$$\phi^{-1}: Y \to X$$

also exists and is C^∞.

R^n and C^n, n-dimensional real and complex Euclidean spaces, are considered as manifolds in the usual way.

If Y is an open subset of a manifold X, it inherits a manifold structure.

Let X continue to denote a manifold. A set

$$(f_1,\dots,f_n)$$

of elements of F(X) is a coordinate system for X if the map

$$x \to (f_1(x),\dots,f_n(x))$$

is a diffeomorphism between X and an open subset of R^n. Of course, the very definition of "manifold" implies that X can be covered by open subsets with such coordinate systems, but X itself does not necessarily have a global coordinate system. However, to keep in contact with classical ideas, we usually work with coordinate systems, but in a way that is independent of the chosen coordinate system. This guarantees that concepts have a "global" meaning.

For notational convenience, we shall denote a coordinate system of functions on X by

$$(x^1,\dots,x^n) \quad . \tag{2.1}$$

(It is important to keep in mind that $x^1,\dots,x^n$ are real-valued functions on X.)

Adopt the summation convention, and the following range of indices:

$$1 \le i,j \le n \quad .$$

Thus, a coordinate system may be denoted by

$$(x^i) \quad .$$

We can now define vector fields

$$\frac{\partial}{\partial x^i} \in V(X) \quad ,$$

such that:

$$df = \frac{\partial f}{\partial x^i} dx^i \quad , \qquad (2.2)$$

for each $f \in F(X)$.

These vector fields define a basis for the $F(X)$-module $V(X)$.

If (y^i) is another coordinate system for X, the vector field basis $\partial/\partial y^i$ they determine is related to $\partial/\partial x^i$ via the transformation law:

$$\frac{\partial}{\partial x^i} = \frac{\partial y^j}{\partial x^i} \frac{\partial}{\partial y^j} \qquad (2.3)$$

(Notice that 2.3 conforms to the "chain rule" of calculus.)

A given, fixed, coordinate system (x^i) serves to identify X with R^n, i.e., a point $x \in X$ is identified with the point

$$(x^1(x),\ldots,x^n(x)) \in R^n \quad .$$

Thus, geometric objects and operations on X may be referred back, via this identification, to R^n. For example, in Lie's work, the mappings

$$\phi\colon X \to X$$

play a prominent role. In terms of R^n, such a mapping is determined by functions of the form:

$$\phi(x) = (x'^1, \ldots, x'^n)$$

with

$$x'^1 = \phi^1(x), \ldots, x'^n = \phi^n(x) \quad .$$

Alternately, the component functions $\phi^1, \ldots, \phi^n$ defining the map may be defined via the relation:

$$\phi^*(x^i) = \phi^i \quad ,$$

where $\phi^*: F(X) \to F(X)$ is the dual pull-back map on functions determined by ϕ.

3. FLOWS, INFINITESIMAL GENERATORS, AND ORBITS

Let X be a manifold, with a coordinate system (x^i). Let t be a real variable, varying, say, over

$$-\infty < t < \infty \quad .$$

A <u>flow</u> on X is a map

$$t \to \phi_t$$

of R into the set of <u>diffeomorphisms</u> of X, satisfying the following conditions:

$\phi_0 \;=\;$ identity map

The map

$$(t,x) \to \phi_t(x)$$

of $R \times X \to X$

is C^∞ .

Given such a flow, define a map

$$t \to A_t$$

of $R \to V(X)$ via the following formula:

$$A_t(f) = (\phi_t^{-1})^* \frac{\partial}{\partial t} (\phi_t^*(f)) \tag{3.1}$$

for $f \in F(X)$.

The one-parameter family

$$t \to A_t$$

of vector fields is called the <u>infinitesimal generator</u> of the flow $t \to \phi_t$.

<u>Remark</u>. From a modern point of view, Lie's work may largely be considered as the study of this correspondence between flows and their infinitesimal generators.

This relation between flows and generators has its physical origin in hydrodynamics. If

$$X = R^n ,$$

and the flow ϕ_t is that determined by a fluid flow, the vector field

$$t \to A_t$$

is called the <u>Eulerian velocity field</u>. See DGCV and GPS.

<u>Definition</u>. Let (ϕ_t) be a flow on X. A curve $t \to x(t)$ in X is an <u>orbit of the flow</u> if there is a point $x_0 \in X$ such that:

$$x(t) = \phi_t(x_0) \tag{3.2}$$

The orbits can be reconstructed from the infinitesimal generator $t \to A_t$ of the flow. Suppose that:

$$A_t = A_t^i \frac{\partial}{\partial x^i} , \tag{3.3}$$

where, for each t, $A_t^i \in F(X)$.

<u>Theorem 3.1</u>. If $t \to x(t)$ is an orbit of the flow, and if

$$x^i(t) \equiv x^i(x(t)) , \tag{3.4}$$

then:

$$\frac{dx^i}{dt} = A_t^i(x(t)) \tag{3.5}$$

<u>Proof</u>. From 3.4,

$$x^i(t) = x^i(\phi_t(x_0)) = \phi_t^*(x^i)(x_0) .$$

Hence,

$$\frac{dx^i}{dt} = \frac{\partial}{\partial t}\phi_t^*(x^i)(x_0)$$

$$= \text{, using 3.1, } \phi_t^*(A_t(x^i))(x_0)$$

$$= \text{, using 3.3, } \phi_t^*(A_t^i)(x_0)$$

$$= A_t^i(\phi_t(x_0))$$

$$= \text{, using 3.2, } A_t^i(x(t)) \quad ,$$

which proves 3.5.

Remark. Formula 3.5 is a key to understanding Lie's ideas in modern terms. For, the steps used to derive it are reversible (modulo "global" complications). There is a one-one correspondence between systems of ordinary differential equations of the form 3.5, and flows.

However, this was not precisely Lie's viewpoint. He thought more intuitively in terms of infinitesimal transformations. In the next few sections I will try to describe this concept in a way that will aid the modern reader in understanding Lie's work.

4. INFINITESIMAL TRANSFORMATIONS

Let X be a manifold. Define a equivalence relation on the set of flows in X, as follows:

Definition. Two flows (ϕ_t), (α_t) in X are said to be infinitesimally equivalent if the following condition is satisfied:

For each point $x_0 \in X$, the orbit curves

$$t \to \phi_t(x_0)$$

$$t \to \alpha_t(x_0)$$

which begin at x_0 have the same first order contact at $t = 0$, i.e., have the same tangent vector.

Theorem 4.1. Let (A_t), (B_t) be the infinitesimal generators of the flows (ϕ_t), (α_t). Then, the flows are infinitesimally equivalent if and only if:

$$A_0 = B_0 \tag{4.1}$$

The proof of 4.1 follow readily from formula 3.5.

Definition. An infinitesimal transformation on X is an equivalence class of flows, for the equivalence relation defined above.

Remark. By Theorem 4.1, the set of infinitesimal transformations may be identified with the set of vector fields on X. (Map the flow (ϕ_t) onto the vector field A_0, where (A_t) is the infinitesimal generator of (ϕ_t)). However, there is a difference in the geometric thought that goes into the two concepts of "infinitesimal transformation" and "vector field", and this difference is a key feature in Lie's work.

We can readily describe these ideas in pictures in case X is 2-dimensional, e.g., X is a subset of R^2. A flow and its orbits can be drawn as follows

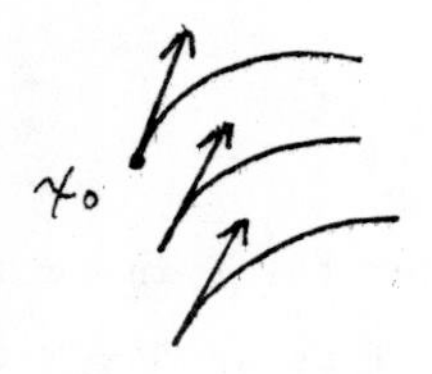

The infinitesimal transformation is then the vector field described by the arrows. Think of a point x_0 going "approximately" into the point

$$x_0 + A(x_0)\,\Delta t$$

in time Δt, where $A(x_0)$ is the value of the vector field-infinitesimal generator at $x = x_0$, $t = 0$.

Here is another way of thinking of these intuitive ideas, for the case where X is a general manifold.

Given f F(X), set:

$$f_t = \phi_t^*(f) \quad .$$

Then, using 3.1,

$$\frac{\partial f_t}{\partial t} = \phi_t^*(A_t(f)) \quad ,$$

or

$$\left.\frac{\partial f_t}{\partial t}\right|_{t=0} = A(f) \quad ,$$

where: $A = A_0$.

Hence, using Taylor's formula:

$$f_t = f + A(f)t + \cdots \qquad (4.2)$$

(The terms $\cdots$ denote higher order terms in it.)

This formula can be thought of as sending f <u>approximately</u> (for small t) into

$$f + (\delta f)t \quad ,$$

where

$$\delta(f) = A(f) \quad .$$

The correspondence

$$f \rightarrow \delta f$$

is then thought of as the "infinitesimal transformation". In particular, it may be applied to the coordinate functions

$$x^i \to x^i + (\delta x^i)t \quad .$$

Now, if

$$A = A^i \frac{\partial}{\partial x^i} \quad ,$$

i.e., if (A^i) are the components (contravariant, in terms of classical tensor analysis) of the vector field A in the given coordinate system, then the infinitesimal transformation takes the form:

$$x^i \to x^i + A^i t + \cdots$$

Warning. Keep in mind that this transformation on functions is dual to the transformation on points. This is the source of a certain amount of notational confusion. In the next section we will pursue this further in the case that X is a vector space.

5. FLOWS AND INFINITESIMAL TRANSFORMATION ON VECTOR SPACES

Vector spaces are manifolds where points can be added. This additional structure is the source of both simplifications and confusions. Since many of the formulas in Lie's work may be most readily interpreted under the assumption that X is a vector space, I will now review and reinterpret in this context the formulas presented in previous sections for the case where X was only assumed to be a manifold, with no additive structure.

Warning. Many of the formulas to be derived in this section should only be interpreted "symbolically", or "locally", since in real life flows will not be defined globally on vector spaces. However, usually everything can be reinterpreted satisfactorily in terms of manifold theory, or some alternate formulism (for example, algebraic geometry, formal power series, sheaves of analytic functions, etc.)

Let X be a real vector space. X^d denotes its dual sapce, i.e., the space of linear maps: $X \to R$. Then, X^d is a linear subspace of $F(X)$.

Given $f \in F(X)$, $y \in X$, $x \in X$, define an R-linear map

$$v(x,y) \colon F(X) \to R$$

via the following formula:

$$v(x,y)(f) = \frac{\partial}{\partial t} f(x+ty)\Big|_{t=0} \tag{5.1}$$

for $f \in F(X)$

For fixed $x \varepsilon X$, this defines a linear isomorphism map

$$y \to v(x,y)$$

of $X \to X_x$.

As x varies, it defines an isomorphism

$$X \times X \to T(X) \quad .$$

Hence:

> <u>For a vector space</u> X, <u>the tangent bundle</u> $T(X)$ <u>may be identified with the product</u> $X \times X$.

<u>Remark</u>. We can also write formula 5.1 as follows:

$$v(x,dx)(f) = df(x,dx) \tag{5.2}$$

This interprets "dx" as an <u>element of</u> X. This is essentially what is called the <u>total differential</u> in advanced calculus.

With this identification of $T(X)$ with $X \times X$, the space of cross-section maps

$$A: X \to T(X)$$

i.e., the space of <u>vector fields on the manifold</u> X, becomes identified with the space of maps

$$A: X \to X \quad .$$

For example, in ordinary, 3-dimensional vector analysis

$$X = R^3$$

and a "vector field" (e.g., electric or magnetic field, velocity fields, etc.) is indeed a map

$$R^3 \to R^3 \quad .$$

With the identification of a vector field A as a map $X \to X$, the action of A on functions $f \in F(X)$ can be written as follows:

$$A(f)(x) = \frac{\partial}{\partial t} f(x+tA(x))\Big|_{t=0} \tag{5.3}$$

This formula--which is a key one linking the modern differential-geometric viewpoint and the formalism used by Lie--can also be written as follows:

$$f(x+tA(x)) = f(x) + tA(f)(x) + \cdots \tag{5.4}$$

(The terms $\cdots$ involve t^2.)

Now let

$$t \to \phi_t$$

be a flow on X. Let

$$t \to A_t$$

be its infinitesimal generator defined by formula 3.1. Then,

$$\text{for } x \in X, \quad f \in F(X) \ ,$$

$$A_t(f)(x) = \frac{\partial}{\partial t}(\phi_t^*(f))(\phi_t^{-1}x)$$

$$= \frac{\partial}{\partial s}(\phi_s^*(f))(\phi_t^{-1}(x))\Big|_{s=t}$$

$$= \frac{\partial}{\partial s}(f(\phi_s(\phi_t^{-1}(x)))\Big|_{s=t}$$

Hence, we have (identifying each vector field with a map $X \to X$):

$$A_t(x) = \frac{\partial}{\partial s}\phi_s(\phi_t^{-1}(x)))\Big|_{s=t} \qquad (5.5)$$

Here is a fluid mechanics interpretation of this formula. Suppose

$$X = R^3 ,$$

and that the orbits

$$t \to \phi_t(x)$$

of the flow represent stream-lines of a fluid flow. Then:

> $A_t(x)$, given by 5.5, is the velocity vector of the particle which is <u>at</u> x <u>at time</u> t. $\{A_t\}$ then represents the <u>Eulerian velocity vector field</u>.

Another intuitive insight into formula 5.5 is obtained by writing the right hand side of 5.5 as <u>approximately</u>

$$\frac{\phi_{t+\Delta t}(\phi_t^{-1}(x)) - \phi_t(\phi_t^{-1}(x))}{\Delta t} \tag{5.6}$$

Thus, we have:

$$\phi_{t+\Delta t}(x) \approx \phi_t(x) + A_t(\phi_t(x))\,\Delta t \tag{5.7}$$

In particular, set $t = 0$, and

$$A = A_0 \quad .$$

Consider the flow

$$x \to x + A(x)t \quad .$$

It may be considered as the infinitesimal transformation of the flow $\{\phi_t\}$.

Here is another way Lie sometimes denotes this. Write 5.6 as:

$$\phi_{t+\Delta t}(\phi_t^{-1}(x)) \approx x + A_t(x)\,\Delta t \quad . \tag{5.8}$$

This suggests that the "infinitesimal transformation" associated with the flow be denoted as:

$$\phi_{t+\Delta t}\,\phi_t^{-1}$$

or

$$\phi_{t+dt}\,\phi_t^{-1} \tag{5.9}$$

Chapter B

SOME ALGEBRAIC AND GEOMETRIC STRUCTURES INHERENT IN LIE'S WORK

In this chapter I briefly review general concepts of present-day mathematics which are useful in understanding Lie's work.

In general, Lie deals with a variety of binary algebraic operations, defining various algebraic structures, such as group, semigroup, Lie algebra, etc. See IM, vol. I, for a brief introduction to these concepts. In addition, certain ideas of category theory also occur implicitly in Lie's work.

1. BINARY AND PARTIALLY DEFINED BINARY OPERATIONS

Let X be a set. A binary operation on X is a mapping

$$X \times X \to X \quad .$$

We usually denote the image of

$$(x_1,x_2) \in X \times X$$

under this mapping by

$$x_1 x_2 \quad .$$

(This is called the <u>multiplicative notation</u> for the binary operation.)

A <u>partially defined binary operation</u> on a set X is defined as a subset

$$Y \subset X \times X \quad ,$$

together with a mapping

$$Y \to X \quad .$$

The image of

$$(x_1, x_2) \in Y$$

under this mapping is again usually denoted, multiplicatively, as

$$x_1 x_2 \quad .$$

Y is called the <u>distinguished subset</u> of $X \times X$.

Let X and X' be two sets with binary operation. A map

$$\phi: X \to X'$$

is a <u>homomorphism</u> if

$$\phi(x_1 x_2) = \phi(x_1)\phi(x_2)$$

$$\text{for} \quad x_1, x_2 \in X \quad .$$

If X, X' have partially defined binary operations, with distinguished subsets Y, Y', a map

$$\phi: X \to X'$$

is a <u>homomorphism</u> if:

$$\phi(Y) \subset Y'$$

$$\phi(x_1 x_2) = \phi(x_1)\phi(x_2)$$

for $(x_1, x_2) \in Y$.

Many examples of such binary operations should be familiar to the reader, and I will not go into detail. The main examples of "partially defined" operations are <u>categories</u>, which we consider later.

2. SEMIGROUPS, MONOIDS AND GROUPS

Let X be a set. A binary operation on X defines X as a <u>semigroup</u> if the following rule is satisfied:

$$x_1(x_2 x_3) = (x_1 x_2)x_3 \qquad (2.1)$$

for x_1, x_2, x_3 X .

It is called the <u>associative law</u>.

A <u>monoid</u> is a semigroup X, with an element 1 such that:

$$1x = x1 = x$$

for all $x \in X$.

1 is called the unit element. In words:

A monoid is a semigroup with a unit.

Let X be a monoid. Given $x \in X$, an element $x^{-1} \in X$ is called an inverse to x if:

$$xx^{-1} = 1 = x^{-1}x$$

for all $x \in X$.

A group is a monoid where every element has an inverse.

3. CATEGORIES AND FUNCTIONS

Of course, one can put in a "partially defined" condition on the algebraic structures defined in Section 2. (For example, a partially defined group is called a pseudo-group.) A category is a partially defined monoid, where a special emphasis is put on how the distinguished subset

$$Y \subset X \times X$$

is defined.

Suppose given two sets, $\underset{\sim}{A}$, $\underset{\sim}{\Phi}$. The elements of $\underset{\sim}{A}$ are called objects, and the elements of $\underset{\sim}{\Phi}$ are called morphisms. Set:

$$X = \underset{\sim}{A} \times \underset{\sim}{\Phi} \times \underset{\sim}{A} . \tag{3.1}$$

Given $x \varepsilon X$, of the form

$$x = (A_1, \phi, A_2) ,$$

with A_1, $A_2 \varepsilon \underset{\sim}{A}$, $\phi \varepsilon \underset{\sim}{\Phi}$, the element A_1 is called the <u>domain</u> of x, and the element A_2 is called the <u>codomain</u>. Set:

$$Y = \{(x_1, x_2) \varepsilon X \times X: \text{codomain } x_2 = \text{domain } x_1\} \tag{3.2}$$

<u>Definition</u>. A <u>semi-category</u> structure on X is a partially defined semigroup structure on X, with Y defined by 3.2 as the distinguished subset of $X \times X$. For $(x_1, x_2) \varepsilon Y$,

$$\text{domain } (x_1 x_2) = \text{domain } x_2$$

$$\text{codomain } (x_1 x_2) = \text{codomain } x_1$$

<u>Remark</u>. Do not be confused by the use of the terms domain, morphism, etc. Elements of categories are <u>not</u> mappings. As we shall see in a moment, mappings form one type of category, and the terminology has been chosen to reflect this special case. Thus, categories are algebraic abstractions, just as groups, fields, etc. are abstractions from familiar algebraic examples. Of course, a category is a good deal more general! I will not use or prove any theorems about categories. They will occasionally be useful as forming a convenient language.

Here is the notation used for the product in such a semi-category structure. Given

$$x_1 = (A_1,\phi_1,A_2), \qquad x_2 = (A_3,\phi_2,A_1)$$

with $(x_1,x_2) \in Y$, then

$$x_1 x_2 = (A_3,\phi_1\phi_2,A_1)$$

Definition. A semi-category is a category if, to each object $A \in \underset{\sim}{A}$, there is a unit element

$$(A,1_A,A)$$

such that:

$$(A,1_A,A)x = x$$

if domain $x = A$.

$$x(A,1_A,A) = x$$

if domain $x = A$.

Definition. A category is a set-mapping category, if, for $x \in (A_1,\phi,A_2)$, A_1 and A_2 are sets, ϕ is actually a mapping from A_1 to A_2, the multiplication is defined as mapping composition, and the unit element is the identity map.

Remark. Because the categories of mappings are the main example, one often denotes

$$(A_1,\phi,A_2)$$

by

$$A_1 \xrightarrow{\phi} A_2 \quad \text{or} \quad \phi: A_1 \to A_2 \quad ,$$

even for more general categories.

Definition. Let $X = \underset{\sim}{A} \times \underset{\sim}{\Phi} \times A$, $X' = \underset{\sim}{A} \times \Phi' \times \underset{\sim}{A}'$ be two categories.

Let

$$\alpha: \underset{\sim}{A} \to \underset{\sim}{A}'$$

$$\beta: \underset{\sim}{\Phi} \to \underset{\sim}{\Phi}'$$

be mappings of objects and morphisms.

If

$$x = (A_1,\phi,A_2) \quad ,$$

define:

$$(\alpha,\beta): X \to X'$$

as follows:

$$(\alpha,\beta)(A_1,\phi,A_2) = (\alpha A_1,\beta\phi,\alpha A_2) \quad .$$

(α,β) is called a functor if

$$(\alpha,\beta)(x_1x_2) = (\alpha,\beta)(x_1)(\alpha,\beta)(x_2) \quad ,$$

i.e., if (α,β) is a "homomorphism" of the partially defined algebraic structure associated with the category.

<u>Remark</u>. A "functor" stands to "category" as "homomorphism" stands to "group".

The original examples of "functors" came from topology. The topologists' notion of "commutative diagram" is often very useful as a notation for a functor. It would go something like this:

$$\begin{array}{ccc} A_1 & \xrightarrow{\phi} & A_2 \\ \alpha\downarrow & & \downarrow\alpha \\ A_1' & \xrightarrow{\phi'} & A_2 \end{array}$$

Lie's examples will give us a typical kind of category. Let M be a topological space, (typically, a differential manifold). The objects of the category are certain families of open subsets of M. The mappings defined between such open subsets, the "morphism" of the categories, are certain homeomorphisms of these open subsets. (Typically, the homeomorphisms will be defined by Lie algebras of vector fields on the manifold M.)

Another example of a category that appears naturally in Lie's work is the category whose "objects" are transformation groups, and whose "morphisms" are what are called "intertwining maps" or "prolongations". (The latter term is used by Lie and Cartan.) We now consider this situation.

4. TRANSFORMATION SEMIGROUPS AND GROUPS

Let G be a semigroup, and let M be a set. A <u>transformation group action of</u> G <u>on</u> M is defined as a map

$$G \times M \to M \quad ,$$

denoted by

$$(g,p) \to gp \quad ,$$

such that:

$$g_1(g_2 p) = (g_1 g_2)p$$

for $g_1, g_2 \in G$, $p \in M$.

Given two transformation semigroups (G,M), (G',M'), an <u>intertwining map</u> is a pair of maps (ϕ,π),

$$\phi: G \to G'$$

$$\pi: M \to M'$$

such that:

ϕ is a <u>homomorphism</u> of the semigroup structure on G

π is a map between M and M' such that

$$\pi(gp) = \phi(g)\pi(p)$$

for $g \in G$, $p \in M$.

In other words, the following diagram of maps is commutative:

$$\begin{array}{ccc} G\times M & \longrightarrow & M \\ \phi\downarrow\ \downarrow\pi & & \downarrow\pi \\ G'\times M & \longrightarrow & M \end{array}$$

If G is a group, with unit element 1, and if

$$1p = p \qquad \text{for all } p \in M ,$$

i.e., if the unit element of G acts as the identity mapping on M, then (G,M) is called a transformation group. It is the basic algebraic structure investigated by Lie. (Of course, he always dealt with the case where G was a "continuous" group, M a manifold.)

Definition. An intertwining map $\phi: G \to G'$, $\pi: M \to M'$ is called an isomorphism if:

a) ϕ is a homomorphism between groups G, G',

b) π is one-one and onto.

Remark. The notion of "isomorphism" roughly coincides with the use of the term "ähnlich" or "similar" by Lie. (Often, he also thought of isomorphisms in a local sense.) A good deal of the paper is concerned with classifying certain types of transformation group actions up to isomorphism!

Consider the intertwining maps as morphisms, the transformation groups as objects. This defines a category.

5. LIE ALGEBRAS AND THEIR ACTION AS INFINITESIMAL TRANSFORMATIONS

Let K be a field. A Lie algebra (with K as field of scalars) is a vector space (over K), with an algebraic multiplication which satisfies the Jacobi identity. Denote one by $\underset{\sim}{G}$. Then, elements $A,B \in \underset{\sim}{G}$ can be added,

$$A,B \to A + B \quad ;$$

and multiplied by scalars in K:

$$k,A \to kA \quad ;$$

and multiplied together:

$$A,B \to [A,B] \quad .$$

The Jacobi identity is:

$$[A,[B,C]] = [[A,B],C] + [B,[A,C]] \tag{5.1}$$

Here is a typical example of a Lie algebra. Let V be a vector space, with K field of scalars. Let

$$L(V)$$

denote the space of K linear maps

$$A: V \to V \quad .$$

Make L(V) into a vector space, by adding linear maps in the obvious way. The Lie algebra structure on L(V) is defined by defining the Lie multiplication as the commutator:

$$[A,B] = AB - BA \tag{5.2}$$

for $A,B \in L(V)$.

In differential geometry, Lie albegras are obtained from vector fields. Let M be a manifold, with F(M) the C^∞, real-valued functions on M. F(M) becomes--under pointwise addition and multiplication--a commutative associative algebra over the real numbers.

Set:

$$V(M) = \text{space of derivations of } F(M) .$$

Thus, $A \in V(M)$ is an R-linear mapping

$$f \to A(f)$$

of functions such that:

$$A(f_1 f_2) = A(f_1)f_2 + f_1 A(f_2)$$

for $f_1, f_2 \in F(M)$.

(One proves that such an A is, in local coordinates, a first order linear, homogeneous differential operator.)

Make V(M) into a Lie algebra using 5.2, i.e., the Lie algebra bracket of $(A,B) \to [A,B]$ (called the Jacobi bracket, in this case) is the commutator of the differential operators.

Definition. Let G be an abstractly given Lie algebra, with the real numbers as field of scalars.

An (infinitesimal) action of $\underset{\sim}{G}$ on the manifold M is defined as an R-bilinear map

$$\underset{\sim}{G} \times F(M) \to F(M) \quad ,$$

denoted by

$$(A,f) \to A(f)$$

such that:

$$[A,B](f) = A(B(f)) - B(A(f))$$

$$\text{for } A,B \in \underset{\sim}{G} \,, \quad f \in F(M) \quad .$$

Another way of putting it is to say that an action of $\underset{\sim}{G}$ on M is determined by a Lie algebra homomorphism:

$$\underset{\sim}{G} \to V(M) \quad .$$

Definition. Let $(\underset{\sim}{G},M)$, $(\underset{\sim}{G}',M')$ be two Lie algebras of vector fields. An intertwining map

$$(\underset{\sim}{G},M) \to (\underset{\sim}{G}',M')$$

is defined as a pair of maps

$$\pi\colon M \to M'$$

$$\phi\colon \underset{\sim}{G} \to \underset{\sim}{G}'$$

such that:

a) ϕ is a Lie algebraic homomorphism

b) $\pi^*(\phi_*(A)(f)) = A(\pi^*(f'))$

for $A \in \underset{\sim}{G}$, $f' \in F(M')$.

Remarks. Such an intertwining map is also called a prolongation, in Lie's work.

The Lie algebraic actions on manifolds, with the intertwining maps as "morphisms", form a category. A basic idea in Lie's work is that there is a functor from certain categories of transformation groups to categories of Lie algebras acting on manifolds.

6. LIE GROUPS AND LIE TRANSFORMATION GROUPS

Now we come to the type of mathematical structure which is most readily abstracted from Lie's works. In fact, what I will now define as a "Lie group" is one of the main concepts of modern mathematics!

Definition. A Lie group structure is defined on a set G by the following data:

a) G carries a manifold structure

b) G carries a group structure, usually denoted multiplicatively,

$$(g_1, g_2) \to g_1 g_2 \quad .$$

These structures satisfy the following condition which links them:

c) The mapping $(g_1, g_2) \to g_1 g_2^{-1}$ of $G \times G \to G$ is a (C^∞) differentiable map.

To make the set of Lie groups into a category, it is necessary to define a notion of "homomorphism".

Definition: A homomorphism between Lie groups G, G' is a map

$$\phi: G \to G'$$

such that:

ϕ is a (C^∞) differentiable map in the sense of manifold theory.

ϕ is a homomorphism in the sense of algebra.

Now, it is trivial to observe that the "Lie groups" and "homomorphisms" defined in this way form a category. Notice how nicely this leads to a view of Lie group theory as a combination of "algebraic" and "differentiable" ideas!

Remark. Chevalley's book [1] is the first modern treatment of Lie group theory. He restricts the manifold structure to be real-analytic (denoted by C^ω). We might denote this as the C^ω Lie group category. Since a C^ω manifold determines a C^∞ manifold there is a functor

$$C^\omega \text{ Lie group} \to C^\infty \text{ Lie group} .$$

One can prove that this functor is one-one and onto.

Another possibility is to allow the manifolds only to be manifolds in the sense of topology (also called C^0

or _continuous_ manifolds). There is then a functor

$$C^{\infty} \text{ Lie group } \to C^{0} \text{ Lie group } .$$

It is relatively easy (but non-trivial) to show that this map is one-one, i.e., each C^0 Lie group arises from _at most one_ C^{∞} Lie group. _Hilbert's Fifth Problem_ says, in modern language, that it is _onto_, i.e., each C^0 Lie group _can be made into a_ C^{∞} _mapping_, by appropriate restrictions on the coordinate neighborhoods, _so that the group operators are given by_ C^{∞} _functions_. This is highly non-trivial and was only proved by a massive effort. See the book by Montgomery and Zippin [1], which is the definitive reference for much of the purely topological side of group theory.

One may think of this as an "abstract" Lie group. Many examples can be considered immediately, such as groups of matrices, the Galilean and Lorentz groups of physics, the groups of various geometries (affine, projective, conformal, and so forth). These groups all appear "naturally" as transformation groups on various spaces. Lie himself always thought of them in this way. Here is the appropriate

Definition. A _Lie transformation group_ is a pair (G,M), satisfying the following conditions:

a) G is a (C^{∞}) Lie group.

b) M is a (C^∞) manifold.

c) There is a (C^∞) map

$$G \times M \to M$$

which defines G as a transformation group on M in the algebraic sense. (See Section 4 and Volume I.)

Remark. Again, this concept is constructed by starting with the purely algebraic notion and adding "differentiability" conditions.

Now, add the appropriate notion of "homomorphism", to make the objects into a category.

Definition. Let (G,M), (G',M') be two transformation groups. A homomorphism

$$\phi: (G,M) \to (G',M')$$

is defined as a pair (α,β) of maps

$$\alpha: G \to G'$$
$$\beta: M \to M'$$

such that:

a) α is a homomorphism in the Lie group sense.

b) β is a differentiable mapping in the manifold sense.

c) The following diagram of mappings is commutative:

$$\begin{array}{ccc} G\times M & \longrightarrow & M \\ {\scriptstyle \alpha\times\beta}\downarrow & & \downarrow{\scriptstyle \beta} \\ G'\times M' & \longrightarrow & M' \end{array}$$

Remark. Such a ϕ is also called an intertwining map or a prolongation.

Of course, a homomorphism such that:

α and β are diffeomorphisms,

is called an isomorphism. One might say that the goal of much of Lie's work is to classify Lie transformation groups up to isomorphisms. Unfortunately, the problem in this global form is too hard, and has never been attacked in this generality. (Instead, it is usually split up into more tractable subproblems, e.g., classify transitive transformation groups, compact transformation groups, etc.) However, a local version is more tractable, and is basically the problem that Lie did attempt to solve. (In fact, he carried it to completion for the case where the manifolds M are of dimension 1, 2 or 3.) Accordingly, we now briefly discuss this notion.

7. LOCAL LIE GROUPS AND TRANSFORMATION GROUPS

The idea of a "local Lie group" is intricate to state precisely, but it is intuitively simple--it should be an object that "locally" (in the neighborhood of the identity element) looks like a "global" Lie group (as defined in Section 6), but with the property that the group operation is not necessarily defined when one is far from the identity element. Then, it is, algebraically, a sort of <u>partially defined semigroup</u> structure, as defined in Section 1.

Suppose that LG is a manifold, with O an open subset of LG.

We suppose further that there is a (differentiable) mapping

$$O \times O \to LG \quad ,$$

denoted multiplicatively by

$$(g_1, g_2) \to g_1 g_2$$

such that:

$$1 \in O$$

$$g1 = g = 1g$$

for all $g \in O$, and for some element 1 of LG.

$$g_1(g_2g_3) = (g_1g_2)g_3$$

$$\text{for } g_1,g_2,g_3 \in O$$

<u>whenever these products are defined</u>, i.e.,

$$g_2g_3 \in O \; ; \quad g_1,g_2 \in O \quad .$$

There is a (differentiable) mapping $O \to O$ denoted by

$$g \to g^{-1} \quad ,$$

such that:

$$gg^{-1} = 1 = g^{-1}g \quad .$$

We then say that (LG,O) is a <u>local Lie group</u>, with O as the <u>distinguished neighborhood of the identity</u>. Two such structures, $(LG,O),(LG,O')$, are <u>equivalent</u> if the multiplication operation and inverse <u>agree</u> on the open set

$$O \cap O' \quad .$$

Usually, we are only interested in local Lie groups up to equivalence.

The local analogue of a "transformation group" can be developed. Here is one way to do it. Let LG be a local Lie group, and let M be a manifold. Let U be an open subset of

$$(LG) \times M$$

which contains the subset

$1 \times M$.

A local Lie group action of (LG) on M is determined by a mapping

$$U \to M ,$$

denoted multiplicatively as

$$(g,p) \to gp ,$$

which satisfies all the usual algebraic laws for transformation group actions whenever the action is defined.

I do not want to discuss here the finer details of these structures. (This should probably be done in the general context of category theory.) In practice, what one usually does is to reduce everything to Lie algebras. Thus, a local Lie group has a Lie algebra "functorally" attached to it (for example, defined as the tangent space to the identity element), and a local Lie group action defines this Lie algebra as a Lie algebra of vector fields on M. At this point, the real work starts.

The ideas of algebraic geometry offer another way of approaching these concepts. In the typical example, the group action is only defined "locally" because the multiplicative action develops "singularities". Thus, in certain situations it might be useful to define actions as "maps"

$$G \times M \to M$$

with certain types of "singularities". Here again, the appropriate general language for a general foundational discussion is category theory, particularly as it is used in modern algebraic geometry.

8. LINEAR FRACTIONAL TRANSFORMATIONS AS LOCAL TRANSFORMATION GROUPS

A typical example of a "local" transformation group action is the "group" of linear fractional transformations on the real numbers, R. Let us consider this example in detail in order to distinguish the various concepts we have briefly discussed in Section 7.

Denote the manifold of the real numbers by R, and a point of R by

$$x \; .$$

A linear fractional transformation on R is a "mapping" $R \to R$ of the following form:

$$x \to \frac{ax + b}{cx + d} \; , \tag{8.1}$$

with a, b, c, d real constants.

Now, this is not a mapping in the set theory sense (at least with domain and range R) because for certain values of x the right hand side of 8.1 is infinite.

Let G be the group of 2×2 real matrices of <u>determinant one</u>. (It is usually denoted as $SL(2,R)$) Denote $g \in G$ as

$$g = \begin{pmatrix} a & b \\ c & d \end{pmatrix}$$

with $a,b,c,d \in R$. Write 8.1 as:

$$gx = \frac{ax + b}{cx + d} \qquad (8.2)$$

Let O be any neighborhood of the identity in G whose closure is compact, and such that

$$O^{-1} \subset O$$

(G,O), with the usual product of matrices, forms a local group. (It is, of course, a "localization" of a <u>global</u> Lie group.

Let:

$$U = \{(g,x) \colon g \in G,\ x \in R \text{ such that } gx \neq \infty\} .$$

U is indeed an open subset of $G \times R$ which contains the subset

$$(1,R) .$$

Define the map

$$U \to X$$

via formula 8.2. It defines the required <u>local transformation group</u>.

An alternate approach is to consider G and X as algebraic manifolds, and the map 8.2 as defining a <u>rational map</u>

$$G \times R \to R \quad ,$$

satisfying the appropriate algebraic rules.

9. LIE'S IDEA OF THE GROUP-TRANSFORMATION GROUP NOTION

Of course, this article itself is one of the key papers in which Lie explains his ideas. However, several preliminary comments might be useful to the modern reader.

First, Lie did not put the "abstract" notion of a group in the foreground. For him, a "group" means a transformation group. (He did develop the idea of "isomorphism", in its modern sense, so presumably he appreciated that it is useful to consider groups as "abstract" objects.)

Here is the context in which he often worked. Let

$$x = (x_1,\ldots,x_n)$$

denote a point of R^n. Introduce another Euclidean space, R^m, with coordinates

$$a = (a_1,\ldots,a_m) \quad .$$

Lie often assumes that a "transformation group" is given as a set of mappings

$$x' = f(x;a) \quad , \tag{9.1}$$

parameterized by points $a \in R^m$, such that:

For $a,b \in R^m$, there is a $c \in R^m$ with:

$$f(f(x,b),a) = f(x,c) \tag{9.2}$$

for all $x \in R^n$

He is not usually precise about the domain and range of the mappings 9.1. Let us say, for the moment, that they are the entire R^n. Let us also say that the parameters a vary over R^m.

Thus, 9.2 says that the composition of two maps $R^n \to R^n$ of the form 9.1 is again of the form 9.1, i.e., that we are given a set of maps which forms a semigroup. This semigroup is not the most general, since it is "parameterized" by a. However, Lie, at this stage, doesn't usually discuss in detail the correspondence

$$(a,b) \to c$$

implicitly defined by 9.2. It is, of course, reasonable to assume that this is a bona fide differentiable mapping

$$R^m \times R^m \to R^m \quad ,$$

which defines R^m as a Lie semigroup. The formula 9.1

then may be supposed to define a transformation action of the semigroup on R^n.

It is now appropriate, in the context of modern mathematics, to replace R^n and R^m by manifolds M and G. Then, the modern reader who sees Lie's definition might be tempted to define the basic object of Lie's work as the study of the pair

$$(G,M) \quad ,$$

with the following structures:

a) G is a manifold and a semigroup, such that the map $G \times G \to G$ defined by the group-multiplication is differentiable.

b) M is a manifold.

c) A differentiable transformation semigroup action $G \times M \to M$.

Unfortunately, Lie found that his methods (which were "analytic", rather than geometric and topological, and usually invoked application of the theory of differential equations) gave very few non-trivial results without the group property of G. As the reader will see in the text, he didn't like to assume this, but hoped to be able to deduce it. In fact, it is impossible to do this, i.e., there are examples of Lie transformation semigroups which are not groups. C. Loewner has been the mathematician in modern times who has most extensively studied transformation semigroups in the way envisaged by Lie. Their general theory seems to be very hard, but important!

Accordingly, Lie reluctantly assumed that G contained inverses, which leads to a "Lie group" in the modern definition. Now, return to formula 9.1. It is clear in the context of Lie's work that he did not want or need (to deduce his main result) to assume that the functions on the right hand side of 9.1 were defined everywhere. Basically, he was interested in the "infinitesimal" form of the transformation 9.1. Thus, one appropriate modern context in which to consider Lie's work is the theory of local groups and their local actions. Another possible context is the case where the transformations are only defined on open subsets of a manifold M. Thus, we might deal with a category of maps, whose domain and range are open subsets of M, such that the maps themselves are diffeomorphisms between their domain and range. (Such objects are also called pseudogroups.) If this framework were to be pursued, it would also be necessary to formulate a notion of "parameterization" of these mappings by means of finite dimensional manifolds. Again, the theory of categories would be the appropriate tool for studying these objects. However, in practice, Lie usually reduced problems to their "infinitesimal" version, usually involving the theory of Lie algebras of vector fields in manifolds.

10. COMPLETELY INTEGRABLE VECTOR FIELD SYSTEMS AND FOLIATIONS

The theory of Lie groups is closely related to a part of the theory of differential equations called (recently) the theory of foliations. In fact, the first modern book on the theory of Lie groups--by Chevalley [1]--also contains the first comprehensive "global" treatment of basic existence theorems of foliation theory. Chevalley's treatment (of the "non-singular" case) was later extended (see Hermann [,], Sussman [1]) to the "singular" case. In this section I will briefly review the needed material.

Let M be a manifold. Recall that $V(M)$ and $F(M)$ denote the (differentiable) vector fields and real-valued functions on M. $V(M)$ is an $F(M)$-module.

If V is a subset of $V(M)$, $x \in M$, set:

$$V(x) = \{A(x): A \in V\} .$$

$V(x)$ is then a subset of M_x, the tangent space to M at x. It is a linear subspace if V is an R-linear subspace of $V(M)$.

Definition. A vector field system on M is defined by an $F(M)$-submodule V of $V(M)$. It is said to be non-singular if the following condition is satisfied:

$$\dim (V(x)) \text{ is constant, as } x \text{ ranges over } M . \tag{10.1}$$

It is completely integrable if:

$$[V,V] \subset V . \tag{10.2}$$

($[\ ,\ ]$ denotes the Jacobi bracket operation on vector fields.)

Recall that a submanifold of a manifold M is a pair

$$(\phi,N) ,$$

where N is a manifold,

$$\phi: N \to M$$

is a (differentiable) map such that the differential map

$$\phi_*: T(N) \to T(M)$$

is one-one. For $x \in N$, this enables us to identify:

x with $\phi(x)$

N_x with the linear subspace $\phi_*(N_x) \subset M_{\phi(x)}$.

Usually, we suppress explicit mention of ϕ, and write

$$x \in N \subset M$$

$$N_x \subset M_x .$$

Definition. Let $V \subset V(M)$ be a vector field system. A submanifold

$$N \subset M$$

is an orbit of V if the following condition is satisfied:

$$N_x = V(x) \tag{10.3}$$

for all $x \in N$

Definition. An orbit curve of the vector field system V is a map

$$\sigma\colon [a,b] \to M$$

of an interval $a \le t \le b$ of real numbers such that:

a) σ is continuous

b) σ is piecewise C^∞

c) For $a \le t \le b$, $\sigma'(t)$ (the tangent vector to σ at t) is an element of

$$V_{\sigma(t)} .$$

Definition. Let V be a vector field system on a manifold M. A submanifold $N \subset M$ is said to be a maximal orbit of V if it satisfies the following conditions:

N is an orbit, i.e., it satisfies 10.3

N is a connected manifold

If $t \to \sigma(t)$; $a \leq t \leq b$, is an orbit curve of V such that $\sigma(a) \in N$, then

$$\sigma(t) \in N \quad \text{for } a \leq t \leq b \quad ,$$

i.e., N contains all orbits starting from each point of N.

So far, we have only quoted definitions. (However, in this subject it is important to understand the basic concepts in a general way. These definitions are designed to lead the reader in this direction.) Here are some basic results. (See the references by Chevalley, Hermann, Sussman quoted above, and DGCV.)

Theorem 10.1. Suppose V is a completely integrable vector field system on M satisfying the following condition:

For *each* orbit curve $t \to \sigma(t)$, $a \leq t \leq b$, the dimension of $V(\sigma(t))$ is constant as t varies (10.4)

Then, through each point of M there passes a unique maximal orbit.

Theorem 10.2. Suppose V is a completely integrable vector field system, and that *either* of the following two conditions are satisfied:

a) The manifold M, and V, are *real analytic*.

b) V is locally finitely generated, as an $F(M)$-module.

Then, condition 10.4 is satisfied, and through each point of M there is a maximal orbit.

Remark. Sussman has investigated [1] what happens if these conditions are not satisfied. He shows that the set of points that can be joined to a given point of M by an orbit curve is indeed a submanifold of M. It is an orbit submanifold of *another* vector field system on M, which may be larger than V.

Condition 10.4 is, of course, automatically satisfied if V is non-singular. This is the case covered by Chevalley [1]. (He calls a non-singular vector field system a *distribution*, and a completely integrable non-singular vector field system an *involutive distribution*.)

Definition. A *foliation* of a manifold M is defined as an *equivalence relation* on M, such that each equivalence class is a *submanifold* of M. If x is a point of M, the equivalence class to which x belongs is called the *leaf of the foliation passing through the point* x.

Thus, a vector field system satisfying the hypotheses of Theorem 10.1 defines a foliation, whose leaves are the maximal orbits.

11. SUBMERSIONS AND FOLIATIONS

Definition. A (differential) mapping

$$\pi: M \to N$$

between manifolds is said to be a submersion if the following condition is satisfied:

The differential of π

$$\pi_*: T(M) \to T(N) \tag{11.1}$$

is onto.

Remark. Recall that a map

$$\phi: N \to M$$

is a submanifold mapping if

$$\phi_*: T(N) \to T(M)$$

is one-one. Hence, a "submersion" is in a sense dual to a submanifold mapping.

The implicit function theorem enables one to describe the local structure of submersion maps. In fact, they are locally equivalent to Cartesian projection maps of Euclidean spaces. (See DGCV, Chapter 5.) Here is one useful way of thinking of this.

Choose indices as follows:

$$1 \leq a,b \leq n = \dim N$$

$$1 \leq i,j \leq m = \dim M \quad .$$

Let (y^a) be a coordinate system of functions on N. Condition 11.1 then means that:

> The functions $\pi^*(y^a)$ are functionally independent, i.e., $\pi^*(dy^a)$ are linearly independent in the F(M)-module sense.

The Implicit Function Theorem then asserts that there is, locally, a coordinate system

$$(x^i)$$

for M such that:

$$x^a = \pi^*(y^a) \quad .$$

Return to the global setting: If π is a map satisfying 11.1, say that a vector field A on M is <u>vertical</u> if:

$$\pi_*(A(x)) = 0 \tag{11.2}$$

for all $x \in M$.

Let V be the set of all such vertical vector fields. Here are some properties of V which follow readily from 11.1:

> V is an F(M)-submodule of V(M), i.e., V defines a <u>vector field system</u> on M.

V is non-singular, i.e., $\dim(V(x))$ is constant as x ranges over M.

$[V,V] \subset V$, i.e., V is completely integrable.

The connected component of the fiber of π,

$$\pi^{-1}(\pi(x)) \quad ,$$

passing through x, is the leaf of the foliation defined by V.

Definition. The map π is said to be the quotient map associated with the foliation V.

Remark. A foliation does not necessarily have a global quotient map. See Palais [1] and Sussman [1] for what can be said about its existence. However, the Frobenius theorem (see DGCV) asserts that such quotient maps exist locally.

Chapter C

FILTRATIONS AND PROLONGATIONS OF VECTOR FIELDS

1. INTRODUCTION

One of the key features of Lie's work, from the modern point of view, is the emphasis on the study of what we would now call the "structure" (both algebraic and geometric) of Lie algebras (both finite and infinite dimensional) of vector fields on manifolds. Indeed, one of my main reasons for preparing this version of Lie's work is that many of his ideas and results have never been adequately assimilated into modern mathematics.

In this chapter I will develop certain elementary (but not necessarily well-known) properties of such Lie algebras. In particular, I will emphasize (following my own paper "Cartan connections and the equivalence problem for geometric structures") the role of certain abstract types of Lie algebras called _filtered Lie algebras_. We shall see that these algebraic objects are well-suited to expressing many of Lie's results in terms of modern mathematics.

2. LIE ALGEBRAS OF VECTOR FIELDS AND THEIR FILTRATIONS

Let X be a manifold, and let $\underset{\sim}{G}$ be a Lie algebra of vector fields on X. A vector field $A \in \underset{\sim}{G}$ is <u>said to vanish at a point</u> $x \in X$ if:

$$A(f)(x) = 0$$

for all $f \in F(X)$.

It is said to <u>vanish to the second order</u> if:

$$[B,A] \text{ vanishes at } x$$

$$\text{for } \underline{\text{all}} \quad B \in V(X) \quad . \tag{2.1}$$

Set:

$$\underset{\sim}{G}^0(x) = \underset{\sim}{G}$$

$$\underset{\sim}{G}^1(x) = \{A \in \underset{\sim}{G}: A \text{ vanishes at } x\}$$

$$\underset{\sim}{G}^2(x) = \{A \in \underset{\sim}{G}: A \text{ vanishes to second order at } x\}$$

and so forth,

i.e., inductively on n:

$$\underset{\sim}{G}^n(x) = \{A \in \underset{\sim}{G}: [B,A] \text{ vanishes to } (n-1)\text{-st order at } x \text{ for all } B \in V(X)\} \quad .$$

Usually, we consider x fixed. Suppress explicit mention of "x" in the notation, and denote the subspace of $\underset{\sim}{G}$ by $\underset{\sim}{G}^0, \underset{\sim}{G}^1, \underset{\sim}{G}^2, \ldots$

Here are some basic properties:

$$\underset{\sim}{G} = \underset{\sim}{G}^0 \supset \underset{\sim}{G}^1 \supset \underset{\sim}{G}^2 \supset \cdots \qquad (2.2)$$

$$[\underset{\sim}{G}^j, \underset{\sim}{G}^k] \subset \underset{\sim}{G}^{j+k-1} \qquad (2.3)$$

$$\underset{\sim}{G}^1 \text{ is a Lie subalgebra of } \underset{\sim}{G} . \qquad (2.4)$$

$$\underset{\sim}{G}^1, \text{ for } j \geq 2, \text{ is a Lie } \underline{\text{ideal}} \text{ of } \underset{\sim}{G} . \qquad (2.5)$$

We can now abstract from these properties a general:

<u>Definition</u>. Let $\underset{\sim}{G}$ be an abstract Lie algebra. A (descending) <u>filtration</u> of $\underset{\sim}{G}$ is defined by a sequence $\underset{\sim}{G}^0, \underset{\sim}{G}^1, \cdots$ of linear subspaces of $\underset{\sim}{G}$ which satisfies conditions 2.2-2.3.

<u>Warning</u>. This is not the only sort of natural filtration Lie algebras can have. The <u>linear differential operations</u> on a manifold are filtered by their <u>order</u>. This is an <u>ascending filtration</u>, i.e., an element of filtration j is contained in the set of filtrations k, for $j<k$. Such a filtration is important in quantum mechanics.

<u>Remark</u>. The descending filtration of Lie algebras of vector fields can be described more concretely in terms of coordinates. Let

$$(x^i) \quad , \qquad 1 \leq i,j \leq n \quad ,$$

be a coordinate system for X. Suppose coordinates are chosen so that the given point "x" has coordinates 0. A vector field A can be written as:

$$A = A^j \frac{\partial}{\partial x^i} .$$

One sees readily that:

A vanishes to the r-th order at x if and only if the components A^j vanish to the r-th order, i.e., if they are of the following form:

$$A^i = a^j_{j_1 \cdots j_r}(x)\, x^{j_1} \cdots x^{j_r}$$

3. TRANSITIVE LIE ALGEBRA ACTIONS

Let $\underset{\sim}{G}$ be a Lie algebra of vector fields on a manifold X. Each $A \in \underset{\sim}{G}$ has a <u>value</u> at x, which is a tangent vector, i.e., an element of X_x. The value is denoted by:

$$A(x) .$$

As definition,

$$A(x)(f) = A(f)(x)$$
$$\text{for all } f \in F(X) .$$

Set:

$$\underset{\sim}{G}(x) = \{A(x) : A \in \underset{\sim}{G}\}$$

$\underset{\sim}{G}(x)$ is then a linear subspace of X_x.

<u>Definition</u>. $\underset{\sim}{G}$ is said to act transitively on X if:

$$\underset{\sim}{G}(x) = X_x \quad (3.1)$$

for all $x \in X$.

If 3.1 is satisfied, the evaluation map

$$A \to A(x)$$

is an R-linear map

$$\underset{\sim}{G} \to X_x ,$$

whose kernel is $\underset{\sim}{G}^1(x)$. This results in the following:

$\underset{\sim}{G}/\underset{\sim}{G}^1(x)$ is vector-space isomorphic to X_x .

Set:

$$\underset{\sim}{L} = \underset{\sim}{G}^1(x) .$$

$\underset{\sim}{L}$ is a Lie subalgebra of $\underset{\sim}{G}$.

We can give an abstract definition:

<u>Definition</u>. Let $\underset{\sim}{G}$ be a Lie algebra, $\underset{\sim}{L}$ a subalgebra. The linear subspaces $\underset{\sim}{G}^0, \underset{\sim}{G}^1, \cdots$ defined as follows,

$$\begin{aligned}
\underset{\sim}{G}^0 &= \underset{\sim}{G} \\
\underset{\sim}{G}^1 &= \underset{\sim}{L} \\
\underset{\sim}{G}^2 &= \{A \in \underset{\sim}{G}: [A,\underset{\sim}{G}] \subset \underset{\sim}{G}^1\} \qquad (3.2) \\
&\vdots \\
\underset{\sim}{G}^n &= \{A \in \underset{\sim}{G}: [A,\underset{\sim}{G}] \subset \underset{\sim}{G}^{n-1}\} ,
\end{aligned}$$

define a filtration of $\underset{\sim}{G}$, called the filtration of $\underset{\sim}{G}$ defined by the subalgebra $\underset{\sim}{L}$.

We see that:

> If $\underset{\sim}{G}$ acts transitively on X, and if $\underset{\sim}{L} = \underset{\sim}{G}^1(x)$, then the filtration of $\underset{\sim}{G}$ determined by the action on X is identical with the filtration of $\underset{\sim}{G}$ defined, via formulas 3.2, purely algebraically by the subalgebra $\underset{\sim}{L}$.

Much of E. Cartan's work on what he called "infinite dimensional Lie groups" can be viewed as a structure theory of such filtered Lie algebras of vector fields which act transitively. Cartan also states that his methods work to classify the intransitive cases also, but this is much less worked out. In fact, the most interesting part of this paper by Lie is the classification of intransitive Lie algebras of vector fields on 2-dimensional manifolds. Not much is known about higher-dimensional generalizations.

4. IMPRIMITIVE LIE ALGEBRAS OF VECTOR FIELDS AND PROLONGATIONS

Let X be a manifold. Let

$$V \subset V(X)$$

be a <u>completely integrable</u>, <u>non-singular vector field system on</u> X. This means that the following conditions are satisfied:

a) V is an F(X)-submodule of V(X)

b) dim V(x) is constant, as x ranges over X

c) $[V,V] \subset V$.

Such a system determines, by the Frobenius-Chevalley global integrability theorem (see DGCV, Chapter 8), a decomposition of X into submanifolds, called <u>leaves</u>. Each leaf Y is a <u>maximal connected</u> integral submanifold of V, and satisfies:

$$Y_y = V(y) \qquad \text{for all } y \in Y .$$

Each point of X belongs to precisely one leaf of V. The "leaved" structure is called a <u>foliation</u>. Usually, I will not be so precise, and say that the foliation <u>is</u> V.

<u>Definition</u>. Let $\underset{\sim}{G}$ be a Lie algebra of vector fields on a manifold X, and let $V \subset V(X)$ be a foliation on X. $\underset{\sim}{G}$ is <u>said to leave the foliation invariant</u> if the following condition is satisfied.

$$[\underset{\sim}{G},V] \subset V . \tag{4.1}$$

<u>Definition</u>. A given Lie algebra $\underset{\sim}{G}$ of vector fields on X is said to act <u>imprimitively</u> if each point $x \in X$ is

contained in an open subset U, which carries a foliation V which is left invariant by $\underset{\sim}{G}$.

$\underset{\sim}{G}$ is said to act <u>primitively</u> if it "does not act imprimitively". In other words, "$\underset{\sim}{G}$ acts primitively" means that it leaves invariant no foliation, <u>even locally</u>.

A good deal of Cartan's work on what are called "infinite dimensional Lie groups" (in Vol. II, Part 2 of his Collected Works) involves the classification of <u>primitive transitive</u> Lie algebras. It is also a key concept in Lie's work, although he does not use the term explicitly. (It is used in his treatise "Transformations-gruppen".) Instead, he speaks of a "group which does not leave any differential equation invariant."

We can examine the geometric meaning of condition 4.1.

<u>Definition</u>. Let V be a foliation on X. A mapping

$$\pi: X \to Y$$

between manifolds is a <u>decomposition submersion</u> for the foliation V if the following conditions are satisfied:

a) π is a submersion mapping in the sense of differential topology, i.e.,

$$\pi(X) = Y$$

$$\pi_*(X_x) = Y_{\pi(x)} \qquad \text{for all } x \in X .$$

b) For $x \in X$,

$$V(x) = \pi_*^{-1}(0) \tag{4.2}$$

For $f \in F(Y)$, condition 4.2 means that:

$$A(\pi^*(f)) = 0 , \tag{4.3}$$

i.e., $\pi^*(f)$ is <u>constant along the leaves of</u> V, or is a <u>conserved function</u> of V. (The old terminology is "integral function of V," but I am trying to eliminate the many confusing uses of the term "integral" in the classical literature.)

<u>Theorem</u>. $\underset{\sim}{G}$ satisfies 4.1 if and only if

For each conserved function f, $B(f)$ is again conserved, for all $B \in \underset{\sim}{G}$. (4.4)

In particular, $\underset{\sim}{G}$ acts as a Lie algebra of vector fields, denoted by

$$\pi_*(\underset{\sim}{G}) ,$$

on the base space Y of the decomposition submersion.

<u>Proof</u>. For $A \in V$, $B \in \underset{\sim}{G}$, $f \in F(Y)$,

$$\begin{aligned} A(B(\pi^*(f)) &= [A,B](\pi^*(f)) + BA(\pi^*(f)) \\ &= 0 , \end{aligned}$$

since $[A,B] \in V$.

This proves that $B(\pi^*(f))$ is again a conserved function of V.

We can define

$$\pi_*(B)$$

as a vector field on Y via the rule:

$$\pi^*(\pi_*(B)(f)) = B(\pi^*(f)) \tag{4.5}$$

for all $f \in F(Y)$.

Remark. The assignment

$$B \to \pi_*(B),$$

$$\underset{\sim}{G} \to \pi_*(\underset{\sim}{G})$$

is also a Lie algebra homomorphism. It is an example of what Lie and Cartan call a prolongation, i.e., an "abstract" Lie algebra homomorphism which is realized via a geometric mapping.

Much of the work Lie does in classifying Lie algebras of vector fields involves putting together arguments involving filtrations and prolongations.

Of course, if $\underset{\sim}{G}$ acts intransitively on X, the orbits of $\underset{\sim}{G}$ determine a foliation which is left invariant by $\underset{\sim}{G}$. For simplicity, let us assume that $\underset{\sim}{G}$ acts "non-singularly" in the following sense:

<u>Definition</u>. $\underset{\sim}{G}$ acts on X in a <u>non-singular way</u> if:

$$\dim \underset{\sim}{G}(x) \text{ is constant}$$
$$\text{for all } x \in X \;.$$

We can then define a foliation V by requiring that:

$$V(x) = \underset{\sim}{G}(x)$$
$$\text{for all } x \in X \;,$$

i.e., that V is the smallest $F(X)$-submodule of $V(X)$ determined by $\underset{\sim}{G}$. The Lie algebra condition for $\underset{\sim}{G}$ implies that

$$[V,V] \subset V \;,$$

i.e., V is completely integrable. Hence, a foliation of X is determined. The leaves of the foliation are called the <u>orbits</u> of $\underset{\sim}{G}$.

If $\pi: X \to Y$ is a decomposition submersion for this foliation, then

$$\pi_*: (\underset{\sim}{G}) = 0 \;,$$

i.e., $\underset{\sim}{G}$ acts trivially on Y. Geometrically, this means that $\underset{\sim}{G}$ is <u>tangent</u> to the leaves, so that each $B \in \underset{\sim}{G}$ generates a group preserving each leaf.

Chapter D

SOME GENERAL CONCEPTS OF LIE ALGEBRA THEORY THAT ARE USEFUL IN THE INTERPRETATION OF LIE'S WORK

Many of Lie's proofs can be simplified using ideas of modern Lie algebra theory. In this chapter, I present in outline those concepts that I have found to be most useful and that I will use in the text.

The basic reference for Lie algebra theory is now Samelson's book [1]. The treatises by Jacobson [1], Sagle and Walde [1], Wallach [1], F. Warner [1], Serre [1], and Hochschild [1] contain much useful material. The material of greater importance for the applications to physics is developed (some only sketchily) in LGP, VB, Vol. II, and LMP, Vol. II.

1. BASIC DEFINITIONS

All Lie algebras will be vector spaces, with the real or complex numbers as field of scalars, and finite dimensional, unless mentioned otherwise. A certain amount carries over to more general situations.

A Lie algebra, typically denoted by $\underset{\sim}{G}$, is defined by giving two algebraic structures on the set $\underset{\sim}{G}$:

a) A vector space structure

b) A <u>bilinear</u> map

$$\underset{\sim}{G} \times \underset{\sim}{G} \to \underset{\sim}{G} \quad ,$$

denoted by $(A,B) \to [A,B]$, called the <u>bracket</u>, satisfying the following two conditions:

$$[A,B] = -[B,A] \tag{1.1}$$

$$[A,[B,C]] = [[A,B],C] + [B,[A,C]] \tag{1.2}$$

for $A,B,C \in \underset{\sim}{G}$.

1.1 is called the <u>skew-symmetry</u> of the bracket. 1.2 is called the <u>Jacobi identity</u>. (Lie often describes it by the term "the familiar relations".)

If $\underset{\sim}{H},\underset{\sim}{K}$ are two linear subspaces of a Lie algebra $\underset{\sim}{G}$,

$$[\underset{\sim}{H},\underset{\sim}{K}]$$

denotes the space of all linear combinations of elements of the form $[A,B]$, with

$$A \in \underset{\sim}{H}, \quad B \in \underset{\sim}{K} \quad .$$

Let $\underset{\sim}{G}$ be a Lie algebra. A linear subspace $\underset{\sim}{H} \subset \underset{\sim}{G}$ is a <u>(Lie)</u> <u>subalgebra</u> if

$$[\underset{\sim}{H},\underset{\sim}{H}] \subset \underset{\sim}{H} \quad , \tag{1.3}$$

i.e., if $\underset{\sim}{H}$ inherits the bracket structure. If, further,

$$[\underset{\sim}{G},\underset{\sim}{H}] \subset \underset{\sim}{H} \quad , \tag{1.4}$$

then $\underset{\sim}{H}$ is called a (Lie) ideal of $\underset{\sim}{G}$.

A homomorphism between two Lie algebras $\underset{\sim}{G}, \underset{\sim}{G}'$ is a linear map

$$\phi: \underset{\sim}{G} \to \underset{\sim}{G}'$$

such that:

$$\phi([A,B]) = [\phi(A),\phi(B)]$$

$$\text{for } A,B \in \underset{\sim}{G} .$$

Thus, with the "Lie algebras" as objects, and the "homomorphisms" as morphisms, there is defined a category, called the Lie algebra category.

If $\underset{\sim}{H}$ is an ideal of $\underset{\sim}{G}$, a new Lie algebra structure can be defined on the quotient vector space

$$\underset{\sim}{G}/\underset{\sim}{H} .$$

The quotient map

$$\underset{\sim}{G} \to \underset{\sim}{G}/\underset{\sim}{H}$$

is then a Lie algebra homomorphism.

Here is the simplest way to form a Lie algebra. Let V be a vector space. Define $L(V)$ as the space of linear maps

$$\alpha: V \to V$$

Make $L(V)$ into a Lie algebra by defining the bracket as the commutator:

$$[A,B] = AB - BA$$

If $\underset{\sim}{G}$ is a Lie algebra, V is a vector space, a <u>linear representation of</u> $\underset{\sim}{G}$ <u>in</u> V is a homomorphism

$$\rho: \underset{\sim}{G} \to L(V) \quad .$$

Another way of putting this is to say that a linear representation is a bilinear map

$$\underset{\sim}{G} \times V \to V \quad ,$$

$$(A,v) \to \rho(A)(v) \quad ,$$

such that:

$$\rho([A,B])(v) = \rho(A)(\rho(B)(v)) - \rho(B)(\rho(A)(v))$$

$$\text{for } A,B \in \underset{\sim}{G}, \quad v \in V \quad .$$

A linear representation

$$\rho: \underset{\sim}{G} \to L(V)$$

is said to be <u>reducible</u> if there is a linear subspace $V' \subset V$ (different from the "trivial" cases $V' = (0)$ or V) such that

$$\rho(G)(V') \subset V' \quad .$$

If there is no such subspace, it is said to be <u>irreducible</u>.

A <u>reducible representation</u>

$$\rho: \underset{\sim}{G} \to L(V) \quad ,$$

such that $\rho(\underset{\sim}{G})(V') \subset V'$,

defines two other representations, defined as follows.

$$\rho': \underset{\sim}{G} \to L(V') \quad ,$$

with:

$$\rho'(A)(v') = \rho(A)(v) \quad ,$$

i.e., $\rho'(A)$ is $\rho(A)$ restricted to V'.

$$\rho'': \underset{\sim}{G} \to L(V/V') \quad ,$$

with $\rho''(A)$ = quotient map of $\rho(A)$,

i.e., $\rho''(A)$ is defined by the following commutative diagram:

$$\begin{array}{ccc} V' & \xrightarrow{\rho'(A)} & V' \\ \downarrow & & \downarrow \\ V & \xrightarrow{\rho(A)} & V \\ \downarrow & & \downarrow \\ V/V' & \xrightarrow{\rho''(A)} & V/V' \end{array}$$

ρ' is called the subrepresentation, ρ'' is called the quotient representation.

A representation ρ of $G \to L(V)$ is the direct sum of representations

$$\rho': \underset{\sim}{G} \to L(V')$$

$$\rho'': \underset{\sim}{G} \to L(V'')$$

if:

$$V = V' \oplus V'' \quad ,$$

i.e., V is the direct sum of the vector spaces V', V'', and:

$$\rho(A)(v' \oplus v'') = \rho'(A)(v') \oplus \rho''(A)(v'')$$

$$\text{for} \quad A \in \underset{\sim}{G}; \quad v' \in V'; \quad v'' \in V'' \quad .$$

Two representations

$$\rho: \underset{\sim}{G} \to L(V)$$

$$\rho': \underset{\sim}{G} \to L(V')$$

are _equivalent_ if there is a vector space automorphism

$$\phi: V \to V'$$

such that:

$$\rho'(A)(v') = \phi(\rho(A)(v')) \quad ,$$

i.e., ϕ _intertwines_ the action of $\underset{\sim}{G}$.

A representation is said to be _completely reducible_ if it is equivalent to a direct sum of irreducible ones.

Every Lie algebra defines--by its own structure--one linear representation called the _adjoint representation_,

$$\text{Ad}: \underset{\sim}{G} \to L(\underset{\sim}{G})$$

$$\text{Ad}(A)(B) = [A,B]$$

$$\text{for} \quad A,B \in \underset{\sim}{G} \quad .$$

(Notice that it is the Jacobi identity which guarantees that Ad really is a Lie algebra representation.)

2. INNER AUTOMORPHISMS AND DERIVATIONS

Let $\underset{\sim}{G}$ be a Lie algebra. (In order to have everything said in this chapter hold rigorously, it will be necessary to assume that $\underset{\sim}{G}$ is a finite dimensional real or complex Lie algebra. However, the spirit of the ideas carries over to certain infinite dimensional situations--and as we shall see, Lie's own work often requires infinite dimensional Lie algebras.)

Definition. An automorphism of $\underset{\sim}{G}$ is a one-one linear map

$$\alpha: \underset{\sim}{G} \to \underset{\sim}{G}$$

such that

$$\alpha([A,B]) = [\alpha(A),\alpha(B)] \quad (2.1)$$

for $A,B \in \underset{\sim}{G}$.

The set of automorphisms form a group, called Aut $(\underset{\sim}{G})$. Consider a curve

$$t \to \alpha_t$$

in Aut $\underset{\sim}{G}$, such that:

$$\alpha_0 = \text{identity} .$$

It defines a <u>linear flow</u> on $\underset{\sim}{G}$. (See Chapter A and Volume III of IM.) The <u>infinitesimal generator</u> of the flow is the curve

$$t \to \gamma_t$$

in $L(\underset{\sim}{G})$ defined by the following formula:

$$\gamma_t = \frac{d}{dt}(\alpha_t)\alpha_t^{-1} \tag{2.2}$$

<u>Definition</u>. A <u>derivation</u> of $\underset{\sim}{G}$ is a linear map $\gamma: \underset{\sim}{G} \to \underset{\sim}{G}$ such that

$$\gamma([A,B]) = [\gamma(A),B] + [A,\gamma(B)] \tag{2.3}$$

for all $A,B \in \underset{\sim}{G}$

<u>Theorem 2.1</u>. A flow $t \to \alpha_t$ in $GL(\underset{\sim}{G})$ lies in $\mathrm{Aut}\,\underset{\sim}{G}$ if and only if its infinitesimal generator curve $t \to \gamma_t$ satisfies:

γ_t is a derivation of $\underset{\sim}{G}$ for each t.

<u>Proof</u>. Express the fact that each α_t is an automorphism:

$$\alpha_t([A,B]) = [\alpha_t(A), \alpha_t(B)] \ . \tag{2.4}$$

Differentiate each side of 2.4:

$$\frac{d}{dt}(\alpha_t)([A,B]) = \left[\frac{d}{dt}\alpha_t(A), \alpha_t(B)\right] + \left[\alpha_t(A), \frac{d}{dt}\alpha_t(B)\right] \tag{2.5}$$

Since this holds for <u>each</u> $A,B \in \underset{\sim}{G}$,

$$\frac{d}{dt}(\alpha_t)\alpha_t^{-1}([A,B]) = \frac{d}{dt}(\alpha_t)[\alpha_t^{-1}A, \alpha_t^{-1}B]$$

$$= \text{using 2.5,}$$

$$[\gamma_t(A),B] + [A,\gamma_t(B)] \quad ,$$

which proves one-half of Theorem 2.1.

To prove the other, converse half is left as an exercise, i.e., to prove that if $t \to \alpha_t$ is a one parameter family of convertible linear maps: $\underset{\sim}{G} \to \underset{\sim}{G}$ such that each

$$\alpha_t = \frac{d}{dt}(\alpha_t)\alpha_t^{-1}$$

is a derivation, then each α_t is an automorphism.

Now, the <u>derivations</u> of $\underset{\sim}{G}$ form a Lie algebra, with the bracket the <u>commutator</u>. This Lie algebra is denoted by

$$\text{Der}\,(\underset{\sim}{G}) \quad .$$

Recall the <u>adjoint representation</u>

$$\text{Ad}: \underset{\sim}{G} \to L(\underset{\sim}{G}) \quad .$$

The elements of $\text{Ad}\,(\underset{\sim}{G})$ are derivations; they are called <u>inner derivations</u>. In other words, an inner derivation is a map $\gamma: \underset{\sim}{G} \to \underset{\sim}{G}$ of the form

$$B \to \gamma(B) = [A,B]$$

for <u>some</u> $A \in \underset{\sim}{G}$.

Each derivation $\gamma \in \text{DER}(\underset{\sim}{G})$ determines an <u>automorphism</u>, denoted by

$$\exp(\gamma) \quad ,$$

and defined by the formula:

$$\exp(\gamma)(B) = B + \gamma(B) + \frac{\gamma^2(B)}{2!} + \cdots$$
$$= \sum_{j=0}^{\infty} \frac{\gamma^j(B)}{j!} \tag{2.6}$$

In particular, if this formula is applied to an inner derivation $\gamma = \text{Ad } A$, it takes the following form:

$$\exp(\text{Ad } A)(B) = B + [A,B] + \frac{1}{2}[A,[A,B]] + \cdots \tag{2.7}$$

<u>Definition</u>. The <u>inner automorphisms</u> of $\underset{\sim}{G}$ consist of the smallest subgroup of Aut (G) containing all automorphism of form 2.7, where A runs through all elements of $\underset{\sim}{G}$. This group is denoted by

$$\text{IN AUT}(\underset{\sim}{G})$$

<u>Remark</u>. One can prove--using Lie group theory--that IN AUT $(\underset{\sim}{G})$ is a <u>connected Lie group</u>, and that its Lie algebra is the subalgebra of DER $(\underset{\sim}{G})$ consisting of the inner automorphisms. It is also important to notice that not <u>every</u> $\gamma \in$ IN AUT $(\underset{\sim}{G})$ is of the form 2.6, since the

set of γ of form 2.6 does not necessarily form a group, (although it may do so.)

The action of IN AUT ($\underset{\sim}{G}$) on $\underset{\sim}{G}$ as a transformation group is very important for Lie's work. In particular, the following notion plays a key role:

Definition. Two Lie subalgebras

$$\underset{\sim}{H}, \underset{\sim}{H}' \subset \underset{\sim}{G}$$

are said to be conjugate or transformable into each other via inner automorphism if there is a $\alpha \in$ IN AUT ($\underset{\sim}{G}$) such that:

$$\underset{\sim}{H}' = \alpha(\underset{\sim}{H}) \tag{2.8}$$

3. SOLVABLE AND SEMISIMPLE LIE ALGEBRAS THE LEVI-MALCEV AND LIE-MOROSOV THEOREMS

Let $\underset{\sim}{G}$ be a Lie algebra. Set

$$\underset{\sim}{G}_1 = [\underset{\sim}{G}, \underset{\sim}{G}] ,$$

$\underset{\sim}{G}_1$ is an ideal of $\underset{\sim}{G}$. (This follows from the Jacobi identity.) It is called the first derived subalgebra. The quotient

$$\underset{\sim}{G}/\underset{\sim}{G}_1$$

is abelian.

Now, set:

$$G_2 = (G_1)_1 \equiv [[G,G],[G,G]]$$

It is also an ideal of G, called the <u>second derived subalgebra</u>.

Continue in this way to define a sequence

$$G \supset G_1 \supset G_2 \supset \cdots$$

of ideals.

<u>Definition</u>. G is a <u>solvable Lie algebra</u> if:

$$G_n = 0 \qquad \text{for } n \text{ sufficiently large.}$$

<u>Remark</u>. Lie calls such an algebra an <u>integrable</u> one. It was found to be the abstract algebraic situation which mirrored (in what we would now call a "functorial" sense) what the 19<u>th</u> century mathematicians called "<u>integrability of differential equations by quadratures</u>."

<u>Definition</u>. A Lie algebra G is said to be <u>semi-simple</u> if it has no non-zero solvable ideals.

General (finite dimensional) Lie algebras are built up from solvable and semisimple ones. Here is how.

The sum of two solvable ideas is again a solvable ideal. Hence the sum of <u>all</u> solvable ideals is a solvable ideal of G called the <u>radical</u> R. It may also be characterized as the unique <u>maximal solvable ideal</u>.

$\underset{\sim}{G}/\underset{\sim}{R}$ is <u>semisimple</u>. The <u>Levi Theorem</u> asserts that there is a semisimple <u>subalgebra</u> $\underset{\sim}{S}$ of $\underset{\sim}{G}$ such that the quotient map

$$\underset{\sim}{S} \to \underset{\sim}{G} \to \underset{\sim}{G}/\underset{\sim}{R}$$

of $\underset{\sim}{S} \to \underset{\sim}{G}/\underset{\sim}{R}$ is an isomorphism. (Alternatively,

$\underset{\sim}{G}$ = vector space direct sum of $\underset{\sim}{R}$ and $\underset{\sim}{S}$.)

The <u>Malcev Theorem</u> asserts that, for any semisimple subalgebra $\underset{\sim}{S}' \to \underset{\sim}{G}$, there is an inner automorphism $\alpha\colon \underset{\sim}{G} \to \underset{\sim}{G}$ such that

$$\alpha(\underset{\sim}{S}') \subset \underset{\sim}{S} \quad .$$

(Both the Levi and Malcev Theorems are now proved using Lie algebra cohomology theory. See VB, Vol. II, Chapter 3.)

<u>Theorem</u> (<u>Lie</u>). Let V be a vector space, and

$$\underset{\sim}{G} \subset L(V)$$

a solvable Lie subalgebra of $L(V)$. Suppose that V is a vector space whose scalar field is algebraically closed (e.g., the complex numbers). Then, there is a basis for V with respect to which the elements of $\underset{\sim}{G}$ are represented by matrices in <u>triangular form</u>. Alternatively, one may say that $L(V)$ has a maximal solvable subalgebra MS (namely, the Lie algebra of matrices in triangular form with respect to a given basis of V) such that any other solvable Lie subalgebra of $L(V)$ is transformable by an inner automorphism of $L(V)$ to a subalgebra of MS.

To add to the usefulness of Lie's theorem, here are some elementary properties of solvable Lie algebras.

Theorem. If $\mathbf{G}$ is a solvable Lie algebra, any subalgebra or quotient algebra of $\mathbf{G}$ is also solvable.

In particular, if $\rho\colon \mathbf{G} \to L(V)$ is a linear rep. of a solvable $\mathbf{G}$, then the Lie Theorem applies to $\rho(\mathbf{G})$.

The following generalization of Lie's Theorem is also very useful.

Theorem (_Morosov_). Let $\mathbf{H}$ be a semisimple Lie algebra over the complex numbers. Then, any two maximal solvable subalgebras of $\mathbf{H}$ are transformable into each other by an inner automorphism of $\mathbf{H}$.

Finally, here is another basic result:

Theorem (_Cartan_). Let $\mathbf{G}$ be a semisimple Lie algebra, $\gamma\colon \mathbf{G} \to \mathbf{G}$ a derivation. Then, γ is _inner_. In symbols,

$$\text{IN DER}\,(\mathbf{G}) = \text{DER}\,(\mathbf{G}) \quad .$$

4. NILPOTENT LIE ALGEBRAS AND THE ENGEL THEOREM

Let V be a vector space, and

$$\alpha: V \to V$$

a linear transformation. α is said to be <u>nilpotent</u> if:

$$\alpha^n = 0 \quad \text{for } n \text{ sufficiently large} \quad (3.1)$$

A subset of $L(V)$ is said to be <u>nilpotent</u> if every element of the subset is nilpotent.

<u>Definition</u>. Let $\underset{\sim}{G}$ be a Lie algebra,

$$\rho: \underset{\sim}{G} \to L(V)$$

a linear representation. ρ is said to be <u>nilpotent</u> if $\rho(\underset{\sim}{G})$ is a nilpotent subset of $L(V)$.

<u>Definition</u>. Let $\underset{\sim}{G}$ be a Lie algebra. $\underset{\sim}{G}$ is said to be <u>nilpotent</u> if the adjoint representation is nilpotent.

<u>Theorem 4.1</u>. If $\underset{\sim}{G}$ is nilpotent then it is solvable. If $\underset{\sim}{G}$ is solvable then $\underset{\sim}{G}_1 = [\underset{\sim}{G},\underset{\sim}{G}]$, the first derived algebra, is nilpotent.

<u>Theorem 4.2</u> (<u>Engel</u>). If $\rho: \underset{\sim}{G} \to L(V)$ is a linear representation such that each linear map $A \in \rho(\underset{\sim}{G})$ is nilpotent, then $\underset{\sim}{G}$ is nilpotent as a Lie algebra.

5. THE KILLING FORM AND THE CARTAN CRITERION

Let $\underset{\sim}{G}$ be a finite dimensional Lie algebra, (with a field K, of zero characteristic, as field of scalars.) For $A,B \in \underset{\sim}{G}$, consider

$$(\text{Ad } A)(\text{Ad } B)$$

It is a linear map: $\underset{\sim}{G} \to \underset{\sim}{G}$, hence has a <u>trace</u>, as does any linear map. (Recall that the trace is the sum of the diagonal elements in a matrix representation of the linear map.) Set:

$$\beta(A,B) = \text{trace }((\text{Ad } A)(\text{Ad } B)) \tag{5.1}$$

It is readily provable that β so defined is a symmetric, bilinear form

$$\underset{\sim}{G} \times \underset{\sim}{G} \to K \quad ,$$

called the <u>Killing form</u>. It is a key tool in the algebraic study of Lie algebras.

<u>Theorem</u> (<u>Cartan</u>). $\underset{\sim}{G}$ is semisimple if and only if the Killing form is non-degenerate, i.e., iff:

$$\beta(A,\underset{\sim}{G}) = 0 \qquad \text{implies} \quad A = 0 \quad .$$

If β is <u>identically</u> zero, then $\underset{\sim}{G}$ is solvable.

If $\underset{\sim}{G}$ is nilpotent, then β is identically zero.

A Lie algebra is said to be simple if it contains no ideals at all, except for (0) and the algebra itself. The following result is proved using these properties of the Killing form.

Theorem. $\underset{\sim}{G}$ is semisimple if and only if it is the direct sum of its simple ideals.

6. LIE ALGEBRA COHOMOLOGY AND THE J. H. C. WHITEHEAD THEOREMS

Let $\underset{\sim}{G}$ be a Lie algebra (not necessarily finite dimensional) and let

$$\rho: \underset{\sim}{G} \to L(V)$$

be a linear representation. It will be convenient to suppress explicit mention of ρ, and write

$$A(v) \equiv \rho(A)(v) \qquad (6.1)$$

$$\text{for} \quad A \in \underset{\sim}{G}, \quad v \in V \quad .$$

For non-negative integer n, we define three vector spaces, denoted by

$$C^n(\underset{\sim}{G},\rho), \quad Z^n(\underset{\sim}{G},\rho), \quad B^n(\underset{\sim}{G},\rho)$$

such that

$$B^n \subset Z^n \subset C^n \quad .$$

The elements of C^n are called the n-cochains; the elements of Z^n are called the n-cocycles; and the elements of B^n are called the n-coboundaries.

The representation ρ defines representations of $\underset{\sim}{G}$ on each C^n, Z^n, B^n, denoted in a way similar to 6.1, with suppression of explicit use of ρ.

There will also be linear maps defined:

$$d: C^n \to C^{n+1} \quad ,$$

called the coboundary map, such that:

$$Z^n = \text{kernel } d$$

$$B^{n+1} = d(C^n)$$

$$d(d) = 0$$

Once these objects are defined, we will define $H^n(\underset{\sim}{G},\rho)$ as follows:

$$H^n(\underset{\sim}{G},\rho) = Z^n/B^n \quad .$$

H^n is called the n-th cohomology group of $\underset{\sim}{G}$, with coefficients in ρ.

Now, to define these objects, for each n. In fact, we shall only do so for

$$n = 0,1,2 \quad .$$

One can proceed to higher values of n by induction on n, as described in my paper "Analytic construction of group representations" and in VB, Vol. II, Chapter 3.

First, $n = 0,1$:

$$C^0 = V, \quad B^0 = (0)$$

$C^1 = L(\underset{\sim}{G},V)$ space of linear maps $\theta: \underset{\sim}{G} \to V$

$d: C^0 \to C^1$ is defined as follows:

$$d(v)(A) = A(v)$$

for $v \in V$, $A \in \underset{\sim}{G}$

$Z^0 = \text{kernel } d \subset C^0$,

$= \text{set of } v \in V \text{ such that } \underset{\sim}{G}(v) = 0$,

i.e., Z^0 consists of the invariants of $\underset{\sim}{G}$.

Hence,

$$H^0 = Z^0 .$$

Next, $n = 1,2$:

$C^2 = L(\underset{\sim}{G} \wedge \underset{\sim}{G},V) \equiv$ skew-symetric, bilinear maps:

$\underset{\sim}{G} \times \underset{\sim}{G} \to V$.

$d: C^1 \to C^2$ is defined as follows:

$$d(\theta)(A,B) = A(\theta(B)) - B(\theta(A)) - \theta([A,B])$$

for $A,B \in \underset{\sim}{G}$, $\theta \in C^1$.

$Z^1 = \text{kernel } (d) \subset C^1$.

$B^1 = d(C^0)$

$H^1(\underset{\sim}{G},\rho) = Z^1/B^1$.

Finally, for $n = 2$:

$$C^3(\underset{\sim}{G},V) = L(\underset{\sim}{G} \wedge \underset{\sim}{G} \wedge \underset{\sim}{G},V)$$

$$\equiv \text{ trilinear, skew-symetric maps: } \underset{\sim}{G} \times \underset{\sim}{G} \times \underset{\sim}{G} \to V \quad .$$

$d: C^2(\underset{\sim}{G},V) \to C^3(\underset{\sim}{G},V)$ is defined as follows:

$$\begin{aligned} d(\theta)(A_1,A_2,A_3) &= A_1(\theta(A_2,A_3)) - A_2(\theta(A_1,A_3)) \\ &+ A_3(\theta(A_1,A_2)) - \theta([A_1,A_2],A_3) \\ &+ \theta(A_1,[A_2,A_3]) - \theta(A_2,[A_1,A_3] \end{aligned}$$

$$Z^2 = \text{kernel } d$$

$$B^2 = d(C^1)$$

$$H^2(\underset{\sim}{G},\rho) \equiv Z^2/B^2 \quad .$$

Theorem 6.1 (J.H.C. Whitehead). If $\underset{\sim}{G}$ is finite dimensional and semisimple, and if also V is finite dimensional, then

$$H^1(\underset{\sim}{G},\rho) = (0) = H^2(\underset{\sim}{G},\rho) \quad .$$

Theorem 6.2. Let $\rho: \underset{\sim}{G} \to L(V)$ be a linear representation, and let

$$V' \subset V$$

be a linear subspace such that

$$\rho(\underset{\sim}{G})(V') \subset V' \quad .$$

Let $V'' = V/V'$, ρ'' = quotient representation of $\underset{\sim}{G}$ in V''. If V is finite dimensional, and if

$$H^1(\underset{\sim}{G},\rho'') = 0 ,$$

then ρ is equivalent to the direct sum of ρ'' and another representation.

Putting together both results, proves the following:

Theorem 6.3 (H. Weyl). If $\underset{\sim}{G}$ is finite dimensional and semisimple, and if $\rho: \underset{\sim}{G} \to L(V)$ is a representation in a finite dimensional vector space V, then ρ is completely reducible, i.e., is a direct sum of irreducible representation.

7. DEFORMATIONS OF LIE ALGEBRA HOMOMORPHISMS

Now we turn to material that is not really part of the standard Lie algebra repetoire, but is very important and useful in all sorts of applications.

Let $\underset{\sim}{G}$ and $\underset{\sim}{L}$ be Lie algebras. For simplicity, we deal with the case where $\underset{\sim}{G}$ and $\underset{\sim}{L}$ are finite dimensional. However, the ideas carry over, to a certain extent, to the infinite dimensional cases. Many of Lie's results can be interpreted as Lie algebra and Lie group deformation theory in an infinite dimensional setting.

Let

$$\text{Hom}\ (\underset{\sim}{G},\underset{\sim}{L})$$

denote the space of Lie algebra homomorphism mappings

$$\phi\colon \underset{\sim}{G} \to \underset{\sim}{L} \ .$$

Let

$$L = \text{group of inner automorphisms of } \underset{\sim}{L}.$$

L acts as a transformation group on $\text{Hom}\ (\underset{\sim}{G},\underset{\sim}{L})$:

$$(\ell\phi)(A) = \ell(\phi(A))$$

$$\text{for } A \in \underset{\sim}{G},\quad \phi \in \text{Hom}\ (\underset{\sim}{G},\underset{\sim}{L}) \ .$$

A major algebraic problem is:

<u>Find the orbits of</u> L <u>on</u> $\text{Hom}\ (\underset{\sim}{G},\underset{\sim}{L})$.

For, two homomorphisms lie on the same orbit of L if and only if they are <u>transformable into each other by an inner automorphism</u>. (One also says, <u>are conjugate by an inner automorphism</u>.)

Now,

$$\text{Hom}\ (\underset{\sim}{G},\underset{\sim}{L})$$

is a subset of $L(\underset{\sim}{G},\underset{\sim}{L})$, which is a manifold. In fact, it is an algebraic variety in the vector space $L(\underset{\sim}{G},\underset{\sim}{L})$, since the condition that a linear map from the vector space $\underset{\sim}{G}$ to the vector space $\underset{\sim}{L}$ is a homomorphism is defined by algebraic conditions on the matrix coordinates of linear maps.

For a brief introduction to the ideas of algebraic geometry, see Volume VIII of IM. Notice we are working with the real or complex numbers as scalar field. Recall that by an algebraic variety I mean any subset which is defined by polynomial equations, whereas in the literature of algebraic geometry they often reserve this term for one which has the "irreducibility" property.

We can define the tangent space to Hom $(\underset{\sim}{G},\underset{\sim}{L})$ at a point ϕ as usual, as the collection of tangent vectors to $L(\underset{\sim}{G},\underset{\sim}{L})$ at ϕ such that:

> The curve
>
> $$t \to \phi + t\theta$$
>
> satisfies the equation defining Hom $(\underset{\sim}{G},L)$ up to the first order.

(Here, $\theta \in L(\underset{\sim}{G},\underset{\sim}{L})$, i.e., θ is a linear map $\underset{\sim}{G} \to L$. As the space of linear maps is a vector space, it is identified with its tangent space.) The condition for this is then that:

$$0 = \frac{d}{dt} (\phi+t\theta)([A,B]) - [(\phi+t\theta)(A),(\phi+t\theta)(B)]\Big|_{t=0} \quad \text{for } A,B \in \underset{\sim}{G} \qquad (7.1)$$

Working out condition 7.1, we can put it into the following equivalent form:

$$0 = \theta([A,B]) - [\theta(A),\phi(B)] - [\phi(A),\theta(B)]$$

or:

$$\rho_\phi(A)(\theta(B)) - \rho(B)(\theta(A)) - \theta([A,B]) = 0 \tag{7.2}$$

where:

$$\rho_\phi(A)(C) = [\phi(A),C] \tag{7.3}$$

for $A \in \underset{\sim}{G}$, $C \in \underset{\sim}{L}$.

Notice that:

ρ_ϕ defines a linear representation

$\underset{\sim}{G} \to L(\underset{\sim}{L})$.

θ defines a 1-cochain of $\underset{\sim}{G}$; with coefficients determined by representation, i.e., an element of $C^1(\underset{\sim}{G},\rho_\phi)$.

Condition 7.2 means that:

$\theta \in Z^1(\underset{\sim}{G},\rho_\phi)$, i.e., its coboundary is zero.

Hence, we have proved:

<u>Theorem 7.1</u>. The tangent space to $\text{Hom}\ (\underset{\sim}{G},\underset{\sim}{L})$ at a point $\phi \in \text{Hom}\ (\underset{\sim}{G},\underset{\sim}{L})$ is

$$Z^1(\underset{\sim}{G},\rho_\phi) \quad ,$$

the space of 1-cocycles defined by the representation ρ_ϕ.

We may think of this result as giving geometric interpretation to the space of 1-cocycles. What is the geometric interpretation of its quotient space, the cohomology group $H^1(\underset{\sim}{G},\rho_\phi)$?

To provide such a geometric interpretation, consider a curve

$$t \to \phi_t$$

in Hom $(\underset{\sim}{G},\underset{\sim}{L})$, with:

$$\phi_0 = \phi \quad ,$$

Such a curve defines a <u>one-parameter deformation of</u> ϕ, <u>with the set of Lie algebra homomorphisms</u>.

Consider two such one-parameter deformations,

$$t \to \phi_t ; \quad \phi_0 = \phi$$

$$t \to \phi'_t ; \quad \phi'_0 = \phi' \quad .$$

They are said to be <u>equivalent</u> (up to <u>inner automorphism</u>) if there is a curve $t \to \ell(t) \in L$ ($\equiv$ group of inner automorphisms of $\underset{\sim}{L}$) such that:

$$\phi'_t = \ell(t)(\phi'_t) \tag{7.4}$$
for all t .

Let us work out the conditions that 7.4 hold. Set,

$$\theta_t = \frac{d}{dt}\phi_t \in L(\underset{\sim}{G},\underset{\sim}{L}) \equiv C^1(\underset{\sim}{G},\rho_{\phi_t})$$

$$\theta'_t = \frac{d}{dt}\phi'_t \in L(\underset{\sim}{G},\underset{\sim}{L}) \equiv C^1(\underset{\sim}{G},\rho_{\phi'_t}) \quad .$$

θ_t and θ'_t are, we know, 1-cocycles with respect to the representations $\rho_{\phi'_t}, \rho_{\phi'_t}$.

Set:

$$C(t) = \frac{d}{dt}(\ell(t))\ell(t)^{-1} \tag{7.5}$$

Now, the right hand side of 7.5 is the infinitesimal generator of the curve $t \to \ell(t)$ in L. We know this may be identified with a curve in $\underset{\sim}{L}$, since:

$$L = \text{group of inner automorphisms of } \underset{\sim}{L}.$$

We can differentiate 7.4 to obtain a basic relation:

$$\theta'_t = \frac{d\ell(t)}{dt}\ell(t)^{-1}\phi'_t + \ell(t)(\theta_t) \tag{7.6}$$

This can now be written as follows:

$$\theta'_t(A) = [C(t), \phi'_t(A)] + \ell(t)(\theta_t)(A)$$

$$\text{for } A \in \underset{\sim}{G},$$

or

$$\theta'_t(A) = -\rho_{\phi'_t}(A)(C(t)) + \ell(t)(\theta_t)(A) \tag{7.7}$$

We can write this in cohomological form. Consider $C(t)$ as a 0-cochain, i.e.,

$$C(t) \in C^0(\underset{\sim}{G}, \rho_{\phi'_t}) \tag{7.8}$$

Then, 7.7 is equivalent to the following formula:

$$\theta'_t = -dC(t) + \ell(t)(\theta_t) \tag{7.9}$$

(Of course, it has to be understood that "d" on the right hand side of 7.9 refers to the coboundary operation with respect to the representation $\rho_{\phi_t'}$.) Then, we see that:

> The cohomology class in $H^1(\underset{\sim}{G},\rho_{\phi_t})$ to which θ_t, the 1-cocycle θ_t belongs, is an <u>invariant</u> of the equivalence relation among 1-parameter deformations.

In certain situations, one can show that the cohomology groups <u>parameterize</u> the 1-parameter deformations. I will not go any further into this general aspect of the theory at this point--I hope I have done enough to convince the reader that Lie algebra cohomology is the correct algebraic tool to study deformation classes of homomorphisms.

<u>Remark</u>. There seems to be a general principle here, "cohomology" of an algebraic structure serves to "parameterize" the "deformations" of that algebraic structure. Presumably, this should be studied in the context of category theory and the theory of general algebraic structures.

I will now turn to a more specific computational way of looking at these matters, in terms of power series.

8. DEFORMATIONS OF LIE ALGEBRA HOMOMORPHISMS IN THE CASE OF VANISHING COHOMOLOGY

Let $\underset{\sim}{G}$ and $\underset{\sim}{L}$ be Lie algebras. Let

$$t \to \phi_t : \underset{\sim}{G} \to \underset{\sim}{L}$$

be a one-parameter family of Lie algebra homomorphisms.

Let

$$\rho(A)(C) = [\phi_0(A), C] \quad \text{for } A \in \underset{\sim}{G}, \; C \in \underset{\sim}{L} \tag{8.1}$$

be the representation

$$\underset{\sim}{G} \to L(\underset{\sim}{L})$$

defined by formula 8.1.

Set:

$$\phi_t = \sum_{j=0}^{\infty} \theta_j t^j \tag{8.2}$$

<u>Remark</u>. By = in 8.2, I mean that the power series on the right hand side of 8.2 is the <u>Taylor series</u> of the function of t on the left hand side. Of course, if $t \to \phi_t$ depends <u>analytically</u> on the parameter t, this will mean equality in the function sense--but this is not necessarily required.

What are the θ_j on the right hand side of 8.2? Note that 8.2 means that:

$$\phi_t(A) = \sum_{j=0}^{\infty} \theta_j(A)t^j \tag{8.3}$$

for all $A \varepsilon \underset{\sim}{G}$.

Thus, each θ_j is a linear map

$\underset{\sim}{G} \to \underset{\sim}{L}$.

Interpret this as a 1-cochain of $\underset{\sim}{G}$, with coefficients determined by the representation ρ, i.e.,

$\theta_j \varepsilon C^1(\underset{\sim}{G},\rho)$

for $j = 0,1,2,...$

Of course,

$\theta_0 \equiv \phi_0$.

Let us determine the conditions imposed on the θ_j by the requirement that each ϕ_t be a homomorphism:

$$\phi_t([A,B]) = [\phi_t(A),\phi_t(B)] .$$

Then,

$$\sum_{i=0}^{\infty} \theta_i([A,B])t^i = \sum_{j,k=0}^{\infty} [\theta_j(A),\theta_k(B)]t^{j+k}$$

Here, the conditions can be written in the following form:

$$\theta_i([A,B]) = \sum_{j+k=i} [\theta_j(A),\theta_k(B)] \tag{8.4}$$

for $A,B \varepsilon \underset{\sim}{G}$, $i = 0,1,2,...$

Let us write down explicitly the first few of these conditions:

<u>For i = 0</u>:

$$\theta_0[A,B] = [\theta_0(A),\theta_0(B)] \quad ,$$

which is, of course, the condition that

$\theta_0 \equiv \phi_0$ is a Lie algebra homomorphism.

<u>For i = 1</u>:

$$\theta_1([A,B]) = [\phi_0(A),\theta_1(B)] + [\theta_1(A),\phi_0(B)]$$

$$= \rho(A)(\theta_1(B)) - \rho(B)(\theta_1(A))$$

Hence, condition 8.4 means, in this case, that:

$$d\theta_1 = 0 \tag{8.5}$$

where "d" is the coboundary operator of the cohomology defined by ρ, i.e., that:

$$\theta_1 \in A^1(\underset{\sim}{G},\rho) \quad . \tag{8.6}$$

<u>For i = 2</u>:

$$\theta_2([A,B]) = [\phi_0(A),\theta_2(B)] + [\theta_1(A),\theta_1(B)] + [\theta_2(A),\phi_0(B)]$$

$$= \rho(A)(\theta_2(B)) - \rho(B)(\tau_2(A)) + [\theta_1(A),\theta_1(B)] \quad .$$

We can write this as:

$$d\theta_2 = [\theta_1,\theta_1] \quad , \tag{8.6}$$

where $[\theta_1,\theta_1]$ is the 2-cochain defined as follows:

$$[\theta_1,\theta_1](A,B) = [\theta_1(A),\theta_1(A)] \quad .$$

<u>Remark</u>. The mapping

$$\theta_1 \to [\theta_1,\theta_1]$$

is an example of what topologists call a <u>cohomology operation</u>. It passes to the quotient to define an algebraic operation on <u>cohomology classes</u>.

Thus, the one-parameter deformation $t \to \phi_t$ of $\phi \equiv \phi_0$ may be represented by the sequence

$$\{\theta_j\} \equiv \{\theta_0,\theta_1,\theta_2,\ldots\}$$

of 1-cochains, satisfying relations 8.4. Let us examine how these cochains change when the deformation is replaced by one which is equivalent with respect to a one-parameter <u>group</u> of inner automorphisms of L. For $C \in \underset{\sim}{L}$, set:

$$\phi'_t(t) = \exp(\mathrm{Ad}\ tC)(\phi_t(t)) \tag{8.7}$$

$$= \sum_{j=0}^{\infty} \frac{(\mathrm{Ad}\ C)^j t^j}{j!} \phi_t(t)$$

$$= \sum_{j,k} \frac{(\mathrm{Ad}\ C)^j \theta_k}{j!} t^{j+k}$$

Set:

$$\phi'_t = \sum_i \theta'_i t^i \tag{8.8}$$

Comparing 8.7 and 8.8, we have:

$$\theta'_i = \sum_{j+k=i} \frac{(\mathrm{Ad}\ C)^j \theta_k}{j!} \tag{8.9}$$

Again, we can readily write out these relations for the first few values of i, and consider the cohomology interpretation of the resulting relations:

$$\theta'_0 = \theta_0 \ , \tag{8.10}$$

i.e., the initial homomorphism does not change. This is, of course, evident from 8.7.

<u>For $i = 1$</u>:

$$\begin{aligned} \theta'_1(A) &= [C,\phi(A)] + \theta_1(A) \\ &= -dC(A) + \theta_1(A) \end{aligned} \tag{8.11}$$

Here "C" means "interpret the element C of $\underset{\sim}{L}$ as an element of

$$C^0(\underset{\sim}{G},\rho) \ ,$$

and interpret

$$dC$$

as its image under the cohomology operation:

$$d\colon C^0(\underset{\sim}{G},\rho) \to C^1(\underset{\sim}{G},\rho) \ .$$

Then, we can write relation 8.11 in the following form:

$$\theta_1 - \theta_1' = dC , \tag{8.12}$$

i.e., θ_1 and θ_1' cobound, i.e., determine the same element of $H^1(\underset{\sim}{G},\rho)$.

Conversely, if θ_1 cobounds, i.e.,

$$\theta_1 = dC , \tag{8.13}$$

then we can define ϕ_t' by formula 8.7, and read off the following relations:

$$\phi_0' = \phi_0$$

$$\theta_1' = 0 ,$$

i.e., the Taylor series expansion of the deformation ϕ_t' contains no term of order t^2.

Theorem 8.1. Let $\underset{\sim}{G},\underset{\sim}{L}$ be Lie algebras and let $\phi\colon \underset{\sim}{G} \to \underset{\sim}{L}$ be a Lie algebra homomorphism. Let

$$\rho\colon \underset{\sim}{G} \to L(\underset{\sim}{L})$$

be the representation defined by the following formula:

$$\rho(A)(C) = [\phi(A),C]$$

$$\text{for } A \in \underset{\sim}{G}, \quad A \in \underset{\sim}{L} .$$

Suppose that:

$$H^1(\underset{\sim}{G},\rho) = 0 \tag{8.14}$$

Suppose that:

$$t \to \phi_t$$

is a one-parameter deformation of ϕ, i.e., a curve in Hom $(\underset{\sim}{G},\underset{\sim}{L})$ which is equal to ϕ for $t = 0$.

Then, there is a sequence of one parameter subgroup of L ($\equiv$ group of inner automorphisms of $\underset{\sim}{L}$) such that the deformation

$$\phi_t^{(0)} \equiv \ell_1(t)\ell_2(t) \cdots \ell_n(t)\phi_t$$

have the terms in their Taylor series vanish for $j=1,2,\ldots,n$.

Proof. We have seen above that we can choose C_1 so that

$$\phi_t' \equiv \exp\,(\mathrm{Ad}\; C_1)(\phi_t)$$

has vanishing first term in its Taylor series. Hence, its second order term

$$\theta_j' \in C^1(\underset{\sim}{G},\rho)$$

is a 2-cocycle, hence--by our hypothesis 8.14--cobounds, i.e., is the coboundary of $C_2 \in \underset{\sim}{L}$. Set

$$\phi_t^{(2)} = \exp\,(\mathrm{Ad}\; C_2)(\phi_t') \quad ,$$

which has terms $j = 1,2$ equal to zero in its Taylor series. Continue in this way for arbitrary n.

<u>Remarks</u>. a) Suppose that groups $\ell_1(t), \ell_2(t), \ldots$ can be chosen so that there is a curve $t \to \ell(t)$ in L so that:

$$\ell(t) = \lim_{n\to\infty} \ell_n(t) \tag{8.15}$$

Then set:

$$\phi_t^\infty = \ell(t)(\phi_t) \quad .$$

Since

$$\phi_t^\infty = \lim_{n\to\infty} \ell_n(t)(\phi_t) \quad ,$$

we know that <u>all</u> coefficients (beyond the constant term) in the Taylor series for ϕ_t^∞ vanish. In particular, <u>if all the data is analytic</u>, ϕ_t^∞ is constant in t, equal to ϕ. In particular:

> ϕ_t <u>is equivalent under</u> L <u>to</u> ϕ,
> <u>for each</u> t.

In words, <u>all homomorphism deformations of</u> ϕ <u>are trivial if</u> $H^1(\underset{\sim}{G}, \rho_\phi) = 0$.

The method we have chosen to prove this is direct but crude. More subtle differential and/or algebraic methods have been developed to handle this situation. For example, see the work of R. Richardson listed in the Bibliography.

b) Here is one typical situation to which this argument may be applied. Suppose that $\underset{\sim}{G}$ is a finite dimensional semisimple Lie algebra, L is any finite dimensional Lie algebra. We see then that, by the J.H.C. Whitehead Theorem quoted in a previous section,

$$H^1(\underset{\sim}{G},\rho) = 0$$

This means, geometrically, that a Lie algebra homomorphism

$$\phi: \underset{\sim}{G} \to \underset{\sim}{L}$$

is "rigid", in the sense that, for any other homomorphism

$$\phi': \underset{\sim}{G} \to \underset{\sim}{L}$$

which is <u>sufficiently close to</u> ϕ <u>in the vector space topology of</u> $L(\underset{\sim}{G},\underset{\sim}{L})$, there is an $\ell \in L \equiv$ group of inner automorphisms of $\underset{\sim}{L}$ such that:

$$\phi' = \ell(\phi') .$$

(In words, ϕ' is <u>equivalent to</u> or <u>congruent to</u> ϕ)

In particular, this may be applied to the case where:

V = finite dimensional vector space

$\underset{\sim}{L} = L(V)$

$\phi: \underset{\sim}{G} \to \underset{\sim}{L} \equiv L(V)$ is an irreducible linear representation of $\underset{\sim}{G}$.

This means that:

Any other linear representation of $\underset{\sim}{G}$ which is "close" to ϕ is equivalent to ϕ. In particular, this is a qualitative explanation why the equivalence classes of irreducible representations of semisimple Lie algebras form a discrete set. (In fact, they are parameterized by points on a "lattice", determined by their maximal weights.)

c) A typical situation encountered while interpreting Lie's work usually requires that $\underset{\sim}{L}$ be infinite dimensional. In this case, one will typically obtain by these arguments the result that deformations are trivial in the formal power series sense. One knows very little about when these formal power series can be made to converge. (A hint of an idea for the case where $\underset{\sim}{G}$ is finite dimensional and semisimple might be Guillemin and Sternberg's convergence proof [1] of a formal power series argument of mine [1], using the "unitary trick", passing from non-compact to compact groups by analytic continuation.)

Here is one important speculation. Suppose

$\underset{\sim}{G}$ = an abstract Lie algebra

$\underset{\sim}{L} = V(X) \equiv$ Lie algebra of C^∞ vector fields on a finite dimensional manifold X.

Then,

$$L = \text{group of diffeomorphisms of } X.$$

A homomorphism

$$\phi\colon \underset{\sim}{G} \to \underset{\sim}{L} \equiv V(X)$$

determines an action of $\underset{\sim}{G}$ on X, i.e., an "infinitesimal transformation group." Two such are said to be <u>equivalent</u> if they can be transformed into each other by a diffeomorphism of X. One of Lie's main problems (which he "solves" by brute force calculation for $\dim X = 1,2$, $\underset{\sim}{G}$ finite dimensional, in this paper) is to find these equivalence classes. (Usually, under the assumption that the problem is "local" in the sense that classification up to <u>local</u> diffeomorphism is required.)

Then, if

$$H^1(\underset{\sim}{G},\rho_\phi) = 0 \quad ,$$

where ϕ is a given homomorphism: $\underset{\sim}{G} \to \underset{\sim}{L}$, i.e., geometrically, a given infinitesimal transformation group action of $\underset{\sim}{G}$ on X, then ϕ is equivalent to any ϕ' which can be connected to ϕ via a curve in $\text{Hom}\,(\underset{\sim}{G},\underset{\sim}{L})$, <u>at least in the formal power series sense</u>. (Of course, much of 19<u>th</u> century analysis we would now characterize as "what one can do with formal power series.") When these formal power series can be replaced by actual diffeomorphisms (or local diffeomorphisms) is a key question to which I do not know the answer, at least in adequate generality.

More generally,

$$H^1(\underset{\sim}{G},\rho_\phi)$$

may be considered as the <u>tangent space</u> to the <u>space of deformation of a given homomorphism</u> ϕ. Its <u>dimension</u> may often be expected to coincide with the number of "independent parameters" found in Lie's calculations of equivalence classes of infinitesimal transformation group action.

9. DEFORMATIONS OF SUBALGEBRAS OF LIE ALGEBRAS

There are many interesting deformation problems connected with Lie algebra theory that are both mathematically interesting and that have important ramifications in geometry or physics. We have just discussed one--the deformation of homomorphism problem. Here is another.

Let $\underset{\sim}{L}$ be a Lie algebra, with L the group of inner automorphisms of $\underset{\sim}{L}$. Recall that a Lie subalgebra $\underset{\sim}{S}$ of $\underset{\sim}{L}$ is defined as a subset such that:

a) $\underset{\sim}{S}$ is a sub-vector space of $\underset{\sim}{L}$.

b) $[\underset{\sim}{S},\underset{\sim}{S}]$.

Let $\mathcal{S}$ denote the <u>set</u> of all such subalgebras.

Let L act on $\mathcal{S}$ in the natural way.

Two subalgebras $\underset{\sim}{S}_1,\underset{\sim}{S}_2 \in \mathcal{S}$ are <u>equivalent</u>

(with respect to L) if there is an element $\ell \in L$ such that

$$S_2 = \ell(S_1) \quad .$$

Most generally, the "deformation" problem is to parameterize the orbits of L acting on $\mathcal{S}$. More specifically, it may be thought of as the problem of deciding when two "smooth" curves

$$t \to S_1(t)$$

$$t \to S(t)$$

in $\mathcal{S}$ differ by a "smooth" curve in L.

There are certain complications due to the fact that $\mathcal{S}$ is not necessarily a manifold, or even an algebraic variety. One difficulty is that the dimension of the elements of $\mathcal{S}$ may vary, and the isomorphism class of the Lie algebra structure on S may also vary. I will not go into the full story at this point, but will only treat one-parameter deformations which involve <u>isomorphic</u> Lie algebra structures. In other words, we consider one-parameter formulas

$$t \to S(t) \subset L \quad ,$$

such that there is a Lie algebra G such that:

$$S(t) \text{ is isomorphic to } G \quad .$$

We suppose that this isomorphism can be chosen as

$$\phi_t: \underset{\sim}{G} \to \underset{\sim}{S}(t) \subset \underset{\sim}{L}$$

We also <u>suppose</u> that it depends smoothly on t.

Suppose

$$\phi'_t: \underset{\sim}{G} \to \underset{\sim}{S}'(t)$$

is another such family of isomorphic subalgebras. When are they equivalent under L? Such equivalence requires the existence of a curve

$$t \to \ell(t)$$

in L such that:

$$\ell(t)(\underset{\sim}{S}'(t)) = \underset{\sim}{S}(t)$$

for all t.

Define:

$$\alpha(t) = \phi_t^{-1}\ell(t)^{-1}\phi'_t$$

Then,

$\alpha(t)$ is a map: $\underset{\sim}{G} \to \underset{\sim}{G}$ which is a Lie algebra isomorphism.

Further,

$$\ell(t)\phi_t\alpha = \ell(t)\ell(t)^{-1}\phi'_t$$

$$= \phi'_t \ .$$

Here is how to interpret this formula. Let:

G = group of automorphisms of $\underset{\sim}{G}$.

Construct the direct product group

$$L \times G \quad .$$

Let it act on $\mathrm{Hom}\,(\underset{\sim}{G},\underset{\sim}{L})$ as follows:

$$(\ell,g)(\phi) = \ell\phi g^{-1} \tag{9.1}$$

for $(\ell,g) \in L \times G$, $\phi \in \mathrm{Hom}\,(\underset{\sim}{G},\underset{\sim}{L})$.

Here is the problem:

Given two curves

$$t \to \phi_t, \phi'_t$$

in $\mathrm{Hom}\,(\underset{\sim}{G},\underset{\sim}{L})$, find the "invariants" required for equivalence, i.e., the existence of a curve $t \to \ell(t), g(t)$ in L and G such that

$$\phi'_t = \ell(t)\phi_t g(t)^{-1} \tag{9.2}$$

Remark. In this form, we clearly have encountered a very general problem of differential geometry. Let H be a group, M a manifold on which H acts as a transformation group. The problem: Given two curves $t \to p(t), p'(t)$ in M, find the "invariants" that determine when there is a curve $t \to h(t)$ in H such that:

$$p'(t) = h(t)p(t) \quad .$$

Much of Lie's geometric work (to be covered partially in later volumes) invokes special cases of this problem in quite different circumstances.

Return to the study of deformations of Lie subalgebras of $\underset{\sim}{L}$. Given two curves

$$t \to \phi_t$$

$$t \to \phi'_t$$

in $\text{Hom}\,(\underset{\sim}{G},\underset{\sim}{L})$ with

$$\phi_0 = \phi = \phi'_0 \quad ,$$

we know from previous work how to write down a sequence of 1-cochains $\theta_1,\theta_2,\ldots,\theta'_1,\theta'_2,\ldots \in C^1(\underset{\sim}{G},\rho)$, where $\rho\colon \underset{\sim}{G} \to L(\underset{\sim}{L})$ is the representation defined as follows:

$$\rho(A)(C) = [\phi(A),C]$$

for $A \in \underset{\sim}{G}$, $C \in \underset{\sim}{L}$.

In fact, these 1-cochains are just the coefficients in the Taylor series expansion:

$$\phi_t = \phi + \sum_{j=1}^{\infty} \theta_j t^j \tag{9.3}$$

$$\phi'_t = \phi + \sum_{j=1}^{\infty} \theta'_j t^j \tag{9.4}$$

We can then find the relations between the (θ_j) and the (θ'_j) by substituting 9.2 into this relation. Alternately, we can differentiate 9.2 repeatedly with respect to t. For variety, I will choose the latter approach.

Set:

$$\begin{aligned} \theta_1(t) &= \frac{d}{dt}\phi_t \\ \theta'_1(t) &= \frac{d}{dt}\phi'_t \\ \theta_2(t) &= \frac{1}{2!}\frac{d^2}{dt^2}\phi_t \\ \theta'_2(t) &= \frac{1}{2!}\frac{d^2}{dt^2}\phi'_t \end{aligned} \tag{9.5}$$

Set:

$$\frac{d\ell}{dt}\ell(t)^{-1} = \beta(t) \tag{9.6}$$

$$\gamma(t) = \frac{dg}{dt}g(t)^{-1} \tag{9.7}$$

$t \to \beta(t)$ and $t \to \gamma(t)$ are then the _infinitesimal generators_ of the flows $t \to \ell(t)$ and $t \to g(t)$. Let us now combine relations 9.2, 9.5, 9.6 and 9.7.

$$\theta_1' = \frac{d\ell}{dt}\phi_t g(t)^{-1} + \ell(t)\theta_1 g(t)^{-1} - \ell(t)\phi_t g(t)^{-1}\frac{dg}{dt}g(t)^{-1}$$

$$= \beta(t)\phi_t' + \ell(t)\theta_1 g(t)^{-1} - \phi_t'\gamma(t) \qquad (9.8)$$

Consider the consequences, say for:

$$t = 0 \quad .$$

We know that:

$$\theta_1(0), \theta_1'(0) \in Z^1(\underset{\sim}{G}, \rho) \quad .$$

Let us pass to the quotient under the map

$$Z^1 \to H^1(\underset{\sim}{G}, \rho) \quad .$$

<u>It can be proved</u> that the groups G and L act trivially on the cohomology classes.

Now,

$$\beta(0) = \text{Ad } C \quad ,$$

for some $C \in \underset{\sim}{L}$,

since L is the group of <u>inner</u> automorphisms of $\underset{\sim}{L}$.
Denote by

$$\overline{\theta_1(0)}, \overline{\theta_1'(0)} \in H^1(\underset{\sim}{G}, \rho)$$

the cohomology classes to which these cocycles belong.
Then, we have from 9.8:

$$\overline{\theta_1(0)} - \overline{\theta_1'(0)} = \overline{\phi\gamma(0)} \qquad (9.9)$$

__Remark.__ It follows __implicitly__ from this argument that

$$\phi\gamma(0) \quad ,$$

considered as an element of

$$L(\underset{\sim}{G},\underset{\sim}{L}) \equiv C^1(\underset{\sim}{G},\rho) \quad ,$$

is a cocycle. Let us prove it, in general. Suppose then that

$$\phi\colon \underset{\sim}{G} \to \underset{\sim}{L}$$

is an arbitrary Lie algebra homomorphism. Define $\rho\colon \underset{\sim}{G} \to L(\underset{\sim}{L})$ as usual:

$$\rho(A)(C) = [\phi(A),C] \quad .$$

Let $\gamma\colon \underset{\sim}{G} \to \underset{\sim}{G}$ be a __derivation__, i.e., γ satisfies the following relation:

$$\gamma([A,B]) = [\gamma(A),B] + [A,\gamma(B)] \quad .$$

Set:

$$\eta = \phi\gamma,$$

a linear map: $\underset{\sim}{G} \to \underset{\sim}{L}$, i.e., an element of

$$C^1(\underset{\sim}{G},\rho) \quad .$$

Let us compute the coboundary of η.

$$\begin{aligned} d\eta(A,B) &= \rho(A)(\eta(B)) - \rho(B)(\eta(A)) - \eta([A,B]) \\ &= [\phi(A),\phi\gamma(B)] - [\phi(B),\phi\gamma(A)] - \phi\gamma([A,B]) \\ &= \phi([A,\gamma(B)] - \gamma([A,B]) + [\gamma(A),B]) \\ &= 0 \quad , \end{aligned}$$

since γ is a derivation. This proves, independently, that η is a 1-cocycle. Its cohomology class defines a mapping

$$\gamma \to \bar{\eta} = \delta(\gamma)$$

of

$$\text{Der}\,(\underset{\sim}{G}) \to H^1(\underset{\sim}{G},\rho) \quad .$$

Let us examine what happens to this cohomology class if γ is an <u>inner</u> derivation, i.e., if there is an $A \in \underset{\sim}{G}$ such that

$$\gamma(B) = [A,B] \quad .$$

Then,

$$\begin{aligned} \eta(B) &= \phi([A,B]) \\ &= [\phi(A),\phi(B)] \\ &= -[\phi(B),\phi(A)] \\ &= -\rho(B)(\phi(A)) = d(\phi(A))(B) \quad , \end{aligned}$$

where $\phi(A)$ is considered as an element of $C^0(\underset{\sim}{G},\rho)$. In particular, we see that:

> The 1-cocycle η cobounds, if γ is an inner derivation of $\underset{\sim}{G}$.

We can restate what we have proved as follows:

> If all derivations of $\underset{\sim}{G}$ are inner (e.g., if $\underset{\sim}{G}$ is finite dimensional

and semisimple) then the map

$$\delta: \text{Der}\ (\underset{\sim}{G}) \to H^1(\underset{\sim}{G},\rho)$$

defined above is zero.

Return to our basic deformation problem. Given a curve

$$t \to \phi_t$$

in $\text{Hom}\ (\underset{\sim}{G},\underset{\sim}{L})$ with

$$\phi_0 = \phi \ ,$$

we see that the first invariant (or first obstruction, in topological language) is the image of

$$\overline{\theta}_1 \quad \text{in} \quad H^1(\underset{\sim}{G},\rho)/\delta(\text{Der}\ (\underset{\sim}{G})) \quad .$$

In particular, if

$$\delta(\text{Der}\ (\underset{\sim}{G})) = H^1(\underset{\sim}{G},\rho) \quad ,$$

the invariant vanishes for "a priori" reasons, and we expect "rigidity" of the deformation problem, i.e., that two Lie subalgebras

$$\underset{\sim}{S}_1,\underset{\sim}{S}_2 \subset \underset{\sim}{L}$$

of $\underset{\sim}{L}$ which are both isomorphic to $\underset{\sim}{G}$ and which are "sufficiently close" (say, in the natural topology of the Grassman manifolds of the vector space $\underset{\sim}{L}$) are equivalent under inner automorphism of $\underset{\sim}{L}$.

THEORY OF TRANSFORMATION GROUPS

By Sophus Lie

INTRODUCTION

In a series of papers, of which the following is the first,* I mean to develop a new theory, which I call the _theory of transformation groups_. The reader will note that these investigations, with which I have zealously occupied myself since 1873,** have many points of contact with several mathematical disciplines, especially with the theory of substitutions, with geometry and the modern theory of manifolds, and finally also with the theory of differential equations. At the end of this work I will enumerate all the investigations known to me which are more or less related to my theory of transformation groups.

This first paper is divided into two parts, of which the first provides the determination of all transformation groups of a 1-dimensional manifold, while the second determines all the groups of a 2-dimensional manifold. Later

* It is the only surviving one. [Ed. note.]

** Göttinger Nachrichten No. 22, 1874; Archiv for Mathematik og Naturvidenskab, vol. I, III, IV, Christiania 1876, 1878, 1879. [Collected Papers, vol. V, papers I-VI.] The first note cited contains a resume of all the results of this paper.

papers will treat on the one hand the general theory of an n-dimensional manifold, and, on the other hand, in connection with these theories they will develop new points of view for the general theory of differential equations.

Part I.

DETERMINATION OF ALL TRANSFORMATION GROUPS OF A 1-DIMENSIONAL MANIFOLD

Chapter 1

THE GENERAL PROBLEM

In this part I settle the following general problem.

Problem. To determine the most general function f of x and r parameters $a_1, a_2, \ldots, a_r$ satisfying an equation of the form

$$f(f(x,a_1,\ldots,a_r),b_1,\ldots,b_r) = f(x,c_1,\ldots,c_r) \qquad (1.1)$$

in which it is assumed that the c_i depend only on the a's and the b's.

This problem can perhaps be more clearly formulated by using the concept of a transformation group, which we now define.

Definition. A family of transformations

$$x' = f(x,a_1,\ldots,a_r) \quad ,$$

where x' denotes the original variable, x the new one, and the a_i parameters, forms a transformation group if the composition of two transformations of the family is a transformation of the family, i.e., when from the equations

$$x' = f(x, a_1, \ldots, a_r) \; ,$$

$$x'' = f(x', b_1, \ldots, b_r) \; ,$$

there follows

$$x'' = f(x, c_1, \ldots, c_r) \; ,$$

where the c_i are functions of the a's and b's alone.

As usual we consider the unknown function $f(x, a_1, \ldots, a_r)$ as a power series in x and the a_i, which converges on some domain of these quantities. Conseequently f is a differentiable function of its arguments. The form of the condition (1.1) imposes the obvious additional requirement that $x, a_1, \ldots, a_r$ can be so chosen in the domain of convergence of f that $f(x, a_1, \ldots, a_r)$ is in the projection on the first factor of the domain of convergence of f.

If the quantity $f(x, a_1, \ldots, a_r)$ satisfies one or more relations of the form

$$\sum_k \Phi_k(a_1, \ldots, a_r) \frac{\partial f}{\partial a_k} = 0 \; ,$$

then the number of parameters can be diminished. Indeed, if $\alpha_1, \ldots, \alpha_{r-1}$ are independent functions of the a_i satisfying the above partial differential equation, then f

can be put in the form

$$f(x,a_1,\ldots,a_r) = \phi(x,\alpha_1,\ldots,\alpha_{r-1})$$

Definition. A group with r parameters

$$x' = f(x,a_1,\ldots,a_r)$$

is called an r-term group if the number of the parameters cannot be diminished.

Using this terminology we can reformulate our problem as:

> Determine all r-term transformation groups of a 1-dimensional manifold.

From a given r-term group $x' = f(x,a_1,\ldots,a_r)$ one easily constructs new r-term groups. In fact, if ϕ is an arbitrary function, with ϕ^{-1} its inverse, then the equation

$$x' = \phi^{-1}[f(\phi(x),a_1,\ldots,a_r)] = F(x,a_1,\ldots,a_r)$$

again determines an r-term group, as one immediately verifies. Note that the given group can then be put in the form

$$x' = \phi[F(\phi^{-1}(x),a_1,\ldots,a_r)] ,$$

so that the relation between the two groups is symmetric.

Definition. Two groups

$$x' = f(x, a_1, \ldots, a_r) \quad ,$$

$$x' = F(x, a_1, \ldots, a_r)$$

are similar if the second can be put in the form

$$x' = \phi^{-1}[f(\phi(x), a_1, \ldots, a_r)] \quad ,$$

where ϕ and ϕ^{-1} are inverse functions, and hence also the first can be put in the form

$$x' = \phi[F(\phi^{-1}(x), a_1, \ldots, a_r)] \quad .$$

From this definition it follows at once that two groups which are similar to a third are similar to each other. Similar groups can be considered identical, if one wishes.

COMMENTS

Here is one suitably general formulation of Lie's general problem.

Let X *and* A *be manifolds. Suppose, given a mapping*

$$X \times A \to X \quad ,$$

denoted by

$$(x,a) \to f(x,a) \quad .$$

For fixed $a \varepsilon A$ *this defines a map*

$$\phi_a : x \to f(x,a) \equiv \phi_a(x) \quad .$$

Suppose also that the maps

$$\{\phi_a\}$$

form a <u>*transformation semigroup*</u> *on* X.
We shall say that such a structure is a <u>*parameterized transformation semigroup on*</u> X.

Of course, Lie makes "smoothness" assumptions on the map

$$X \times A \to X \quad .$$

From today's point of view, smoothness in the sense of C^∞ *manifold theory is one reasonable assumption. Lie uses local, real-analyticity. Various alternate possibilities involving algebra or analytic geomtric ideas may also be considered.*

In these comments, we shall call such an object a <u>*Lie transformation semigroup*</u> *on* X.

A precise definition of what is meant by <u>*r-term*</u> *will be available when we discuss the notion of the* <u>*infinitesimal transformation*</u>.

Chapter 2

THE INFINITESIMAL TRANSFORMATIONS OF A GROUP

2.1 AN EXAMPLE OF A TRANSFORMATION GROUP ON R

The simplest example of a group is the so-called linear group, defined by the equation

$$x' = \frac{x + a_1}{a_2 x + a_3} \tag{2.1.1}$$

That this equation does determine a group is verified by writing down the second equation

$$x'' = \frac{x' + \alpha_1}{\alpha_2 x' + \alpha_3}$$

and then expressing x'' in terms of x:

$$x'' = \frac{x + \dfrac{a_1 + \alpha_1 a_3}{1 + \alpha_2 a_2}}{\dfrac{\alpha_2 + \alpha_3 a_2}{1 + \alpha_1 a_2} x + \dfrac{\alpha_2 a_1 + \alpha_3 a_3}{1 + \alpha_1 a_2}}$$

showing that x'' is a linear-fractional function of x.

If one puts

$$\alpha_1 = \frac{-a_1}{a_3}, \qquad \alpha_2 = \frac{-a_2}{a_3}, \qquad \alpha_3 = \frac{1}{a_3},$$

then $x'' = x$. From this we see that the composition of the two linear transformations

$$(a_1, a_2, a_3) \quad \text{and} \quad \left(\frac{-a_1}{a_3}, \frac{-a_2}{a_3}, \frac{1}{a_3}\right)$$

is the identity transformation $x'' = x$. Two transformations related in this symmetric way are said to be <u>inverse</u> transformations.

If one puts in (2.1.1),

$$a_1 = 0, \quad a_2 = 0, \quad a_3 = 1,$$

then one obtains the identity transformation $x' = x$, which thus belongs to the linear group. On the other hand, if

$$a_1 = \varepsilon_1, \quad a_2 = \varepsilon_2, \quad a_3 = 1 + \varepsilon_3,$$

where $\varepsilon_1, \varepsilon_2, \varepsilon_3$ are independent infinitesimals, then, by throwing out the infinitesimals of second order, one obtains the infinitesimal transformation

$$x' = x + \varepsilon_1 - \varepsilon_3 x - \varepsilon_2 x^2 .$$

In particular, one obtains, by appropriate choice of the quantities ε_i, the three infinitesimal transformations

$$\begin{aligned} x' &= x + \lambda_1 & \text{or:} \quad \delta x &= \lambda_1 \\ x' &= x + \lambda_2 x & \delta x &= \lambda_2 x \\ x'' &= x + \lambda_3 x^2 & \delta x &= \lambda_3 x^2 \end{aligned} \tag{2.1.2}$$

of which all the other infinitesimal transformations of the group are linear combinations.

The linear group (2.1.1) thus contains the identity transformation and three independent infinitesimal transformations. The finite transformations of the group are ordered into pairs of inverse transformations.

COMMENTS

A map of form 2.1.1 is called a linear fractional transformation. It is not literally a mapping of $R \to R$*, since the formula 2.1.1 shows that for certain values of* x, $x' = \infty$. *As explained in Chapter B, Section 8, it can be considered either as a local Lie transformation group or as a group of birational transformations on* R, *in the sense of algebraic geometry.*

Whatever the precise interpretation, formulas 2.1.1 define three vector fields A_1, A_2, A_3, *on* R:

$$A_1 = \frac{\partial}{\partial x} \; ; \quad A_2 = x \frac{\partial}{\partial x} \; ; \quad A_3 = x^2 \frac{\partial}{\partial x} \; . \qquad (2.1.3)$$

The infinitesimal transformations corresponding to the Lie semigroup action 2.1.1 are then linear combinations of these three vector fields.

The assignment of an "infinitesimal transformation group" to a "transformation group" is a basic element in Lie's theory. I will now present a modern way of looking at this for the reasonably general case of a <u>*Lie transformation group acting on a manifold*</u> X. *(See Chapter B for definitions of these concepts.) In order to keep to the notations used in the text, I will denote the Lie group by*

$$A \quad .$$

Thus, the transformation group action is defined by a map

$$X \times A \to X \quad ,$$

determined by formulas of the type:

$$(x,a) \to f(x,a) \; = \; x' \quad .$$

Another notation we use is:

$$\phi_a(x) \; = \; f(x,a) \quad .$$

For fixed a,

$$\phi_a\colon X \to X$$

is a diffeomorphism.

The group law is denoted by

$$(a_1,a_2) \to a_1 a_2 \quad .$$

Thus, the "transformation group" property means that

$$\phi_{a_1 a_2} \; = \; \phi_{a_1}\phi_{a_2}$$

$$\text{for } a_1,a_2 \in A \quad .$$

We are assuming that A *forms a* <u>*group*</u>, *hence contains the identity element, which we denote by* 1, *and inverses, which we denote by* a^{-1}.

Let

$$A_1$$

denote the tangent vector space to A *and the unit element* 1.

Given a tangent vector $v \in A_1$, *we can define a vector field*

$$A_v \in V(X) \quad ,$$

as follows:

$$\begin{aligned} A_v(x) \quad &= \quad \textit{tangent vector at} \quad t=0 \\ &\textit{of the curve} \quad t \to \phi_{a(t)}(x), \end{aligned} \tag{2.1.4}$$

where $t \to a(t)$ *is any curve in* A *such that:*

$$a(0) = 1$$

Tangent vector to $a(t)$ *at* $t=0$ *is* v.

If A *is a vector space, we can write 2.1.4 in the following more concrete form:*

$$\begin{aligned} A_v(x) &= \frac{\partial}{\partial t} f(x,1+tv)\Big|_{t=0} \\ &\equiv \frac{\partial}{\partial t} (\phi_{1+tv}(x))\Big|_{t=0} \end{aligned} \tag{2.1.5}$$

In words, A_v *is the* *<u>infinitesimal transformation</u>* *of the flows in* X *generated by all curves* $t \to a(t)$ *in the parameter space* A *with the same tangent vector* v *and* $t=0$.

We can also write 2.1.5 somewhat symbolically as:

$$A_V(x) = \lim_{\varepsilon \to 0} \frac{(f(x,a_0+\varepsilon v) - x}{\varepsilon}$$

or

$$f(x,a_0+\varepsilon v) \approx x + \varepsilon A_V(x) \quad ,$$

or

$$f(x,a_0+\varepsilon v) = x + \varepsilon A_V(x) + \textit{(higher order terms in } \varepsilon\textit{)}$$

Of course, the transformations 2.1.1 form a group, i.e., contain inverses. Lie's terminology "linear group", is now replaced by "the groups of linear fractional transformations" or the "projective group."

2.2 THE INFINITESIMAL TRANSFORMATIONS

In the theory of permutations it is shown that the elements of a permutation group can be ordered into pairs of elements, each the inverse of the other. Now, since the distinction between a permutation group and a transformation group lies in the fact that the former contains a finite and the latter an infinite number of operations, it is natural to conjecture that the transformations of a transformation group also are ordered into pairs of inverse transformations. In previous works I came to the conclusion that this is actually the case. I now add explicitly to the definition of a transformation group the requirement that the transformations of the group can be ordered into pairs of inverse transformations. In any case, I conjecture that this is a necessary consequence of my original definition of a transformation group. However, it has been impossible for me to prove this in general.

We therefore restrict ourselves to groups

$$x' = f(x, a_1, \ldots, a_r)$$

having the property that to each set of values $a_1, \ldots, a_r$ there is another set of values $\alpha_1, \ldots, \alpha_r$ such that the equation

$$f(f(x, a_1, \ldots, a_r), \alpha_1, \ldots, \alpha_r) = x$$

holds identically.

From this it follows first of all that the groups we consider always contain the identity transformation. We will now show that they also contain infinitesimal transformations.

Let $a_1, \ldots, a_r$ and $\alpha_1, \ldots, \alpha_r$ be the parameters of two inverse transformations of a group, and let $\omega_1, \ldots, \omega_r$ be independent infinitesimals. Then the expression

$$f(f(x, a_1, \ldots, a_r), \alpha_1 + \omega_1, \ldots, \alpha_r + \omega_r)$$

can, by throwing out the second-order infinitesimals, be put in the form

$$f(f(x,a_1,\ldots,a_r),\alpha_1,\ldots,\alpha_r)$$

$$+ \sum_k \omega_k \left[\frac{\partial f(f(x,a_1,\ldots,a_r),\beta_1,\ldots,\beta_r)}{\partial \beta_k}\right]_{(\beta_k=\alpha_k)}$$

and therefore our group contains the transformation

$$x' = x + \sum_k \omega_k \left[\frac{\partial f(f(x,a_1,\ldots,a_r),\beta_1,\ldots,\beta_r)}{\partial \beta_k}\right]_{(\beta_k=\alpha_k)} ,$$

which is evidently an infinitesimal transformation. If one gives the independent infinitesimals ω_i distinct values, successively, then one obtains ∞^{r-1} distinct infinitesimal transformations, which are linear combinations of the r infinitesimal transformations

$$\delta x = \omega_k \left[\frac{\partial f(f(x,a_1,\ldots,a_r),\beta_1,\ldots,\beta_r)}{\partial \beta_k}\right]_{(\beta_k=\alpha_k)}$$

$$(k = 1,\ldots,r).$$

We shall show that these r infinitesimal transformations are independent in the sense that there is no linear relation of the form

$$\sum_k \phi_k(a_1,\ldots,a_r) \left[\frac{\partial f(f(x,a_1,\ldots,a_r),\beta_1,\ldots,\beta_r)}{\partial \beta_k}\right]_{(\beta_k=\alpha_k)} = 0$$

For if one considers the quantities $f,\alpha_1,\ldots,\alpha_r$ as independent variables, then such a relation would take the form

$$\sum_k \frac{\partial f(f,\alpha_1,\ldots,\alpha_r)}{\partial \alpha_k} \psi_k(\alpha_1,\ldots,\alpha_r) = 0 \quad ,$$

from which it would follow that the number of parameters of our group could be diminished.

We thus obtain the following important theorem:

Theorem 2.2.1. If the transformations of an r-term group can be ordered into pairs of inverse transformations, then the group contains an identity transformation and r independent infinitesimal transformations.

Since the expression of our infinitesimal transformations contains not only x but also $a_1,\ldots,a_r$, it seems at first to be thinkable that an r-term group may contain more than r infinitesimal transformations. But then the group would contain ∞^r infinitesimal transformations, and this is impossible since the group is to contain only ∞^r transformations altogether, and these in general are finite.

In general it is advantageous to put the r infinitesimal transformations in the form

$$\delta x = X(x)\delta t \quad ,$$

where $X(x)$ depends only on x and not on $a_1,\ldots,a_r$.

COMMENTS

Here is my version of these arguments. (It will be done in generality adequate to also cover the situation described in Part II.) As in the previous section, let X *be a manifold,* A *a Lie group,*

$$f\colon X \times A \to X$$

a mapping defining a transformation group action of A *on* X. *(All the arguments generalize readily to the case of a local Lie group action.)*

In order to keep as close as possible to Lie's notations we suppose that:

X *and* A *are vector spaces.*

Standard techniques of manifold theory enable one to extend the formalism to more general situations.

For $a \in A$, *let*

$$\phi_a\colon X \to X$$

be the diffeomorphism defined as follows:

$$\phi_a(x) = f(x,a) .$$

Fix $a \in A$. *For* $b \in A_a$, $x \in X$, *set*

$$A_b(x) = \frac{\partial}{\partial\varepsilon}(\phi_{a+\varepsilon b}\phi_a^{-1}(x))\Big|_{\varepsilon=0} \qquad (2.2.1)$$

(Since we assume that A *is a vector space, the tangent space to* A *at* a *is identified with* A *itself. This is the meaning of the formula on the right hand side of 2.2.1.) As* x *varies,*

$$x \to A_b(x)$$

defines a vector field on X.

The map

$$b \to A_b$$

is a linear map

$$A_a \to V(X) . \qquad (2.2.2)$$

The image set consists of the infinitesimal transformations defined by the Lie transformation group.

What does Lie mean when he says that the "group contains the infinitesimal transformations 2.2.1"? Here is a reasonable interpretation. Write 2.2.1 as:

$$\phi_{a+\ b}\phi_a^{-1} \approx 1 + \varepsilon A_b \qquad (2.2.3)$$

For finite ε, *the map on the left hand side of 2.2.3 belongs to the group. Assuming some sort of closure condition, as* $\varepsilon \to 0$ *it is reasonable to say that the "infinitesimal transformation on the right hand side of 2.2.2 belongs to the group."*

What does Lie mean when he says that "the group contains r-terms"? My interpretation is the following:

Definition. The Lie transformation group is said to be an r-term group, if:

a) The parameter manifold A *is r-dimensional*

b) The map 2.2.2 is one-one.

Lie also says that, if the parameter manifold A *is r-dimensional, but the transformation group is not "r-term", the number of parameters can be reduced. To see what might be involved here, for each* $a \in A$, *let:*

$$W_a \subset A_a$$

be the kernel of the linear map 2.2. Using the hypothesis that the action is that of a group which contains inverses, one can show that:

$\dim(W_a)$ = *constant as* a *ranges over* A.

The vector field system on A *is* $a \to W_a$ *non-singular and completely integrable. (See Chapter B.)*

The vector field system thus defines a <u>*foliation*</u> *of* A.
One proves readily that:

> *If* X *is a connected manifold, and if* a, a' *belongs to the same leaf of the foliation of* A, *then*
>
> $$\phi_a = \phi'_a .$$

Hence, the transformation group can be parameterized by the <u>*space of leaves of the foliation*</u>. *Locally, this means "parameterization by a smaller number of parameters", as described by Lie.*

Chapter 3

RELATIONS AMONG THE INFINITESIMAL TRANSFORMATIONS OF A GROUP

3.1 INTRODUCTION

Let

$$\delta x = X_1(x)\delta t, \ \delta x = X_2(x)\delta t, \ldots, \delta x = X_r(x)\delta t$$

be r infinitesimal transformations of a group

$$x' = f(x, a_1, \ldots, a_r) \ ;$$

we shall show that the r quantities $X_1, \ldots, X_r$, which are certain functions of x, are related in pairs by simple differential relations.

3.2 JACOBI BRACKET RELATIONS AMONG INFINITESIMAL TRANSFORMATIONS

We first carry out the finite transformation

$$x' = f(x, a_1, \ldots, a_r) \ ,$$

and then an infinitesimal transformation

$$\delta x' = (\lambda_1 X_1(x') + \ldots + \lambda_r X_r(x'))\ \delta t \ .$$

The composition

$$x'' = f(x, a_1, \ldots, a_r) + \delta t \sum \lambda_k X_k(x')$$

must, by the definition of a transformation group, be a transformation of the form

$$x'' = f(x, a_1+da_1, \ldots, a_r+da_r) \quad .$$

This gives the equation

$$f(x, a_1+da_1, \ldots) = f(x, a_1, \ldots, a_r) + \delta t \sum_k \lambda_k X_k(f) \quad ,$$

or, by introducing infinitesimals and throwing out those of second order,

$$\sum_k \frac{\partial f}{\partial a_k} da_k = \delta t \sum_k \lambda_k X_k(f) \quad .$$

Such an equation holds for all values of the parameters λ_k under the assumption that the differentials da_k have appropriate corresponding values, and since the ratios $da_k : \delta t$ depend only on the quantities λ_i and a_i, and are independent of x, it follows that in particular there are r equations of the form

$$X_1(f) = A_1 \frac{\partial f}{\partial a_1} + A_2 \frac{\partial f}{\partial a_2} + \cdots + A_r \frac{\partial f}{\partial a_r} \quad ,$$

$$X_2(f) = B_1 \frac{\partial f}{\partial a_1} + B_2 \frac{\partial f}{\partial a_2} + \cdots + B_r \frac{\partial f}{\partial a_r} \quad ,$$

$$\cdot \qquad \cdot \qquad \cdot \quad \cdots \quad \cdot$$

$$X_r(f) = L_1 \frac{\partial f}{\partial a_1} + L_2 \frac{\partial f}{\partial a_2} + \cdots + L_r \frac{\partial f}{\partial a_r} \quad ,$$

where the quantities $A_i, B_i, \ldots, L_i$ depend only on $a_1, a_2, \ldots, a_r$. Here the determinant

$$\det(A_1 B_2 \cdots L_r)$$

must not vanish, since otherwise there would be a relation of the form

$$\phi_1(a_1, \ldots, a_r) X_1(f) + \cdots + \phi_r(a_1, \ldots, a_r) X_r(f) = 0 ,$$

and this is impossible because the r infinitesimal transformations are independent.

Therefore, by solving the preceding system of equations one obtains r equations of the form

$$\left.\begin{aligned} \frac{\partial f}{\partial a_1} &= M_1 X_1(f) + \cdots + M_r X_r(f) , \\ \frac{\partial f}{\partial a_2} &= N_1 X_1(f) + \cdots + N_r X_r(f) , \\ &\cdot \quad \cdot \quad \cdots \quad \cdot \\ \frac{\partial f}{\partial a_t} &= R_1 X_1(f) + \cdots + R_r X_r(f) , \end{aligned}\right\} \qquad (3.2.1)$$

where the quantities $M_i, N_i, \ldots, R_i$ depend on $a_1, a_2, \ldots, a_r$ only. The right side of these equations must satisfy, in pairs, the well-known integrability conditions. In this way we obtain $\frac{1}{2} r(r-1)$ relations, of which we shall develop the first:

$$\frac{\partial}{\partial a_2} \sum_k M_k X_k(f) = \frac{\partial}{\partial a_1} \sum_i N_i X_i(f) .$$

In doing this we remind ourselves that the X_k depend only on the single argument f. Thus, one has

$$0 = \left(\sum_k M_k \frac{dX_k}{df}\right)\left(\sum_i N_i X_i\right) - \left(\sum_i N_i \frac{dX_i}{df}\right)\left(\sum_k M_k X_k\right)$$
$$+ \sum_k \left(\frac{\partial M_k}{\partial a_2} - \frac{\partial N_k}{\partial a_1}\right) X_k ,$$

whence:

$$\sum_k \sum_i M_k N_i \left(X_i \frac{dX_k}{df} - X_k \frac{dX_i}{df}\right) = \sum_k \left(\frac{\partial N_k}{\partial a_1} - \frac{\partial M_k}{\partial a_2}\right) X_k ,$$

in which the indices k and i vary through $1,2,\ldots,r$. In the double sum on the left the term

$$X_i \frac{dX_k}{df} - X_k \frac{dX_i}{df}$$

occurs twice, once with the coefficients $M_k N_i$ and again with the coefficient $-M_i N_k$. Hence, the above equations can also be written as

$$\frac{1}{2} \sum_k \sum_i (M_k N_i - M_i N_k)\left(X_i \frac{dX_k}{df} - X_k \frac{dX_i}{df}\right) = \sum_k \left(\frac{\partial N_k}{\partial a_1} - \frac{\partial M_k}{\partial a_2}\right) X_k .$$

Altogether one obtains $\frac{1}{2} r(r-1)$ such equations, whose left sides are linear in the $\frac{1}{2} r(r-1)$ quantities

$$X_i \frac{dX_k}{df} - X_k \frac{dX_i}{df} .$$

We shall show that these equations can be solved for the $\frac{1}{2} r(r-1)$ quantities referred to. This follows from the well-known proposition:

If the determinant Δ of the $r \times r$ matrix (a_{ij}) is not zero, the same is true of the determinant of the $\frac{1}{2} r(r-1) \times \frac{1}{2} r(r-1)$ matrix $(a_{ij}a_{kq} - a_{kj}a_{iq})$. The new determinant is equal to a power of Δ.

Now since the determinant $(M_1N_2 \cdots R_r)$ is non-zero by the preceding remarks, the same must be true of the determinant whose elements are the $M_iN_k - M_kN_i$, and so one finds, by solving the previous $\frac{1}{2} r(r-1)$ linear equations, $\frac{1}{2} r(r-1)$ new relations of the form

$$X_i \frac{dX_k}{df} - X_k \frac{dX_i}{df} = c_{ik1}X_1 + c_{ik2}X_2 + \cdots + c_{ikr}X_r .$$

Here the c_{iks} are functions of $a_1, \ldots, a_r$, and since the X_k as well as the expression on the left depend on f alone, and the quantities $f, a_1, \ldots, a_r$ can be considered as independent, it follows that the c_{iks} must be absolute constants. Thus, the following fundamental theorem has been proved.

Theorem 3.2.1. If $\delta x = X_1(x)\delta t, \ldots, \delta x = X_r(x)\delta t$ *are* r *independent infinitesimal transformations of an r-term transformation group, then the* X's *satisfy relations of the form*

$$X_i \frac{dX_k}{dx} - X_k \frac{dX_i}{dx} = c_{ik1}X_1 + \cdots + c_{ikr}X_r \quad ,$$

where the c_{iks} *are constants.*

This theorem together with the formulas 3.2.1 suffices for the determination of all transformation groups of a 1-dimensional manifold.

Chapter 4

INFINITESIMAL TRANSFORMATIONS OF VARIOUS ORDERS

4.1 UPPER BOUND ON THE NUMBER OF PARAMETERS

We shall show that <u>the number</u> r <u>of parameters cannot exceed three</u>.

Again let

$$\delta x = X_1(x)\delta t, \ldots, \delta x = X_r(x)\delta t \tag{4.1.1}$$

be r independent infinitesimal transformations of an r-term group. We think of the $X_i(x)$ as power series in $x-x_0$:

$$X_i(x) = X_i(x_0) + \left(\frac{dX_i}{dx}\right)_{x_0}(x-x_0) + \frac{1}{2}\left(\frac{d^2X_i}{dx^2}\right)_{x_0}(x-x_0)^2 + \cdots$$

and form the general infinitesimal transformation

$$\delta x = \delta t \sum \lambda_i X_i$$

of the group. We thus obtain $\sum \lambda_i X_i$ as a power series in $x-x_0$, and indeed it is possible to choose the paramaters λ_i so that $\sum \lambda_i X_i$ has the form

$$\sum \lambda_i X_i = A_{r-1}(x-x_0)^{r-1} + A_r(x-x_0)^r + \cdots$$

In this A_{r-1} must be $\neq 0$ because

$$\det \begin{pmatrix} X_1(x_0) & \left(\frac{dX_1}{dx}\right)_{x_0} & \cdots & \left(\frac{d^{r-1}X_1}{dx^{r-1}}\right)_{x_0} \\ \cdot & \cdot & \cdots & \cdot \\ X_r(x_0) & \left(\frac{dX_r}{dx}\right)_{x_0} & \cdots & \left(\frac{d^{r-1}X_r}{dx^{r-1}}\right)_{x_0} \end{pmatrix}$$

can vanish only for special values of x_0 since the infinitesimal transformations 4.1.1 are independent.

From this it follows at once that one can always choose r independent infinitesimal transformations of the group which have the form

$$\begin{aligned}
\delta x &= (a_0 + a_1(x-x_0) + a_2(x-x_0)^2 + \cdots + a_{r-1}(x-x_0)^{r-1} + \cdots)\delta t \\
\delta x &= (b_1(x-x_0) + b_2(x-x_0)^2 + \cdots + b_{r-1}(x-x_0)^{r-1} + \cdots)\delta t \\
\cdot \quad & \quad \cdot \quad \cdot \quad \cdot \quad \cdots \quad \cdot \quad \cdots \\
\delta x &= (b_{r-2}(x-x_0)^{r-2} + b_{r-1}(x-x_0)^{r-1} + \cdots)\delta t \\
\delta x &= (k_{r-1}(x-x_0)^{r-1} + \cdots)\delta t
\end{aligned}$$

We will sometimes say that the last infinitesimal transformation is of order $r-1$ at the point $x = x_0$, that the next to last is of order $r-2$, etc., and finally that the first transformation is of order zero.

Now if the infinitesimal transformations

$$\delta x = Y_i(x)\delta t$$

and

$$\delta x = Y_k(x)\delta t$$

are of orders i and k respectively, then the expression

$$Y_i \frac{dY_k}{dx} - Y_k \frac{dY_i}{dx}$$

is of order $i+k-1$; and since this expression (Theorem 3.2.1) has the form

$$c_0Y_0 + c_1Y_1 + \cdots + c_{r-1}Y_{r-1} ,$$

we can conclude, if

$$i+k-1 > r-1 ,$$

that all the c's are zero. But the quantity

$$Y_i \frac{dY_k}{dx} - Y_k \frac{dY_i}{dx}$$

must not vanish, since otherwise one would have $Y_i =$ const. Y_k, and this is impossible since these two infinitesimal transformations are to be independent. Thus we see that the number $i+k-1$ cannot be greater than $r-1$. Hence, if we put, as we can,

$$i = r-1, \quad k = r-2,$$

we get

$$2r-4 \leq r-1 \quad ,$$

from which it follows that r is at most 3.

4.2 DETERMINATION OF 1-TERM GROUPS

I therefore separate the problem of determining all groups into the three problems of determining all 1-term, all 2-term, and all 3-term groups.

If $\delta x = X(x)\delta t$ is the infinitesimal transformation of an arbitrary 1-term group $x' = f(x,a)$, then, as we saw in §3, there is a relation of the form

$$\frac{df}{da} = A(a) \cdot X(f) \quad ,$$

whence, by integration,

$$\int \frac{df}{X(f)} = \int A(a)\, da \quad ,$$

or

$$\phi(f) = \psi(a) + \text{const.}$$

To determine the integration constant we recall that there is a value a_0 of the quantity a for which $f(x,a_0) = x$. This gives

$$\phi(x) = \psi(a_0) + \text{const.} \quad ,$$

whence

$$\phi(f) - \phi(x) = \psi(a) - \psi(a_0) = x(a) \quad ,$$

so that

$$f(x,a) = \phi^{-1}(\phi(x)+x(a)) .$$

It follows that every 1-term group is of the form

$$x' = \phi^{-1}(\phi(x)+\chi(a)) ,$$

and on the other hand, it is clear that this equation always determines a 1-term group, for any functions ϕ and χ. This gives:

Theorem 4.2.1. *Every 1-term transformation group of a 1-dimensional manifold is similar to the linear group* $x' = x+a$.

COMMENTS

Here are some results which are proved, or are relevant, in this chapter. Throughout our discussion, X *denotes a connected, 1-dimensional manifold. It is well-known that such a manifold is diffeomorphic to an open interval*

$$a < x < b$$

of real numbers.

Theorem 4.3.1. If A *is a vector field on* X *which does not vanish at any point, then there is a coordinate function* y *for* X *such that:*

$$A = \frac{\partial}{\partial y} \quad . \tag{4.3.2}$$

Proof. Suppose that:

$$A = a(x) \frac{\partial}{\partial x} \quad . \tag{4.3.2}$$

We shall show that there is a real-valued function

$$x \to f(x)$$

on X *such that:*

$$A(f) = 1 \tag{4.3.3}$$

The map $x \to f(x)$ *is a diffeomorphism of* X *with an interval of real numbers,* (4.3.4)

and

$$f(0) = 0 \tag{4.3.5}$$

To do this, suppose, for example, that

$$X = R \; .$$

The condition that A *vanishes nowhere means that*

$$a(x) \neq 0 \qquad \text{for all} \quad x \in R \quad .$$

Let $f \to x(t)$, $-\varepsilon < t < \varepsilon$, *be the solution of the differential equation*

$$\frac{dx}{dt} = a(x(t)) \; ; \qquad x(0) = 0 \quad .$$

(Such a solution exists for ε *sufficiently small.)*

If f *satisfies 4.3.3, we have:*

$$\frac{d}{dt} f(x(t)) = 1 ,$$

By 3.5, we have:

$$f(x(t)) = t \qquad (4.3.6)$$

Now, if ε *is sufficiently small, the map* $t \to x(t)$ *is a diffeomorphism. Hence, we can* <u>*define*</u> f *by relation 4.7.6.*

We have shown that, for ε *sufficiently small, there is a solution of 4.3.3 in the interval*

$$-\varepsilon < x < \varepsilon ,$$

such that

$$f(0) = 0 .$$

We continue to define f *for larger values of* f(x) *by the process of analytic continuation. A standard argument shows that it can be defined in this way over the whole interval*

$$-\infty < x < \infty .$$

Now, 4.3.3 means that

$$\frac{df}{dx}(x) \neq 0 \qquad \textit{for all } x, -\infty < x < \infty$$

Hence, either

$$\frac{df}{dx}(x) > 0 \quad \textit{or}$$

$$\frac{df}{dx}(x) < 0 \quad \textit{for all } x. \qquad (4.3.7)$$

Consider the map

$$x \to f(x) = y$$

of $R \to R$. *Its image is connected, hence is an interval*

$$a < y < b \quad .$$

Let us show that it is one-one. Suppose that

$$f(x_1) = f(x_2) \tag{4.3.8}$$

Condition 4.3.7 means that $x \to f(x)$ *is <u>either</u> strictly increasing or decreasing. In either case, 4.3.8 forces*

$$x_1 = x_2 \quad ,$$

i.e., the map is one-one. The implicit function theorem, together with condition 4.3.7, now implies that the inverse map is C^∞, *i.e., the map is a diffeomorphism. If one now introduces* y *as a new coordinate for* X, *it is readily seen that 4.3.1 is implied by 4.3.3.*

This result gives a more precise version of Lie's Theorem 4.2.1.

<u>Theorem 4.3.2</u>. Let A, B *be two vector fields on* X *such that:*

$$[A,B] = 0 \tag{4.3.9}$$

and

$$A(x) \neq 0 \qquad \textit{for all} \quad x \in X \quad .$$

Then, there is a real constant c *such that:*

$$B = cA \tag{4.3.10}$$

Proof. *By Theorem 4.3.1, we can choose the coordinate function* x *such that*

$$A = \frac{\partial}{\partial x}$$

Suppose

$$B = f(x)\frac{\partial}{\partial x} .$$

Condition 4.3.9 implies that

$$\frac{df}{dx} = 0 ,$$

i.e., f = *constant. This proves 4.3.9.*

Remark. *In the* C^{∞} *case, one can construct two vector fields which satisfy 4.3.9, but not 4.3.10. Of course, this cannot happen with real analyticity, which is the assumption Lie makes.*

Theorem 4.3.3. *Let* $\underset{\sim}{G}$ *be a finite dimensional Lie algebra of vector fields on* X, *and let* x_0 *be a point of such that:*

$$\underset{\sim}{G}(x_0) \neq 0 . \tag{4.3.11}$$

If A *is an element of* $\underset{\sim}{G}$ *which vanishes to an infinite order at* x_0, *then:*

$A \equiv 0$ *in a neighborhood of* x_0.

Proof. *Using hypothesis 4.3.11, there is an element* $B \in \underset{\sim}{G}$ *such that:*

$$B(x_0) \neq 0 .$$

By Theorem 4.3.1, there is a coordinate function for X, *in a neighborhood of* x_0, *which we label* "x", *such that:*

$$B = \frac{\partial}{\partial x} .$$

Suppose:

$$A = f(x) \frac{\partial}{\partial x} .$$

The hypothesis that A *"vanishes to an infinite order at* x_0*" means that:*

$$\frac{d^n}{dx^n} f(x_0) = 0 \qquad \textit{for all } n . \qquad (4.3.12)$$

Let

$$\alpha : \underset{\sim}{G} \to \underset{\sim}{G}$$

be the linear map defined as follows:

$$\alpha(A) = [B,A]$$

for all $A \in \underset{\sim}{G}$

(In the now standard notation of Lie algebra theory,

$$\alpha = \text{Ad } B .)$$

<u>Since</u> $\underset{\sim}{G}$ *<u>is a finite dimensional vector space</u>, there is a <u>polynomial</u>* $P(x)$ *such that:*

$$P(\alpha) = 0 \tag{4.3.13}$$

For example, suppose 4.3.13 takes the following form:

$$\alpha^m + a_{m-1}\alpha^{m-1} + \cdots + a_0 = 0$$

with coefficient $a_0, \ldots, a_{m-1}$ *which are real numbers.*

With the special form we have assumed for the vector fields A *and* B, *we then have:*

$$\frac{d^m f}{dx^m} + a_{m-1} \frac{d^{m-1} f}{dx^{m-1}} + \cdots = 0 , \tag{4.3.14}$$

i.e., f *satisfies a <u>linear differential equation with constant coefficients</u>. By the well-known properties of such equations, if a function* $x \to f(x)$ *solves it, and vanishes to the m-th order at a point, it vanishes identically. (Alternately, the solutions of 4.3.14 are <u>real analytic</u>.) In particular, applying 4.3.12 to the particular function* f, *which is the component of the vector field* A *in the coordinate system* (x), *we see that*

$$f \equiv 0 ,$$

finishing the proof of Theorem 4.3.3.

<u>Theorem 4.3.4</u>. Let $\underset{\sim}{G}$ *be a real Lie algebra of vector fields on* X. *Then,*

$$\text{dimension } \underset{\sim}{G} \leq 3 \tag{4.3.15}$$

Proof. *If the vector fields in* $\underset{\sim}{G}$ *vanish* *identically* *at each point of* X, *the relation 4.3.15 is trivial, since* $\dim \underset{\sim}{G} = 0$.

Let x_0 *be a point of* X *such that*

$$\underset{\sim}{G}(x_0) \neq 0 \tag{4.3.16}$$

Introduce the filtration of $\underset{\sim}{G}$ *determined by the order of vanishing at* x_0. *(See Chapter C.) Let*

$$\underset{\sim}{G}^j \subset \underset{\sim}{G}$$

be the subset of these vector fields which vanish to order j *at* x_0.

A basic relation is then that:

$$[\underset{\sim}{G}^j, \underset{\sim}{G}^k] \subset \underset{\sim}{G}^{j+k-1} \tag{4.3.17}$$

Now,

$$\underset{\sim}{G} \supset \underset{\sim}{G}^1 \supset \underset{\sim}{G}^2 \supset \cdots$$

Since $\underset{\sim}{G}$ *acts on a 1-dimensional manifold,*

$$\dim (\underset{\sim}{G}/G^1) = 1 . \tag{4.3.18}$$

Choose a coordinate function x *for* X *so that* x_0 *has coordinate:*

$$x_0 = 0 ,$$

and such that:

$$B = \frac{\partial}{\partial x} \in \underset{\sim}{G} \tag{4.3.19}$$

(Again, recall that this is possible, by Theorem 4.3.1, because of 4.3.16.)

Let A *be a non-zero element of* $\underset{\sim}{G}$. *By Theorem 4.3.3 there is an integer* j *such that* A *vanishes* <u>*precisely*</u> *to order* j *at* 0, *i.e.,*

$$A = x^j f(x) \frac{\partial}{\partial x} , \qquad (4.3.20)$$

with

$$f(0) \neq 0 .$$

This means that

$A \in \underset{\sim}{G}^j$

A *determines a non-zero element of* $\underset{\sim}{G}^1/\underset{\sim}{G}^{j+1}$.

Note the following general property:

$$\dim (\underset{\sim}{G}^k/\underset{\sim}{G}^{k+1}) \leq 1 \qquad (4.3.21)$$

for all k.

Combining 4.3.19, 4.3.20, we have:

[B,A] *has order* j-1

[B,[B,A]] *has order* j-2 , (4.3.22)

and so forth.

Now, choose j *as the largest integer such that there is an element* $A \in \underset{\sim}{G}$ *which vanishes to order precisely* j *at* x_0. *By 4.3.22,*

$$\dim (\underset{\sim}{G}^k/\underset{\sim}{G}^{k+1}) \doteq 1$$

for $k = j, j-1, \ldots$

Now, apply 4.3.17 and Theorem 4.3.2.

$$[\underset{\sim}{G}^k, \underset{\sim}{G}^j] \subset \underset{\sim}{G}^{j+k-1} \quad .$$

Let us find the values of k *for which:*

$$j+k-1 > j \ ,$$

or

$$j > 1 \quad ,$$

or

$$k \geq 2 \quad .$$

Hence, if $k \geq 2$,

$$[\underset{\sim}{G}^k, \underset{\sim}{G}^j] = 0 \quad .$$

By Theorem 4.3.2, the elements of $\underset{\sim}{G}^k$ *and* $\underset{\sim}{G}^j$ *are linearly dependent. This is only possible if:*

$$k = j \ , \quad \text{or}$$

$$\underset{\sim}{G}^k = 0 \quad .$$

Thus, if j *were* ≥ 3, *there would be a contradiction (since* k *could be taken as 2), hence,*

$$j \leq 2 \quad . \qquad (4.3.23)$$

Now, by a standard argument of linear algebra,

$$\dim \underset{\sim}{G} = \dim (\underset{\sim}{G}/G^1) + \dim (\underset{\sim}{G}^1/\underset{\sim}{G}^2) + \dim (\underset{\sim}{G}^2/\underset{\sim}{G}^3) + \cdots$$

By 4.3.23, the terms $\cdots$ *are zero, whence 4.3.15.*

__*Theorem 4.3.4*__. *Let* $\underset{\sim}{G}$ *be a finite dimensional real Lie algebra of vector fields on* X. *Then, there is a coordinate function* x *such that:*

$$\frac{\partial}{\partial x} \in \underset{\sim}{G} ,$$

and each $A \in \underset{\sim}{G}$ *is of the form:*

$$A = (a_0 + a_1 x + a_2 x^2) \frac{\partial}{\partial x} , \qquad (4.3.24)$$

with $a_0, a_1, a_2 \in R$

__*Proof*__. *We already know (from Theorem 4.3.3) that*

$$\dim \underset{\sim}{G} \leq 3 , \qquad (4.3.25)$$

and that each $A \in \underset{\sim}{G}$ *vanishes to* __*at most*__ *the second order at any point of* X.

__*Case 1*__. *There exists a* $B \in \underset{\sim}{G}$ *such that:*

Ad $\underset{\sim}{B}$ *is nilpotent.*

In this case, we can choose a point of X *at which* B *does not vanish, and a coordinate function* x *valid in a neighborhood of this point such that:*

$$B = \frac{\partial}{\partial x} \quad .$$

"Ad B *nilpotent" means that*

$$(\text{Ad } B)^n = 0$$

for n *sufficiently large. Condition 4.3.25 and the Cayley-Hamilton Theorem imply that:*

$$(\text{Ad } B)^3 = 0$$

This means that:

$$\left[\frac{\partial}{\partial x}, \left[\frac{\partial}{\partial x}, \left[\frac{\partial}{\partial x}, A\right]\right]\right] = 0 \tag{4.3.26}$$

for <u>all</u> $A \in \underset{\sim}{G}$,

which implies that A *is of form 4.3.24.*

Now, let us suppose that Case 1 is not satisfied. Suppose the coordinate function x *is chosen so that*

$$B = \frac{\partial}{\partial x} \in \underset{\sim}{G} \quad .$$

Ad B *is <u>not</u> nilpotent, hence has a non-zero eigenvector. If it has a <u>real</u> eigenvalue* λ, *there is an* $A \in \underset{\sim}{G}$ *such that*

$$[B, A] = \lambda A \quad .$$

In particular, Ad A *<u>is</u> nilpotent, which contradicts our assumption that Case 1 is not satisfied, and the non-zero eigenvalues of* Ad B *are complex numbers.*

Hence, there are elements A_1, A_2 *in* G_c, *the complification of* G, *such that:*

$$[B, A_1] = \lambda A_1 ,$$

$$[B, A_2] = \bar{\lambda} A_2 .$$

Then,

$$[B, [A_1, A_2]] = (\lambda + \bar{\lambda}) [A_1, A_2]$$

Using 4.3.25, we have either:

$$\lambda + \bar{\lambda} = 0 \quad \text{or} \quad [A_1, A_2] = 0 .$$

It is left as an exercise for the reader to show that each of these possibilities leads to a contradiction.

Remark. *Condition 4.3.26 is a* *differential equation* *for* A. *It is a simple example of Lie's general technique for* *defining* *Lie algebras, namely by differential equations. The recent book by Kumpera and Spencer [1] represents a vast development of these ideas in the context of modern mathematics.*

These results are the basic structure theorems for the local study of finite dimensional Lie algebras of vector fields on one-dimensional manifolds. (One might have continued to study what happens in the neighborhood of points x_0 *such that*

$$G(x_0) = 0 ,$$

but it is typical of Lie's work that he avoided such "non-generic" situations.) All of the results presented by Lie in this Part I are easily deducible from them.

Chapter 5

SOLUTION OF THE GIVEN PROBLEM

5.1 2-TERM GROUPS

If $\delta x = X_1(x)\delta t$ and $\delta x = X_2(x)\delta t$ are the infinitesimal transformations of a 2-term group $x' = f(x,a_1,a_2)$, then by previous results there are relations of the form

$$\left.\begin{aligned} \frac{\partial f}{\partial a_1} &= A(a_1,a_2)X_1(f) + B(a_1,a_2)X_2(f) , \\ \frac{\partial f}{\partial a_2} &= C(a_1,a_2)X_1(f) + D(a_1,a_2)X_2(f) , \\ X_1 \frac{dX_2}{dx} - X_2 \frac{dX_1}{dx} &= c_1X_1 + c_2X_2 \end{aligned}\right\} \quad (5.1.1)$$

In this the constants c_1 and c_2 cannot both be zero, since otherwise there would be a relation of the form

$$\text{const. } X_1 + \text{const. } X_2 = 0 .$$

Hence we may assume that, for example, $c_1 \neq 0$. We put

$$c_1X_1 + c_2X_2 = X_1' ,$$

$$\frac{1}{c_1} X_2 = X_2' ;$$

then:

$$X_1' \frac{dX_2'}{dx} - X_2' \frac{dX_1'}{dx} = c_1X_1 + c_2X_2 = X_1' .$$

Hence, it is no restriction to assume $c_1 = 1$, $c_2 = 0$ and so

$$X_1 \frac{dX_2}{dx} - X_2 \frac{dX_1}{dx} = X_1 .$$

If, in this equation, we consider X_1 as known, X_2 as unknown, we find by integration that

$$X_2 = X_1 \int \frac{dx}{X_1} ,$$

and, by substituting this into the first two of the equations (5.1.1):

$$\frac{\partial f}{\partial a_1} = A \cdot X_1(f) + B \cdot X_1(f) \int \frac{df}{X_1(f)} ,$$

$$\frac{\partial f}{\partial a_2} = C \cdot X_1(f) + D \cdot X_1(f) \int \frac{df}{X_1(f)} .$$

If we now introduce

$$\int \frac{df}{X_1(f)} = \phi(f)$$

as an unknown function, we obtain the equations

$$\frac{\partial \phi}{\partial a_1} = A + B\phi , \qquad \frac{\partial \phi}{\partial a_2} = C + D\phi ,$$

whose general integral is of the form

$$\phi(f) = \psi_1(a_1, a_2) + K\psi_2(a_1, a_2) .$$

To determine the integration constant K, which depends on x but not on a_1, a_2, we recall that there are values a_1^0, a_2^0 of the quantities a_1, a_2 for which $f(x, a_1^0, a_2^0) = x$. Hence,

$$\phi(x) = \psi_1(a_1^0, a_2^0) + K\psi_2(a_1^0, a_2^0) \quad ,$$

so that K may be eliminated to give

$$\phi(f) = b_1\phi(x) + b_2 \quad ,$$

where b_1 and b_2 are certain functions of a_1 and a_2. Hence

$$f(x, a_1, a_2) = \phi^{-1}(b_1\phi(x)+b_2) \quad ,$$

so that every 2-term group is of the form

$$x' = \phi^{-1}(b_1\phi(x)+b_2)$$

On the other hand it is clear that this equation always determines a 2-term group, which is similar to the linear group $x' = a_1x + a_2$. This gives:

<u>Theorem 5.1.1</u>. <u>Every 2-term group is similar to the group</u> $x' = a_1x + a_2$.

COMMENTS

A 2-term group "is a 2-dimensional Lie subalgebra of V(R)". *The argument used to prove Theorem 4.3.4 implies that a coordinate* x *may be chosen (at least locally) so that* G *has*

$$\frac{\partial}{\partial x},\ x\frac{\partial}{\partial x}$$

is a basis. Theorem 5.1.1 readily follows.

5.2 3-TERM GROUPS

To determine the most general 3-term group $x' = f(x,a_1,a_2,a_3)$, we choose three independent infinitesimal transformations of the group

$$\delta x = X_1\delta t\ ,\quad \delta x = X_2\delta t\ ,\quad \delta x = X_3\delta t$$

which are of the form

$$X_1 = 1 + a_3(x-x_0)^3 + a_4(x-x_0)^4 + \cdots\ ,$$

$$X_2 = (x-x_0) + b_3(x-x_0)^3 + \cdots\ ,$$

$$X_3 = (x-x_0)^2 + \cdots\ ,$$

We now form the three equations

$$X_2X_3' - X_3X_2' = \alpha_1X_1 + \alpha_2X_2 + \alpha_3X_3\ ,$$

$$X_3X_1' - X_1X_3' = \beta_1X_1 + \beta_2X_2 + \beta_3X_3\ ,$$

$$X_2X_1' - X_1X_2' = \gamma_1X_1 + \gamma_2X_2 + \gamma_3X_3\ ;$$

we have then

$$\alpha_1 = \alpha_2 = 0 , \quad \alpha_3 = 1$$

$$\beta_1 = 0 , \quad \beta_2 = -2 ,$$

$$\gamma_1 = -1 , \quad \gamma_2 = 0 .$$

To determine the remaining constants, we put

$$X_i X_k' - X_k X_i' = [X_i X_k]$$

and we note the identity*

$$[[X_1X_2]X_3] + [[X_2X_3]X_1] + [[X_3X_1]X_2] = 0 .$$

Substituting twice into this the values given above for the $[X_iX_k]$ gives a relation of the form

$$L_1X_1 + L_2X_2 + L_3X_3 = 0 ,$$

which shows $\beta_3 = 0$. Thus one has

$$[X_2X_3] = X_3 , \quad [X_3X_1] = -2X_2 , \quad [X_2X_1] = -X_1$$

Now we consider X_3 as given and seek to determine X_2 and X_1. The equation

---

* This equation is a corollary of the well-known <u>Jacobi</u> identity

$$[[AB]C] + [[BC]A] + [[CA]B] = 0 ,$$

if one puts

$$X_k \frac{df}{dx} = H_k$$

and then forms the equation

$$[[H_1H_2]H_3] + [[H_2H_3]H_1] + [[H_3H_1]H_2] = 0 .$$

$$[X_2X_3] = X_3 = X_2X_3' - X_3X_2'$$

shows that

$$X_2 = -X_3\int\frac{dx}{X_3}.$$

The equation

$$X_3X_1' - X_1X_3' = -2X_2 = 2X_3\int\frac{dx}{X_3}$$

can also be written as

$$\frac{d}{dx}\left(\frac{X_1}{X_3}\right) = 2\cdot\frac{1}{X_3}\int\frac{dx}{X_3}$$

and this gives

$$\frac{X_1}{X_3} = \left(\int\frac{dx}{X_3}\right)^2 + \text{const.} ,$$

where the integration constant must be zero because $[X_2X_1] = -X_1$.

We substitute the values we have found

$$X_2 = -X_3\int\frac{dx}{X_3} , \qquad X_1 = X_3\left(\int\frac{dx}{X_3}\right)^2$$

into the three equations

$$\frac{\partial f}{\partial a_i} = A_iX_1(f) + B_iX_2(f) + C_iX_3(f)$$

and thereby find three equations of the form

$$\frac{\partial f}{\partial a_i} = A_iX_3\left(\int\frac{df}{X_3}\right)^2 - B_iX_3\int\frac{df}{X_3} + C_iX_3(f) ,$$

where the quantities A_i, B_i and C_i are functions of a_1, a_2 and a_3. To simplify the formulas we introduce

$$\phi(f) = \int \frac{df}{X_3(f)}$$

as an unknown instead of f and thus obtain the three equations

$$\frac{\partial}{\partial a_i} = A_i\phi^2 - B_i\phi + C_i \quad ,$$

whose general integral is of the form

$$\phi(f) = \frac{\psi_1(a_1,a_2,a_3) + K\psi_2(a_1,a_2,a_3)}{1 + K\psi_3(a_1,a_2,a_3)}$$

To determine the integration constants K, which are on a_1,a_2,a_3 but not on x, we recall that there are values a_1^0,a_2^0,a_3^0 of a_1,a_2,a_3 for which $f(x,a_1^0,a_2^0,a_3^0) = x$. This gives

$$\phi(x) = \frac{\psi_1(a_1^0,a_2^0,a_3^0) + K\psi_2(a_1^0,a_2^0,a_3^0)}{1 + K\psi_3(a_1^0,a_2^0,a_3^0)}$$

whence, by elimination of K,

$$\phi(f) = \frac{b_1 + b_2\phi(x)}{1 + b_3\phi(x)} \quad ,$$

where b_1,b_2,b_3 are certain functions of a_1,a_2,a_3. This proves the following theorem:

Theorem 5.2.1. Every 3-term group of a 1-dimensional manifold is of the form

$$x' = \phi^{-1}\left(\frac{a_1\phi(x) + a_2}{a_3\phi(x) + 1}\right)$$

for some function ϕ (which may be arbitrary). Every such group is therefore similar to the linear group

$$x' = \frac{a_1 x + a_2}{a_3 x + 1}$$

We thus have solved the problem of this part, for by collecting our results we have

Theorem 5.2.2. Every transformation group of a 1-dimensional manifold is similar to a linear group and hence depends on at most three parameters.

COMMENTS

Again, the assumption that $\underset{\sim}{G}$ *is a 3-dimensional Lie algebra of vector fields on* R *implies (using the argument of Theorem 3.3.4) that it has a basis consisting of*

$$\frac{\partial}{\partial x},\ x\,\frac{\partial}{\partial x},\ x^2\,\frac{\partial}{\partial x} \quad .$$

That the group *it generates takes the form indicated in Theorem 5.2.1 is readily seen by various techniques (most straightforwardly, by solving the differential equation;*

$$\frac{dx}{dt} = (a_0 + a_1 x + a_2 x^2)$$

with real parameters a_0, a_1, a_2.*)*

Part II

DETERMINATION OF ALL TRANSFORMATION GROUPS OF A 2-DIMENSIONAL MANIFOLD

Chapter 6

GENERALITIES

Before we turn to the transformation groups of a 2-dimensional manifold, we make some general observations on the transformation groups of an n-dimensional manifold.

If in the equations

$$x_i' = f_i(x_1,\ldots,x_n,a_1,\ldots,a_r) \qquad (i=1,2,\ldots,n)$$

one considers $x_1',\ldots,x_n'$ as original variables and $x_1,\ldots,x_n$ as new variables and $a_1,\ldots,a_r$ as parameters, then these equations define ∞^r transformations. I say that such a family of transformations forms a group if the composition of two transformations of the family is again a transformation of the family, i.e., when from the equations

$$x_i' = f_i(x_1,\ldots,x_n,a_1,\ldots,a_r) = f_i(a) \quad ,$$

$$x_i'' = f_i(x_1',\ldots,x_n',b_1,\ldots,b_r) \quad ,$$

follows

$$x_i'' = f_i(x_1,\ldots,x_n,c_1,\ldots,c_r) \quad ,$$

where $c_1,\ldots,c_r$ depend only on the a's and the b's, and neither on x or the index i. In other words, we require equations

$$f_i(f_1(a),\ldots,f_n(a),b_1,\ldots,b_r) = f_i(x_1,\ldots,x_n,c_1,\ldots,c_r) .$$

As in the preceding part we restrict ourselves to groups whose transformations can be ordered into pairs of inverse transformations, although we still conjecture that every group has this property.

We think of the unknown functions $f_1,\ldots,f_n$ as power series in x and the a's which converge in some domain of these quantities. Consequently, the f_i are single-valued differentiable functions of their arguments. From the definition of a group it follows that one must be able to choose the quantities $x_1,\ldots,x_n,a_1,\ldots,a_r$ so that $f_i(x,a)$ is in the projection on the first factor of the domain of convergence of f_i.

If in the equations of a transformation group

$$x_i' = f_i(x_1,\ldots,x_n,a_1,\ldots,a_r)$$

one replaces $a_1,\ldots,a_r$ by certain functions of these quantities, say $\alpha_1,\ldots,\alpha_r$, which are to be new parameters, then the equations obtained in this way

$$x_i' = \phi_i(x_1,\ldots,x_n,\alpha_1,\ldots,\alpha_r)$$

again determine a transformation group, which is to be considered equivalent to the original. It may happen that the new equations contain a smaller number of parameters than the old. Such a lowering of the number of parameters can only occur when $f_1,\ldots,f_n$, considered as functions of $a_1,\ldots,a_r$, are all solutions of a linear partial differential equation of the form

$$\sum_k \psi_k(a_1,\ldots,a_r)\frac{\partial f}{\partial a_k} = 0 .$$

If all the f's satisfy this equation and $\alpha_1,\ldots,\alpha_{r-1}$ is a system of solutions depending only on $a_1,\ldots,a_r$, then it is possible to put the f_i in the form

$$\phi_i(x_1,\ldots,x_n,\alpha_1,\ldots,\alpha_{r-1}) .$$

For each group there exists a certain minimum number of parameters. If this number is r, we shall say that the group is an r-term group. This definition can also be expressed as follows:

Definition. A group is an r-term if it contains ∞^r distinct transformations.

If the equations

$$x_i' = f_i(x_1,\ldots,x_n,a_1,\ldots,a_r)$$

determine a transformation group, and if one replaces $x_1,\ldots,x_n$ by new variables $y_1,\ldots,y_n$ by means of the equations

$$x_k = \Theta_k(y_1,\ldots,y_n) = \Theta_k$$

then it is easy to see that the equations

$$\Theta_i(y'_1,\ldots,y'_n) = f_i(\Theta_1,\ldots,\Theta_n,a_1,\ldots,a_r)$$

also determine a transformation group. This is because the new and the old equations determine the same transformation of the x's. Two such groups will be called similar.

Definition. Two r-term groups of n variables are similar when one group becomes the other by the introduction of new variables.

My investigations of transformation groups are meant in the first place to settle the following problem.

Problem. To determine all r-term transformation groups of an n-dimensional manifold.

In treating this problem it is permissible and indeed expedient to consider similar groups as identical.

COMMENTS

Here is one way of interpreting this material. Let X *be a manifold,* G *a local Lie group, and suppose given a local Lie group action of* G *on* X. *Let* $\underset{\sim}{G}$ *be the Lie algebra of* G. *The local transformation group defines a Lie algebra homomorphism*

$$\underset{\sim}{G} \to V(X)$$

We consider local transformation group actions for which this map is one-one. The transformation group G *is then said to be an* <u>*r-term*</u> *one if* $\underset{\sim}{G}$ *is an r-dimensional real Lie algebra.*

Let (G',X') *be another local transformation group of this type.* (G,X) *is said to be* <u>*similar*</u> *to* (G',X') *if there are points*

$$x \in X , \quad x' \in X' ,$$

open neighborhoods

$$x \in U \subset X$$
$$x' \in U' \subset X' ,$$

and diffeomorphism

$$\alpha: U \to U'$$

and a Lie algebra isomorphism

$$\phi: \underset{\sim}{G} \to \underset{\sim}{G}'$$

such that:

$$\alpha_*(X(y)) = \phi(X)(\alpha(y))$$

for all $y \in U$.

(Intuitively, this condition means that α *maps the group action on* U *onto the group action on* U'.*) Lie's problem is to* <u>*classify the similarity classes of local transformation group actions*</u>.

Chapter 7

THE INFINITESIMAL TRANSFORMATIONS OF A GROUP

In the following we shall show that every r-term group in which every transformation has an inverse contains ∞^{r-1} infinitesimal transformations which characterize the group. The only, or at least the simplest, method of determining all transformation groups is based on investigating their infinitesimal transformations.

7.1 LINEAR COMBINATIONS OF INFINITESIMAL TRANSFORMATIONS

A transformation is said to be infinitesimal if it can be put in the form

$$x_i' = x_i + X_i(x_1,\ldots,x_n)\delta t$$

where δt is an infinitesimal. Generally, we shall write such equations as

$$\delta x_i = X_i(x_1,\ldots,x_n)\delta t \quad .$$

If one replaces $x_1,\ldots,x_n$ by new variables, say $y_1,\ldots,y_n$, then our infinitesimal transformation assumes the form

$$\delta y_i = \delta t \sum_k \frac{\partial y_i}{\partial x_k} X \quad .$$

On the other hand if we make the same change of variables in the expression

$$A(F) = X_1 \frac{\partial F}{\partial x_1} + \cdots + X_n \frac{\partial F}{\partial x_n} ,$$

we get

$$A(F) = \frac{\partial F}{\partial y_1} \sum_k \frac{\partial y_1}{\partial x_k} X_k + \cdots + \frac{\partial F}{\partial y_n} \sum_k \frac{\partial y_n}{\partial x_k} X_k .$$

Thus we see that the equations of the infinitesimal transformation and the expression $A(F)$ transform in the same way. Therefore, it is analytically permissible to <u>consider $A(F)$ as the symbol of our infinitesimal transformation</u>.

Let us assume that a transformation group contains the two infinitesimal transformations

$$\delta x_i = X_i \delta t \quad \text{and} \quad \delta x_i = Y_i \delta \tau .$$

Then it must contain the composition of these two transformations, namely

$$\delta x_i = X_i \delta t + Y_i \delta \tau .$$

As the ratio $\delta t : \delta \tau$ runs through all possible values, one obtains from this infinitely many distinct transformations, which belong to the group. This gives the theorem:

<u>Theorem 7.1.1</u>. <u>If a group contains two infinitesimal transformations whose symbols are</u>

$$A = X_1 \frac{\partial}{\partial x_1} + \cdots + X_n \frac{\partial}{\partial x_n} ,$$

$$B = Y_1 \frac{\partial}{\partial y_1} + \cdots + Y_n \frac{\partial}{\partial y_n} ,$$

then it also contains the 1-parameter family of transformations determined by the symbol $\lambda A + \mu B$ and depending on the parameter $\lambda : \mu$.

We say that r infinitesimal transformations $A_1, \ldots, A_r$ are independent if they are linearly independent. If a group contains the r infinitesimal transformations $A_1, \ldots, A_r$ then it also contains the ∞^{r-1} infinitesimal transformations determined by the symbols $\lambda_1 A_1 + \cdots + \lambda_r A_r$, in which $\lambda_1, \ldots, \lambda_r$ are arbitrary constants.

COMMENTS

In Chapter A I have suggested that the differential geometric objects called "vector fields" should be understood in two ways--as "infinitesimal transformations", i.e., as equivalence classes of flows, and as derivations on functions, i.e., as homogeneous, linear first order differential operators.

The material in this section should be interpreted in this way. Lie thinks of an "infinitesimal transformation" as an equivalence class of flows. He then defines their symbols as the operation on functions--now called Lie derivative--associated with the "differential operator" definition.

Let us examine Theorem 7.1.1 in this light. Let $t \to \phi_t$ be a flow on a manifold X, i.e., a one-parameter family of diffeomorphisms of X, such that:

$$\phi_0 = \text{identity map} \quad .$$

For each t, $f \in F(X)$, set

$$A_t(f) = (\phi_t^{-1})^* \frac{\partial}{\partial t} \phi_t^*(f) \tag{7.1.1}$$

$t \to A_t$ is then a curve in $V(X)$, called the infinitesimal generator of the flow. Two flows are said to be infinitesimally equivalent if their infinitesimal generators have the same value at $t = 0$. An infinitesimal transformation is an equivalence class of flows.

Let $t \to \alpha_t, \beta_t$ be two flows on X, with infinitesimal generators

$$t \to A_t, B_y \quad .$$

Set:

$$\gamma_t = \alpha_t \beta_t \quad ,$$

the product flow. Let us compute its infinitesimal generator $\{C_t\}$ by direct application of formula 7.1.1.

$$C_t(f) = (\gamma_t^{-1})^* \frac{\partial}{\partial t} (\gamma_t^*(f))$$

$$= (\beta_t^{-1}\alpha_t^{-1})^* \frac{\partial}{\partial t} ((\alpha_t\beta_t)^*(f))$$

$$= (\alpha_t^{-1})^*(\beta_t^{-1})^* \frac{\partial}{\partial t} (\beta_t^*\alpha_t^*(f))$$

$$= (\alpha_t^{-1})^*(\beta_t^{-1})^*\left(\frac{\partial\beta_t^*}{\partial t} (\alpha_t^*(f)\right) + \beta_t^* \left(\frac{\partial}{\partial t} \alpha_t^*(f)\right)$$

$$= \alpha_t^{-1*}(B_t(\alpha_t^*(f)) + A_t(f) \qquad (7.1.2)$$

This is a basic formula of Lie theory. Specializing it, for $t = 0$, *we have:*

$$C_0 = B_0 + A_0 \qquad (7.1.3)$$

This means that:

> *The infinitesimal transformation of the product of two flows is the sum of the infinitesimal transformation.*

This is the essential content of Theorem 7.1.1.

Here is another way of thinking of this argument involving "infinitesimals" more directly.

Suppose that X *is a vector space. Consider the flows* α_t, β_t *written "infinitesimally" as follows:*

$$\alpha_t(x) \approx x + tA(x) \equiv x + t\delta(x)$$

$$\beta_t(x) \approx x + tB(x)$$

Then,

$$\alpha_t(\beta_t(x)) \approx x + tB(x) + tA(x+tB(x))$$

$$\approx x + t(B(x)+A(x))$$

This shows that the *sum*

$$A + B$$

of two vector fields corresponds "infinitesimally" to the *product* *of two flows.*

Similarly, we can show that $-A$ *corresponds infinitesimally to the* *inverse* *flow* $t \to \alpha_t^{-1}$.

Given

$$\alpha_t(x) \approx x + tA(x)$$

set

$$\beta_t(x) = x - tA(x) \quad .$$

Then,

$$\beta_t(\alpha_t(x)) \approx x + tA(x) - tA(x+tA(x))$$

$$\approx x \quad ,$$

i.e., $\beta_t \approx \alpha_t^{-1}$.

Remark. *The "approximate equality" sign* $\approx$ *means* *up to terms of order* ≥ 2 *in* t. *Notice that this can be interprete in a purely algebraic way (say, in terms of formal power series), independently of limiting operations as they are used in calculus.*

7.2 LINEAR INDEPENDENCE OF INFINITESIMAL TRANSFORMATIONS

Every transformation $a_1,\ldots,a_r$ of the r-term group

$$x_i' = f_i(x_1,\ldots,x_n,a_1,\ldots,a_r)$$

has, by our hypothesis, an inverse transformation in the group. If the parameters of the inverse are $\alpha_1,\ldots,\alpha_r$, where the α's are certain functions of the a's, then one has the n equations

$$f_i(f_1(x,a),\ldots,f_n(x,a),\alpha_1,\ldots,\alpha_r) = x_i \quad .$$

We now consider the transformation

$$x_i' = f_i(f_1(a),\ldots,f_n(a),\alpha_1+\omega_1,\ldots,\alpha_r+\omega_r) \quad ,$$

which is in the group. In this the ω_i denote infinitesimals, and hence the transformation equations can be written as

$$x_i' = f_i(f_1(a),\ldots,f_n(a),\alpha_1,\ldots,\alpha_r) + \\ + \sum_k \omega_k \left[\frac{\partial f_i(f_1(a),\ldots,f_n(a),\beta_1,\ldots,\beta_r)}{\partial\beta_k}\right]_{(\beta_k=\alpha_k)}$$

$$\delta x_i = \sum_k \omega_k \left[\frac{\partial f_i(f_1(a),\ldots,f_n(a),\beta_1,\ldots,\beta_r)}{\partial \beta_k} \right]_{(\beta_k = \alpha_k)}$$

thus determined, is infinitesimal, and depends on the r arbitrary infinitesimals $\omega_1,\ldots,\omega_r$. I claim that the ∞^{r-1} infinitesimal transformations so determined are all distinct.

For, suppose that for each i there were a linear relation of the form

$$\sum_k \phi_k(a_1,\ldots,a_r) \left[\frac{\partial f_i(f_1(a),\ldots,f_n(a),\beta_1,\ldots,\beta_r)}{\partial \beta_k} \right]_{(\beta_k = \alpha_k)} =$$

in which the ϕ_k is independent of the index i. Replacing $x_1,\ldots,x_n,a_1,\ldots,a_r$ by $f_1(a),\ldots,f_n(a),\alpha_1,\ldots,\alpha_r$ would then give n relations of the form

$$\sum_k \psi_k(\alpha_1,\ldots,\alpha_r) \frac{\partial f_i(f_1,\ldots,f_n,\alpha_1,\ldots,\alpha_r)}{\partial \alpha_k} = 0 ,$$

so that the number of parameters of the given group could then be diminished.

Thus, we get the theorem:

<u>Theorem 7.2.1.</u> <u>Every r-term group contains the identity transformation and</u> r <u>independent infinitesimal transformations.</u>

That an r-term group cannot contain more than r independent infinitesimal transformations follows from the fact that otherwise it would contain ∞^r distinct _infinitesimal_ transformations, which is impossible since the group contains only ∞^r transformations altogether, and in general these are _finite_.

Chapter 8

RELATIONS AMONG THE INFINITESIMAL TRANSFORMATIONS OF A GROUP

Let $A_1, \ldots, A_r$ be r independent infinitesimal transformations of an r-term group

$$x_i' = f_i(x_1, \ldots, x_n, a_1, \ldots, a_r) \quad ;$$

we shall show that <u>each expression</u> $A_i \circ A_k - A_k \circ A_i$ <u>is a linear combination of the</u> A_j.

8.1 DIFFERENTIAL EQUATIONS FOR THE TRANSFORMATION GROUP

I take first a finite transformation

$$x_i' = f_i(x_1, \ldots, x_n, a_1, \ldots, a_r) \quad ,$$

and then an infinitesimal transformation

$$\delta x_i' = X_i(x_1', \ldots, x_n')\delta t$$

of the same group; their composition

$$x_i'' = f_i(x_1, \ldots, x_n, a_1, \ldots, a_r) + \delta t X_i(x_1', \ldots, x_n')$$

must be a transformation of the group, say

$$x_i'' = f_i(x_1, \ldots, x_n, a_1 + da_1, \ldots, a_r + da_r) \quad .$$

This gives for each i a relation of the form

$$\delta t \cdot X_i(f_1, \ldots, f_n) = \sum_k \frac{\partial f_i(x_1, \ldots, x_n, a_1, \ldots, a_r)}{\partial a_k} da_k \quad ,$$

where the infinitesimals da_k are independent of the index i. Dividing by δt now gives relations of the form

$$X_i(f_1,\ldots,f_n) = \sum_k B_k(a_1,\ldots,a_r) \frac{\partial f_i(x_1,\ldots,x_n,a_1,\ldots,a_r)}{\partial a_k}$$

where again the B_k are independent of the index i and also of $x_1,\ldots,x_n$.

Let now $A_1,\ldots,A_r$ be r infinitesimal transformations, where

$$A_q = X_{q1}\frac{\partial}{\partial x_1} + \cdots + X_{qn}\frac{\partial}{\partial x_n} \quad ;$$

then for each q there are n relations of the form

$$X_{qi}(f_1,\ldots,f_n) = \sum_k B_{qk}(a_1,\ldots,a_r) \frac{\partial f_i(x_1,\ldots,x_n,a_1,\ldots,a}{\partial a_k}$$

where the B_{qk} are independent of the index i, but do depend on the index q.

I claim that $\det(B_{qk}) \neq 0$; otherwise, for each i there would be a relation

$$\sum \psi_q(a_1,\ldots,a_r)X_{qi}(f_1,\ldots,f_n) = 0 \quad ,$$

with ψ_k independent of i, and this is impossible since our r infinitesimal transformations were assumed to be independent. This shows that $\det(B_{qk}) \neq 0$, and therefor we can solve to get relations of the form

$$\frac{\partial f_i(x_1,\ldots,x_n,a_1,\ldots,a_r)}{\partial a_k} = \sum_q L_{kq}(a_1,\ldots,a_r)X_{qi}(f_1,\ldots,f_n)$$

where again L_{kq} is independent of i. Obviously $\det\,(L_{kq}) \neq 0$.

8.2 LIE ALGEBRA RELATIONS AS INTEGRABILITY CONDITIONS

The right side of the last equations satisfy the well-known integrability conditions. This gives $\frac{1}{2}\,r(r-1)$ relations, of which we develop the first:

$$\frac{\partial}{\partial a_2}\left(\sum_q L_{1q}X_{qi}\right) - \frac{\partial}{\partial a_1}\left(\sum_j L_{2j}X_{ji}\right) = 0 \quad .$$

We recall that X depends only on the arguments $f_1,\ldots,f_r$. Therefore,

$$\sum_q \sum_d L_{1q}\,\frac{\partial X_{qi}}{\partial f_\alpha}\,\frac{\partial f_\alpha}{\partial a_2} - \sum_j \sum_\alpha L_{2j}\,\frac{\partial X_{ij}}{\partial f_\alpha}\,\frac{\partial f_\alpha}{\partial a_1} + \sum_q X_{qi}\left(\frac{\partial L_{1q}}{\partial a_2} - \frac{\partial L_{2q}}{\partial a_1}\right) = 0$$

By substituting the values of $\partial f_\alpha/\partial a_2$ and $\partial f_\alpha/\partial a_1$, this gives

$$\sum_q \sum_\alpha L_{1q}\,\frac{\partial X_{qi}}{\partial f_\alpha}\,\sum_j L_{2j}X_{j\alpha} - \sum_j \sum_\alpha L_{2j}\,\frac{\partial X_{ji}}{\partial f_\alpha}\,\sum_q L_{1q}X_{q\alpha} +$$

$$+ \sum_q X_{qi}\left(\frac{\partial L_{1q}}{\partial a_2} - \frac{\partial L_{2q}}{\partial a_1}\right) = 0$$

or

$$\sum_q \sum_j L_{1q}L_{2j} \sum_\alpha \left(X_{j\alpha} \frac{\partial X_{qi}}{\partial f_\alpha} - X_{q\alpha} \frac{\partial X_{ji}}{\partial f_\alpha} \right) + \sum_q X_{qi} \left(\frac{\partial L_{1q}}{\partial a_2} - \frac{\partial L_{2q}}{\partial a_1} \right) = 0$$

or finally

$$\sum_q \sum_j L_{1q}L_{2j}(A_j(X_{qi})-A_q(X_{ji})) + \sum_q X_{qi} \left(\frac{\partial L_{1q}}{\partial a_2} - \frac{\partial L_{2q}}{\partial a_1} \right) = 0 .$$

In this equation the term $A_j(X_{qi}) - A_q(X_{ji})$ occurs twice: once multiplied by $L_{1q}L_{2q}$ and the other time multiplied by $-L_{1j}L_{2q}$. Hence our equation can be put in the form

$$\frac{1}{2} \sum_q \sum_j (L_{1q}L_{2j}-L_{1j}L_{2q})(A_j(X_{qi})-A_q(X_{ji})) =$$

$$= \sum_q X_{qi} \left(\frac{\partial L_{2q}}{\partial a_1} - \frac{\partial L_{1q}}{\partial a_2} \right) ,$$

in which the left side contains the $\frac{1}{2}\,r(r-1)$ quantities $A_j(X_{qi}) - A_q(X_{ji})$.

In a similar manner one obtains altogether $\frac{1}{2}\,r(r-1)$ equations which are linear in the $\frac{1}{2}\,r(r-1)$ quantities $A_j(X_{qi}) - A_q(X_{ji})$. And since $\det\,(L_{\alpha\beta}) \neq 0$, it follows that the determinant of the matrix whose entries are the determinants of the 2×2 submatrices of $(L_{\alpha\beta})$ is also $\neq 0$; so that by solving the linear equations found one obtains

$\frac{1}{2}r(r-1)$ relations of the form

$$A_j(X_{qi}) - A_q(X_{ji}) = \sum_s \phi_{jqs}(a_1,\ldots,a_r)X_{si} ,$$

where the ϕ_{jqs} are independent of the index i. But now both the left sides and the X_{si} are functions of $f_1,\ldots,f_n$ alone and are independent of $a_1,\ldots,a_r$, from which we conclude that the ϕ_{jqs} are constants. This gives the following fundamental theorem:

Theorem 8.2.1. *Let $A_1,\ldots,A_r$ be r infinitesimal transformations of an r-term group, where*

$$A_q = X_{q1}\frac{\partial}{\partial x_1} + \cdots + X_{qn}\frac{\partial}{\partial x_n}$$

Then for each i there are $\frac{1}{2}r(r-1)$ relations of the form

$$A_j(X_{qi}) - A_q(X_{ji}) = \sum_s c_{jqs}X_{si} ,$$

where the c_{jqs} are absolute constants, independent of the index i. These equations can be put together into the $\frac{1}{2}r(r-1)$ relations

$$A_j \circ A_q - A_q \circ A_j = \sum_s c_{jqs}A_s .$$

On a later occasion we shall show that conversely that if r expressions $A_1,\ldots,A_r$ have the property that each $A_j \circ A_q - A_q \circ A_j$ is a linear combination of the A's with constant coefficients, then they are the infinitesimal transformations of an r-term group.* The correctness of this assertion for the case $n = 2$ will follow from the further developments of this paper.

By means of Theorem 8.2.1 we determine in this paper the infinitesimal transformations of any group of a 2-dimensional manifold. In the next section it is shown how one then determines the finite transformations.

8.2 COMMENTS: THE LIE ALGEBRA OF A TRANSFORMATION GROUP

Let G *be a Lie transformation group on a manifold* X. *(See Chapter B.) I will show how one may define a finite dimensional Lie algebra of vector fields on* X, *which may be considered as the infinitesimal version of the action of* G *on* X. *(This functor--replacing a transformation group by its "infinitesimal" version--is the basic tool in Lie's work.)*

Denote the action of G *on* X, *multiplicatively, as usual:*

$$(g,x) \to gx \quad .$$

---

* I gave a proof of this theorem in vol. III, p. 94 of the Archiv for Math. og Naturv., Christiania [Collected Papers, vol. V, paper IV, p. 78 ff.]

Definition. *A flow*

$$t \to \phi_t$$

in X *is said to arise from the action of* G *if there is a curve*

$$t \to g(t)$$

in G *such that:*

$$\phi_t(x) = g(t)(x) \qquad (8.2.1)$$

$$g(0) = 1 \qquad (8.2.2)$$

Definition. *The collection of infinitesimal transformations associated with the flows on* X *arising from the action of* G *is called the Lie algebra of vector fields on* X *defined by the transformation group structure or the infinitesimal version of the action of* G *on* X. *(We show below that the set of these infinitesimal transformations--when considered as vector fields on* X *--do indeed form a Lie algebra.)*

Let us now work this out analytically. Given $g \in G$, *let* g *also denote the transformation*

$$x \to gx$$

associated with g.

If $t \to g(t)$ *is a curve in* G, *beginning at the identity element, then the infinitesimal transformation* $A \in V(X)$ *determined by the flow on* X,

$$x \to g(x)x \quad ,$$

is given by the following formula:

$$A(f) = \frac{\partial}{\partial t} (g(t)^*(f))\Big|_{t=0} \qquad (8.2.3)$$

Let $\underset{\sim}{G}$ *denote the set of vector fields on* X *defined in this way.*

<u>*Theorem 8.2.1*</u>. $\underset{\sim}{G}$ *forms a real vector subspace of* V(X).

<u>*Proof*</u>. *Let* $t \to g(t)$, $g_1(t)$ *be two curves in* G *satisfying 8.2.2. Let* $A, B \in V(X)$ *be the corresponding infinitesimal transformation. Then,*

$$(A+B)(f) = \frac{\partial}{\partial t} ((g(t)g_1(t))^*(f)\Big|_{t=0}$$

$$-A(f) = \frac{\partial}{\partial t} ((g(t)^{-1})^*(f)\Big|_{t=0}$$

These formulas exhibit A+B *and* -A *as infinitesimal transformations associated with flows.*

To finish the proof that $\underset{\sim}{G}$ *is a real vector subspace of* V(X), *let* $a \in R$. *Let* $t \to g(t)$ *be a curve in* G, *and let* $A \in V(X)$ *be the corresponding infinitesimal transformation. Then,*

$$g(t)^*(f) \approx f + tA(f) \quad .$$

Hence,

$$g(at)^*(f) \approx f + (at)A(f)$$

$$= f + t(aA)(f)$$

Then, aA *is the infinitesimal transformation defined by the flow associated with the curve* $t \to g(at)$. *This proves that scalar multiples of elements of* $\underset{\sim}{G}$ *again lie in* $\underset{\sim}{G}$.

Suppose that $g \in \underset{\sim}{G}$, *and that* $A \in \underset{\sim}{G} \subset V(X)$ *is an infinitesimal transformation of* G. *Let*

$$g_*(A)$$

be the vector field on X *defined as follows.*

$$g_*(A)(f) = g^{-1*}(A(g^*(f)) \qquad (8.2.4)$$

for $f \in F(X)$

In words, $g_*(A)$ *is the vector field which results from transforming* A *via the transformation on* X *defined by* g.

<u>*Theorem 8.2.2*</u>. *If* $A \in \underset{\sim}{G}$, $g \in G$, *then*

$$g_*(A) \in \underset{\sim}{G} \ . \qquad (8.2.5)$$

<u>*Proof*</u>. *Let* $t \to g(t)$ *be a curve on* G, $g(0) = 1$, *such that the corresponding flow has* A *as infinitesimal generator. Set:*

$$g_1(t) = gg(t)g^{-1} \ .$$

$t \to g_1(t)$ *is again a curve in* G, *with* $g_1(0) = 1$. *Let us compute its infinitesimal transformation; denoted by* B.

$$B(f) = \frac{\partial}{\partial t}(gg(t)g^{-1})^*(f)\Big|_{t=0}$$

$$= \frac{\partial}{\partial t}(g^{-1*}g(t)^*g^*(f)\Big|_{t=0}$$

$$= g^{-1*}Ag^*(f)$$

$$= \text{using 8.2.4,} \quad g_*(A)(f) \quad .$$

This proves 8.2.5.

Theorem 8.2.3. $\underset{\sim}{G}$ *is a finite dimensional subspace of* $V(X)$. *Its dimension is no greater than the dimension of* G.

Proof. Let

$$G_1$$

denote the tangent space to G *at the identity element. Its dimension is, of course, equal to the dimension of* G.

Let $t \to g(t)$ *be a curve in* G *such that* $g(0) = 1$. *Let* $A \in \underset{\sim}{G}$ *be the corresponding infinitesimal transformation. Let*

$$v \in G_1$$

be the tangent vector to $t \to g(t)$ *at* $t = 0$.

For fixed $x \in X$, *consider the map* $\phi(x)\colon G \to X$ *defined as follows:*

$$\phi(x)(g) = gx \quad .$$

We see from its definition that:

$$\phi(x)_*(v) = A(x) \in X_x \tag{8.2.6}$$

In particular, the infinitesimal transformation A *is uniquely determined by* v. *There is then a map--essentially defined by formula 8.2.6--of*

$$G_1 \to \underset{\sim}{G} \quad .$$

It is readily seen that this map is *<u>linear</u>* *and* *<u>onto</u>*. *Since* G_1 *is finite dimensional, so is* $\underset{\sim}{G}$.

<u>Theorem 8.2.4</u>. $\underset{\sim}{G}$ *is a Lie subalgebra of* $V(X)$, *i.e., if* $A,B \in \underset{\sim}{G}$, *so does*

$$[A,B] \in \underset{\sim}{G} \quad .$$

<u>Proof</u>. *Let* $t \to g(t)$, $g_1(t)$ *be two curves on* G, *whose corresponding infinitesimal transformations are* A *and* $B \in \underset{\sim}{G} \subset V(X)$.

By Theorem 8.2.2,

$$g_1(s)_*(A) \in \underset{\sim}{G}$$

for all s.

Set:

$$C = \frac{d}{dt} g_1(s)_*(A)\Big|_{s=0} \tag{8.2.7}$$

Since $\underset{\sim}{G}$ *is a finite dimensional vector space, it should be clear that*

$$C \in \underset{\sim}{G} \quad .$$

Let us compute it explicitly, using formula 8.2.4:

$$g_1(s)_*(A)(f) \quad = \quad g_1(s)^{-1}{*}A(g_1(s){*}(f)) \quad ,$$

hence:

$$C(f) \quad = \quad \frac{\partial}{\partial s}\left(g_1(s)^{-1}{*}Ag_1(s){*}(f)\right)\Big|_{s=0}$$

$$= \quad -BA(f) + AB(f)$$

$$= \quad [A,B](f) \quad .$$

Hence,

$$[A,B] \quad = \quad C \in \underset{\sim}{G} \quad .$$

These results cover the main facts about the relation between Lie transformation groups and Lie algebras of vector fields. (They readily extend also to "local" Lie transformation groups.)

Here is a more intuitive way of determining the Lie algebra $\underset{\sim}{G}$. *Suppose that* G *and* X *are vector* *spaces*. *Let* "g" *and* "dg" *denote elements of* G. *Then, the curve*

$$t \to (g+tdg)g^{-1} \quad = \quad g(t)$$

is a curve in G, *whose tangent vector at* $t = 0$ *is* dg. *The vector field* A *is obtained as follows*

$$(g+dg)g^{-1}(x) \quad \approx \quad x + A(x) \quad . \qquad (8.2.8)$$

Another way of writing this is:

$$(g+dg)g^{-1}(x) = x + A(x,dg) + \cdots \tag{8.2.9}$$

where $\cdots$ *denotes terms of higher order in* dg *and*

$$(x,dg) \to A(x,dg) \in X_x$$

is a map

$$X \times G_1 \to T(X)$$

which is *linear* *in* G_1.

Example.

a) X = *real vector space,*

$G = GL(X)$ = *group of invertible linear maps:* $X \to X$.

Since $x \to gx$ *is* *linear* *in* x, *8.2.9 can be rewritten as:*

$$dgg^{-1}(x) = A(x,dg) \quad . \tag{8.2.10}$$

b) X = *real vector space,*

G = *group of* *affine* *automorphisms of* X.

Thus, a $g \in G$ *is a map of the following form:*

$$g(x) = \alpha(x) + y$$

with: $\alpha \in GL(X)$, $y \in X$.

Then,

$$g^{-1}(x) = \alpha^{-1}(x-y)$$

Set:

$$dg(x) = d\alpha(x) + dy$$

Then,

$$dgg^{-1}(x) = dg(\alpha^{-1}(x-y))$$

$$= d\alpha(\alpha^{-1}(x-y)) + dy$$

$$= (d\alpha\alpha^{-1})(x) + dy - (d\alpha\alpha^{-1})(y) \qquad (8.2.11)$$

Here is another less "symbolic" way of interpreting formula 8.2.11. Let:

$$L(X) = \text{space of linear maps: } X \to X.$$

To each $\beta \in L(X)$, $y \in X$, *define a map*

$$A(\beta,y): X \to X$$

via the following formula:

$$A(\beta,y)(x) = \beta(x) + y \quad .$$

Interpret $A(\beta,y)$ *as an element of* $V(X)$ *(identifying* $V(X)$ *with the space of maps* $X \to X$*). Then, as* (β,y) *varies over* $L(X) \times X$, *the* $A(\beta,y)$ *vary over a Lie algebra of vector fields on* X, *which is the* <u>*infinitesimal version of the action of the affine group*</u>.

Chapter 9

A GROUP IS DETERMINED BY ITS INFINITESIMAL TRANSFORMATIONS

In this section it is shown that r <u>independent infinitesimal transformations</u> $A_1,\ldots,A_r$ <u>can belong to at most one r-term group</u>

$$x_i' = f_i(x_1,\ldots,x_n,a_1,\ldots,a_r)$$

9.1 ONE PARAMETER GROUPS

I show first that if an infinitesimal transformation

$$\delta x_i = X_i(x_1,\ldots,x_n)\delta t$$

of an arbitrary group is known, one can always derive from it a 1-parameter family of transformations of this group.

We have previously found the n relations

$$X_i(f_1,\ldots,f_n)\delta t = \sum_k \frac{\partial f_i}{\partial a_k} da_k = df_i \quad ,$$

in which the da_k were independent of the index i and satisfy a certain simultaneous system

$$da_k = \psi_k(a_1,\ldots,a_r)\delta t \quad .$$

Let

$$W_k(f_1,\ldots,f_n,t) = W_k(f_1^{(0)},\ldots,f_n^{(0)},t^{(0)})$$

be the integral equations of the simultaneous system $df_i = X_i(f_1,\ldots,f_n)\delta t$, and let

$$\Omega_k(a_1,\ldots,a_r,t) = \Omega_k(a_1^{(0)},\ldots,a_r^{(0)},t^{(0)})$$

be the integral equations of the system $da_k = \psi_k(a_1,\ldots,a_r)\delta t$. The initial values $f_k^{(0)}$ and $a_k^{(0)}$ can be assumed connected by the equations

$$f_k^{(0)} = f_k(x_1,\ldots,x_n,a_1^{(0)},\ldots,a_r^{(0)}) \quad ,$$

and I choose the $a_k^{(0)}$ so that the equations

$$f_k(x_1,\ldots,x_n,a_1^{(0)},\ldots,a_r^{(0)}) = x_k$$

hold. Then one finds the n equations

$$W_k(f_1,\ldots,f_n,t) = W_k(x_1,\ldots,x_n,t^{(0)}) \quad ,$$

which have a solution

$$f_i(x_1,\ldots,x_n,a_1,\ldots,a_r) = \Phi_i(x_1,\ldots,x_n,t,t^0) \; .$$

In this, t is an arbitrary quantity, while the a_k are known functions of the definite quantities $a_k^{(0)}$ and the parameter t.

Thus we have shown that <u>the 1-parameter family of transformations</u>

$$x_k' = \Phi_k(x_1,\ldots,x_n,t)$$

<u>belong to the group</u>.

9.2 UNIQUENESS OF THE INFINITESIMAL LIE ALGEBRA OF VECTOR FIELDS

Let now $A_1,\ldots,A_r$ be r independent infinitesimal transformations of a group, where

$$A_k = X_{k1}\frac{\partial}{\partial x_1}+\ldots+X_{kn}\frac{\partial}{\partial x_n}$$

Then one finds ∞^r finite transformations of this group as follows.

One forms the general infinitesimal of the group

$$\lambda_1 A_1 +\ldots+ \lambda_r A_r \quad ,$$

and then integrates the simultaneous system

$$\frac{dx_1}{\sum \lambda_k X_{k1}} = \cdots = \frac{dx_n}{\sum \lambda_k X_{kn}} = \delta t \ ,$$

considering the λ_k's as constant. If

$$W_1(x_1,\ldots,x_n,\lambda_1 t,\ldots,\lambda_r t), W_2,\ldots, W$$

are independent solutions of this system, then one solves the equations

$$W_i(x_1',\ldots,x_n',\lambda_1 t,\ldots,\lambda_r t) = W_i(x_1,\ldots,x_n,\lambda_1 t_0,\ldots,\lambda_r t_0)$$

for the x_i'; the transformations

$$x_i' = f_i(x_1,\ldots,x_n),\lambda_1(t-t_0),\ldots,\lambda_r(t-t_0))$$

determined by the preceding equations belong to the group by previous results; moreover, they depend on the r

parameters $\lambda_1(t-t_0),\ldots,\lambda_r(t-t_0)$; therefore if we show that the number of these parameters cannot be decreased, then they in fact provide all the transformations of the group.

The x_i' can be expanded in powers of $(t-t_0)$:

$$x_i' = x_i + (t-t_0)\sum \lambda_k X_{ki}(x_1,\ldots,x_n) + \cdots$$

Hence if for each i there were a relation

$$\sum_q \psi_q(\lambda_1(t-t_0),\ldots,\lambda_r(t-t_0)) \frac{\partial x_i'}{\partial(\lambda_q(t-t_0))} = 0 ,$$

then, by expanding the left side in powers of $(t-t_0)$ and setting the coefficients of $(t-t_0)^0$ equal to zero, one would obtain a relation of the form

$$\sum_q \psi_q^{(0)}\cdot X_{qi}(x_1,\ldots,x_n) = 0 ,$$

where the $\psi_q^{(0)}$ are constants and are independent of the index i. But then our infinitesimal transformations would not be independent, contrary to what has been assumed.*

Therefore, our power series determine ∞^r distinct finite transformations, and so they are actually all the transformations of the given group. Thus we obtain the following fundamental theorem:

---

* In the text it has been tacitly assumed that not all the quantities ψ_q vanish under the substitution $t = t_0$. If this exceptional case does occur then the developments of the text must be modified somewhat.

Theorem 9.2.1. The infinitesimal transformations of an r-term group cannot all belong to a different r-term group.

9.3 COMMENTS: THE EQUATIONS DEFINING A TRANSFORMATION GROUP

Let G be a Lie group, X a manifold on which G acts. In the Comments to Chapter 8, I have shown how G acting on X determines a Lie algebra of vector fields. I will now show how these vector fields determine the transformation group action, at least locally.

Suppose X is a vector space. The transformation group action is determined by a map

$$G \times X \to X \quad .$$

Write this map as

$$(g,x) \to f(g,x) \equiv gx$$

Let G act on $G \times X$ as follows:

> *The transform of a point (g,x) by $g_0 \in G$ is*
> $$(gg_0^{-1}, g_0x) \equiv g_0(g,x) \quad . \qquad (9.3.1)$$

Then,

$$\begin{aligned} f(g_0(g,x)) &= f(gg_0^{-1}, g_0x) = gg_0^{-1}g_0x \\ &= gx \\ &= f(g,x) \quad . \end{aligned} \qquad (9.3.2)$$

Identity 9.3.1 says that:

The X-*valued function* $f: G \times X \to X$ *is* *invariant* *under the action of* G *on* $G \times X$.

Let $t \to g(t)$ *be a curve in* G, *with* $g(0) = 1$. *Its action (via formula 9.3.1) on* $G \times X$ *determines a flow on* $G \times X$, *which has an infinitesimal generator*

$$t \to A''_t \in V(G\times X) \quad .$$

Now, A''_t *can be decomposed as a sum of vector fields on* G *and* X*:*

$$A''_t = A'_t + A_t \quad , \tag{9.3.3}$$

$$A'_t \in V(G), \quad A_t \in V(X) \; .$$

Remark. $t \to A'_t$ *is the infinitesimal generator of the flow*

$$g \to gg(t)^{-1} \quad on \quad G$$

$t \to A_t$ *is the infinitesimal generator of the flow*

$$x \to g(t)x$$

on X.

Hence, the *invariance* *of* f *under the action of* G *on* $G \times X$ *implies it satisfies the following set of differential equations:*

$$A''_t(f) \equiv A'_t(f) + A_t(f) = 0 \quad . \tag{9.3.4}$$

Let us make these equations more explicit, in terms of local coordinates for G *and* X. *(We choose coordinates so as to make contact with Lie's notation.) Choose indices (and the summation convention) as follows:*

$$i \leq i,j \leq n = \dim X$$

$$i \leq u,v \leq m = \dim G$$

Let (x^i) *be coordinates for* X, (a^u) *be coordinates for* G. *Then set:*

$$A''_t = A^i_t \frac{\partial}{\partial x^i} + A^u \frac{\partial}{\partial a^u} .$$

Thus,

$$A'_t = A^u_t \frac{\partial}{\partial a^u}$$

$$A_t = A^i_t \frac{\partial}{\partial x^i}$$

<u>*Remark*</u>. A'_t *is a* <u>*left-invariant vector field on*</u> G. A_t *is the* <u>*infinitesimal generator*</u> *associated with the action of* G *on* X.

Write the map

$$f: G \times X \to X$$

(which, recall, <u>*defines*</u> *the group action of* G *on* X*) in terms of these functions as follows:*

$$f^*(x^i) = f^i(x,a) . \tag{9.3.5}$$

The functions on the right hand side of 9.3.5 then satisfy (making 9.3.4 explicit) the following system of differential equations:

$$A_t^u(a) \frac{\partial f^i}{\partial a^u} + A_t^j \frac{\partial f^i}{\partial x^j} = 0 \tag{9.3.6}$$

Remark. The way Lie proves Theorem 8.2.4 is to show that it is a consequence of the integrability conditions for equations 9.3.6.

Now, the action of G *on* G × X *which gives rise to 9.3.6 (and determines the functions* f^i*) is composed of two separate actions, one on* G *itself via right translations, the other the given action on* X. *Now, the first action is given as soon as* G *is given. Hence, Equations 9.3.6 imply the uniqueness stated in Theorem 9.2.1. Here is a restatement of this result in modern language.*

Theorem 9.3.1. Suppose G *is a connected Lie group,* X *a manifold, and*

$$f: G \times X \to X$$
$$f': G \times X \to X$$

two maps which define G *as Lie transformation groups on* X. *Let* $\underset{\sim}{G}$ *and* $\underset{\sim}{G}'$ *be the Lie algebra of vector fields on* X *determined by the two actions.*

Conclusion: If $\underset{\sim}{G} = \underset{\sim}{G}'$, *then* $f = f'$.

Here is another geometric way of stating these ideas. Let G *be a Lie transformation group on a manifold* X. *Let* M *be the subset of* $G \times X \times X$ *consisting of the points*

$$(g,x,y) \in G \times X \times X$$

such that

$$y = gx \quad .$$

(In words, M *is the* graph *of the map*

$$(g,x) \to gx \qquad G \times X \to X \;)$$

Then, M *is the maximal orbit submanifold (see Chapter B) of a vector field systems on* $G \times X \times X$. *This vector field system is determined by the group structure on* G plus *the Lie algebra of vector fields.*

To understand this result better, consider the special case where:

G = *the* additive *group of the real numbers, denoted by* R.

Denote an element of R *by* t. *Let* f

$$f\colon G \times X \to X$$

be a map determining a transformation group action.

For $t \in G$, *set:*

$$\phi_t(x) = f(t,x) \qquad (9.3.7)$$

For fixed t, ϕ_t *is a map:* $X \to X$. *As* t *varies, it determines a* ***flow*** *on* X. *Let*

$$t \to A_t \in V(X)$$

be its infinitesimal generator.

Now, the ***group property*** *implies that:*

$$\phi_{t_1+t_2} = \phi_{t_1}\phi_{t_2} \tag{9.3.8}$$

for $t_1, t_2 \in R$. *A flow satisfying this property is called a* ***one-parameter group of diffeomorphisms of*** X. *Here is a basic result, whose proof is left to the reader.*

Theorem 9.3.2. *A flow* $t \to \phi_t$ *on a manifold* X *is a one-parameter group of diffeomorphisms of* X *if and only if its infinitesimal generator* $t \to A_t$ *is independent of* t.

Thus, the "infinitesimal generator" of a one-parameter ***group*** *is a single vector field. There is then a mapping*

Infinitesimal generator: (one-parameter group on X*)* $\to$ *(vector fields on* X*)*

As a special case of Theorem 9.3.1, we see that the map is ***one-one***.

If $t \to \phi_t$ *is a one-parameter group of diffeomorphisms of* X, *with* $A \in V(X)$ *its infinitesimal generator,* A *has the following geometric relation to the group.*

> *The orbit curve* $t \to \phi_t(x)$ *of the group are the orbit curves of the vector field* A.

In terms of coordinates (x^i) *for* X, *with:*

$$A = A^i \frac{\partial}{\partial x^i} \quad ,$$

this means that the solutions of the ordinary differential equations

$$\frac{dx^i}{dt} = A^i(x(t))$$

are the orbit curves of the group. Again, this connection between ordinary differential equations and groups is a basic feature of Lie's work.

Return to the general situation. Suppose that G *is a Lie group, which acts on the manifold* X. *Let* $\underset{\sim}{G}$ *be the Lie algebra of vector fields on* X, *which are the infinitesimal transformations of the flows on* X *arising from curves in* G. *Here is another basic result:*

__Theorem 9.3.3__. Let $t \to A_t$, $-\infty < t < \infty$ *be a curve in* $\underset{\sim}{G}$. *Then, there is a curve* $t \to g(t)$, $-\infty < t < \infty$, *in* G, *such that* $g(0) = 1$, *and* A_t *is the infinitesimal generator of the flow in* X *determined by the curve* $t \to g(t)$.

Proof. Let G act on $G \times X$ in the following way:

$$g_0(g,x) = (g_0 g,x) \tag{9.3.9}$$

Remark. This is the product of the *left-action of* G *on itself*, and the trivial action of G on X.

Let

$$f: G \times X \to X \tag{9.3.10}$$

be the map defining the action of G on X, i.e.,

$$f(g,x) = gx \tag{9.3.11}$$

Then, 9.3.9 and 9.3.11 imply that:

$$g_0 f(g,x) = g_0(gx)$$

$=$ *using the transformation group property*,

$$(g_0 g)x = f(g_0(g,x))$$

This means that:

f intertwines the action 9.3.9 of G on $G \times X$ and the given action of G on X. (9.3.12)

Let $\underset{\sim}{G}'$ be the Lie algebra of vector fields on $G \times X$ determined by the action 9.3.9 of G on $G \times X$.

Remark. $\underset{\sim}{G}'$ *consists of the vector fields on* $G \times X$ *which are invariant under right translation by* G, *and do not depend on the X-variables. In other words, they are invariant under the action of the group of all diffeomorphisms of* X, *and the right translation groups of* G *on itself.*

Because of the intertwining property of f, *i.e.*, *9.3.12, we have:*

$$f_*(\underset{\sim}{G}') = \underset{\sim}{G} .$$

Let $t \to A'_t$ *be a curve in* $\underset{\sim}{G}'$ *such that:*

$$f_*(A'_t) = A_t . \qquad (9.3.13)$$

Now, we must show that there is a curve $t \to g(t)$ *in* G, *defined over* $-\infty < t < \infty$, *such that:*

$$g(0) = 1$$

> A'_t *is the infinitesimal generator of the flow on* $G \times X$ *determined by* $g(t)$ *and the action 9.3.9.*

That such a curve exists locally in t *should be clear-- as for any flow, it is just a matter of solving ordinary differential equations. (In this case, in terms of local coordinates* (a^u) *for* G.) *What is remarkable in this case is that the local solution, beginning at* $g = 1$ *for* $t = 0$, *can be analytically continued over* $-\infty < t < \infty$.

I will not go into the full details of this proof--essentially it involves the fact that the vector fields A'_t *are invariant under* right *translation by* G, *acting on* $G \times X$.

That A_t *is the infinitesimal generator of the action of* $t \to g(t)$ *on* X *now follows readily from 9.3.13. This finishes the proof of Theorem 9.3.3.*

An important special case of Theorem 9.3.3 occurs when a single vector field $A \in \underset{\sim}{G}$ *is chosen. The flow it generates is then determined by a curve*

$$t \to g(t)$$

in G *such that:*

$$g(t_1+t_2) = g(t_1)g(t_2) \tag{9.3.14}$$

for $t_1, t_2 \in R$.

Such a curve in G *is called a* <u>one-parameter subgroup</u> *of* G. *They play an important role in the modern and classical theory of Lie groups. Thus, we may say that:*

> *If a Lie group acts on a manifold* X, *the Lie algebra* $\underset{\sim}{G}$ *of vector fields on* X *it determines is precisely the set of infinitesimal generators of one-parameter subgroups of* G.

This correspondence between one-parameter subgroups and Lie algebras can be made more precise and "functorial". In fact, if:

$$t \to g_1(t), g_2(t)$$

are two one-parameter subgroups of G, *set:*

$$g_3(t) = \lim_{n\to\infty} \left[g_1\left(\frac{t}{n}\right) g_2\left(\frac{t}{n}\right) \right]^n$$

$$g_4(t) = \lim_{n\to\infty} \left[g_1\left(\frac{\sqrt{t}}{n}\right) g_2\left(\frac{\sqrt{t}}{n}\right) g_1\left(\frac{-\sqrt{t}}{n}\right) g_2\left(\frac{\sqrt{t}}{n}\right) \right]^{n^2}$$

It can be proved that the formulas determine new one-parameter subgroups. They essentially determine the Lie algebra structure associated with G. *If:*

$A \in V(X)$ *is the infinitesimal generator of* g_1

$B \in V(X)$ *is the infinitesimal generator of* g_2,

then:

$A+B$ *is the infinitesimal generator of* g_3.

$[A,B]$ *is the infinitesimal generator of* g_4.

Here is another important property. Let $t \to A_t$ *be a curve in* $\underset{\sim}{G} \subset V(X)$. *The differential equations*

$$\frac{dx}{dt} = A_t(x(t))$$

determining the orbit curves of the flow generated by A_t *are called a* *Lie system* *associated with the action of* G. *(See* IM, VOLS. *III and IX.) We see that they are determined by curves* $t \to g(t)$ *in* G *which are given by solutions of differential equations* *involving the parameters of* G. *Further, as a consequence of the* *global* *action of* G *on* X, *they have solutions over the full interval* $-\infty < t < \infty$.

Chapter 10

TRANSFORMATION OF LINE-ELEMENTS

10.1 PROLONGATION TO THE CONTACT MANIFOLD

In turning now to the transformation groups of a 2-dimensional manifold x,y, I interpret x and y as Cartesian coordinates of a plane.

I denote the infinitesimal point-transformation

$$\delta x = \xi(x,y)\delta t \ , \quad \delta y = \eta(x,y)\delta t$$

by the symbol

$$A = \xi \frac{\partial}{\partial x} + \eta \frac{\partial}{\partial y}.$$

or

$$A = \xi p + \eta q \ ,$$

where $\partial/\partial x = p$, $\partial/\partial y = q$, and I consider this transformation as an operation taking each point (x,y) of the plane to the neighboring position $(x+\xi\delta t, y+\eta\delta t)$. At the same time the line elements of the plane, whose coordinates are x,y and $dy:dx = y'$ assume certain neighboring positions for whose determination it suffices to compute $\delta y'$. One has:

$$\frac{\delta y'}{\delta t} = \frac{\delta}{\delta t}\frac{dy}{dx} = \frac{dx\,\frac{\delta(dy)}{\delta t} - dy\,\frac{\delta(dx)}{\delta t}}{dx^2}$$

$$= \frac{dx \cdot d\frac{\delta y}{\delta t} - dy \cdot d\frac{\delta x}{\delta t}}{dx^2} = \frac{dxd\eta - dyd\xi}{dx^2}$$

$$= \frac{\partial}{\partial x} + y'\left(\frac{\partial\eta}{\partial y} - \frac{\partial\xi}{\partial x}\right) - y'^2 \frac{\partial\xi}{\partial y}$$

If we want to make it explicit that the transformation $A = \xi p + \eta q$ carries not only the point (x,y) but also the line element (x,y,y') of the plane into a new position, then we denote our transformation by the symbol

$$B = \xi \frac{\partial}{\partial x} + \eta \frac{\partial}{\partial y} + \left[\frac{\partial\eta}{\partial x} + y'\left(\frac{\partial\eta}{\partial y} - \frac{\partial\xi}{\partial x}\right) - y'^2 \frac{\partial\xi}{\partial y}\right]\frac{\partial}{\partial y'}$$

COMMENTS ON SECTION 10.1

The prolongation process is basic to Lie's work. I have described it, in general, in GPS, in terms of "contact manifolds". I will now briefly describe what is involved, in the specific situation dealt with here by Lie.

Let X *be a manifold. Recall that* $T(X)$ *denotes the tangent vector bundel, while* $T^d(X)$ *denotes the cotangent vector bundle. Let*

$$PT^d(X)$$

denote the bundle over X, *whose fiber above a point* $x \in X$ *is the* <u>*projective space*</u> *associated with the vector space*

$$X_x^d \quad ,$$

i.e., the ***projective space of the space of 1-covectors.***

Remark. *The symbol* "$PT^d(X)$" *is read as follows: The* ***projective cotangent bundle of*** X.

Here is another useful way of looking at this. Let

$$R - (0)$$

denote the ***multiplicative group*** *of non-zero real numbers. Let*

$$T^d(X) - (0)$$

denote the bundle of non-zero cotangent vectors. Let $R - (0)$ *act on* $T^d(X) - (0)$ *by* ***dilitation****, i.e.,*

> *The transform of* $\theta \in X_x^d$ *by* $\lambda \in R - (0)$ *is the scalar product* $\lambda\theta$.

$PT^d(X)$ *is then the* ***orbit space*** *of the action of* $R - (0)$ *on* $T^d(X) - (0)$.

Geometrically, $PT^d(X)$ *may be identified with the space of* ***first order contact elements of hypersurfaces of*** X. *In GPS this space was denoted by*

$$C^1(X,n-1) \quad ,$$

where $n = \dim X$.

Here is how this association is set up. Let Y *be an* (n-1)*-dimensional mapping, and let*

$$\phi: Y \to X$$

be a submanifold mapping. For $y \in Y$,

$$\phi_*(Y_y)$$

is an (n-1)*-dimensional linear subspace of* $X_{\phi(y)}$. *This determines a one-dimensional linear subspace of* $X^d_{\phi(y)}$ -- *namely, the set of* $\theta \in X^d_{\phi(y)}$ *such that*

$$\theta(\phi_*(Y_y)) = 0$$

Such a one-dimensional linear subspace defines a *__point__* *of* $PT^d(X)$, *which we denote by*

$$\partial\phi(y) \quad .$$

As y *varies, we obtain a map*

$$\partial\phi: Y \to PT^d(X) \quad ,$$

called the *__prolongation__* *of* ϕ. *It also plays a key role in Lie's work. It is a "lifting" of* ϕ, *in the sense that the following diagram of maps is commutative:*

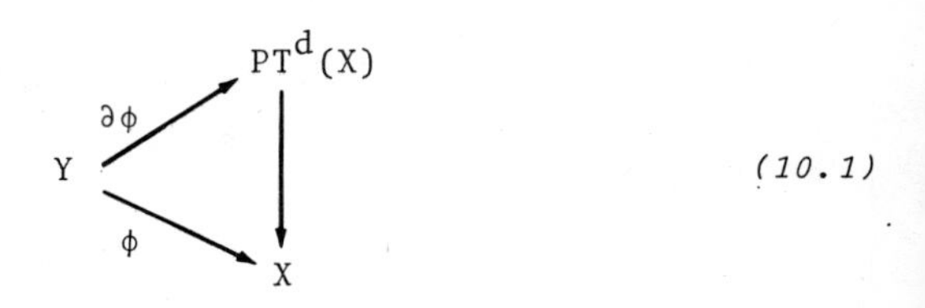

Now, let G *be a group of diffeomorphisms of* X. *Each* $g \in G$ *acts on* $T(X)$:

$$g(v) = g_*(v)$$

for $v \in T(X)$.

Let it act on $T^d(X)$ *as follows:*

$$g(\theta) = (g^{-1})^*(\theta) \qquad (10.2)$$

for $\theta \in T^d(X)$.

This action of G *passes to the quotient to act on*

$$PT^d(X) \quad .$$

It is this action (and its "infinitesimal" equivalent in terms of vector fields) with which Lie is concerned in this section. It is one of the typical examples of "prolongation" of <u>*group actions*</u>*, which again is a key feature in Lie's work.*

The action of $G = \text{diff}\ (X)$ *is also compatible with the diagram 10.1. In words, if*

$$\phi: Y \to X$$

is a submanifold map, if

$$g\phi$$

is its transform by $g \in G$, *then*

$$\partial(g\phi) = g\partial\phi \quad . \qquad (10.3)$$

This indicates the "naturality" or "functoriality" of the prolongation process. (I believe a case could be made that Lie was one of the first mathematicians to appreciate the meaning of "functorial" arguments.)

At the Lie algebra level, the prolongation process leads to a map

$$V(X) \to V(PT^d(X)) \tag{10.4}$$

which is a *linear*, *first order differential operator*, *in the general sense described in Chapter I of GPS.*

Here is the simplest geometric definition of 10.3. Let $t \to g(t)$ *be a one-parameter* *group* *of diffeomorphisms of* X. *(The argument can be modified to cover the case of a one-parameter* *local* *group.) Let* $A \in V(X)$ *be its infinitesimal generator. Let* $t \to g(t)$ *act on* $PT^d(X)$ *via prolongation. Let* B *be the infinitesimal generator of this group. Then*

$$A \to B$$

is the prolongation map 10.3.

B *can also be defined, as explained in GPS, in terms of the* *contact differential form structure* *on* $PT^d(X)$, *defined by a single 1-form* θ *on* $PT^d(X)$. *Let*

$$\pi\colon PT^d(X) \to X$$

be the projection map. Then, B *is* *characterized* *by the following two conditions.*

$$\pi_*(B) = A \tag{10.5}$$

$$B(\theta) = f\theta, \quad \textit{for some} \quad f \in F(PT^d(X))$$

B *is an infinitesimal automorphism of the contact structure.*

To specialize to the case considered by Lie in this section, let:

$$X = R^2 .$$

Denote Cartesian coordinates X, *in the usual analytic-geometry way, by*

$$(x,y) .$$

Let $\hat{x},\hat{y}$ *denote the* ***linear*** *coordinates on* $T^d(R^2)$ *such that:*

$$\hat{x}\left(\frac{\partial}{\partial x}\right) = 1 = \hat{y}\left(\frac{\partial}{\partial y}\right)$$

$$\hat{x}\left(\frac{\partial}{\partial y}\right) = 0 = \hat{y}\left(\frac{\partial}{\partial x}\right)$$

Then,

$$(x,y,\hat{x},\hat{y})$$

forms a coordinate system for $T^d(M)$. *Set:*

$$y' = \frac{\hat{y}}{\hat{x}}$$

(Formally, $\hat{y} = dy/dt$, $\hat{x} = dx/dt$, *hence*

$$y' = \frac{dy}{dx})$$

y' *passes to the quotient, to define a function on* $PT^d(X)$.

$$(x,y,y')$$

thus forms a coordinate system for $PT^d(X)$.

If

$$\phi: x \to (x, y(x))$$

is a one-dimensional submanifold of X, *then*

$\partial\phi$ *is the submanifold:*

$$x \to \left(x, y(x), \frac{dy}{dx} \equiv y'(x)\right) .$$

The contact 1-form θ *on* $PT^d(X)$ *is given in these coordinates by:*

$$\theta = dy - y'dx \tag{10.7}$$

Let

$$A = \xi \frac{\partial}{\partial x} + \eta \frac{\partial}{\partial y} .$$

We can compute its prolongation B *using 10.5 and 10.6. 10.5 requires that:*

$$A(x) = B(x) = \xi$$

$$A(y) = B(y) = \eta$$

Hence,

$$B(\theta) = d\eta - B(y')dx - y'd\xi$$

$$= \frac{\partial \eta}{\partial x} dx + \frac{\partial \eta}{\partial y} dy - B(y')dx - y'\left(\frac{\partial \xi}{\partial x} dx + \frac{\partial \xi}{\partial y} dy\right)$$

$$= f(dy - y'dx)$$

Comparing coefficients, we have:

$$f = \frac{\partial\eta}{\partial y} - y' \frac{\partial\xi}{\partial y}$$

$$B(y') = \left(\frac{\partial\eta}{\partial y} - y' \frac{\partial\xi}{\partial y}\right) y' + \frac{\partial\eta}{\partial x} - y' \frac{\partial\xi}{\partial x}$$

This is precisely the formula for B *given in the text.*

The notation

$$p = \frac{\partial}{\partial x}, \quad q = \frac{\partial}{\partial y}$$

is standard in the 19-th century differential geometry literature. One can, in fact, interpret p *and* q *as functions on the* <u>*cotangent bundle*</u> *to* R^2. *The contact form is then:*

$$\theta = pdx + qdy .$$

10.2 PRESERVATION OF JACOBI BRACKETS UNDER PROLONGATION

Let $A_1,\ldots,A_r$ be r independent infinitesimal transformations (of a group), where

$$A_i = \xi_i p + \eta_i q \qquad (i = 1,\ldots,r)$$

with the relations

$$A_i \circ A_k - A_k \circ A_i = \sum_s c_{iks} A_s \qquad (10.2.1)$$

I then put

$$B_i = \xi_i \frac{\partial}{\partial x} + \eta_i \frac{\partial}{\partial y} + \left[\frac{\partial \eta_i}{\partial x} + y'\left(\frac{\partial \eta_i}{\partial y} - \frac{\partial \xi_i}{\partial x}\right) - y'^2 \frac{\partial \xi_i}{\partial y}\right]\frac{\partial}{\partial y'}$$

or

$$B_i = \xi_i p + \eta_i q + \zeta_i \frac{\partial}{\partial y'} ,$$

and then claim that the B_i satisfy the analogous relations

$$B_i \circ B_k - B_k \circ B_i = \sum_s c_{iks} B_s ,$$

or, equivalently,

$$B_i(\zeta_k) - B_k(\zeta_i) = \sum_s c_{iks} \zeta_s .$$

By direct computation one finds

$$B_i(\zeta_k) - B_k(\zeta_i) = L + My' + Ny'^2 ,$$

where

$$L = A_i\left(\frac{\partial \eta_k}{\partial x}\right) - A_k\left(\frac{\partial \eta_i}{\partial x}\right) + \frac{\partial \eta_i}{\partial x}\left(\frac{\partial \eta_k}{\partial y} - \frac{\partial \xi_k}{\partial x}\right) - \frac{\partial \eta_k}{\partial x}\left(\frac{\partial \eta_i}{\partial y} - \frac{\partial \xi_k}{\partial x}\right) ,$$

$$M = A_i\left(\frac{\partial \eta_k}{\partial y} - \frac{\partial \xi_k}{\partial x}\right) - A_k\left(\frac{\partial \eta_i}{\partial y} - \frac{\partial \xi_i}{\partial x}\right) - 2 \frac{\partial \eta_i}{\partial x} \frac{\partial \xi_k}{\partial y} + 2 \frac{\partial \xi_i}{\partial y} \frac{\partial \eta_k}{\partial x} ,$$

$$N = -A_i\left(\frac{\partial\xi_k}{\partial y}\right) + A_k\left(\frac{\partial\xi_i}{\partial y}\right) - \left(\frac{\partial\eta_i}{\partial y} - \frac{\partial\xi_i}{\partial x}\right)\frac{\partial\xi_k}{\partial y}$$

$$+ \left(\frac{\partial\eta_k}{\partial y} - \frac{\partial\xi_k}{\partial x}\right)\frac{\partial\xi_i}{\partial y}$$

By 10.2.1 one has

$$\xi_i \frac{\partial\eta_k}{\partial x} + \eta_i \frac{\partial\eta_k}{\partial y} - \xi_k \frac{\partial\eta_i}{\partial x} - \eta_k \frac{\partial\eta_i}{\partial y} = \sum_s c_{iks}\eta_s \quad , \tag{10.2.2}$$

and therefore, by differentiating with respect to x:

$$A_i\left(\frac{\partial\eta_k}{\partial x}\right) - A_k\left(\frac{\partial\eta_i}{\partial x}\right) + \frac{\partial\xi_i}{\partial x}\frac{\partial\eta_k}{\partial x} + \frac{\partial\eta_i}{\partial x}\frac{\partial\eta_k}{\partial y} - \frac{\partial\xi_k}{\partial x}\frac{\partial\eta_i}{\partial x} - \frac{\partial\eta_k}{\partial x}\frac{\partial\eta_i}{\partial y}$$

$$= \sum_s c_{iks}\frac{\partial\eta_s}{\partial x} \quad ,$$

or

$$L = \sum_s c_{iks}\frac{\partial\eta_s}{\partial x} \quad .$$

Similarly, if the relation

$$\xi_i \frac{\partial\xi_k}{\partial x} + \eta_i \frac{\partial\xi_k}{\partial y} - \xi_k \frac{\partial\xi_i}{\partial x} - \eta_k \frac{\partial\xi_i}{\partial y} = \sum_s c_{iks}\xi_s \tag{10.2.3}$$

is differentiated with respect to y, one gets

$$N = -\sum_s c_{iks} \frac{\partial \xi_s}{\partial y} .$$

Finally, differentiating the Equations 10.2.2 and 10.2.3 with respect to y and x respectively, and then subtracting these relations, one finds

$$M = \sum_s c_{iks} \left(\frac{\partial \eta_s}{\partial y} - \frac{\partial \xi_s}{\partial x} \right)$$

Therefore,

$$L + My' + Ny'^2 = \sum_s c_{iks} \left[\frac{\partial \eta_s}{\partial x} + y' \left(\frac{\partial \eta_s}{\partial y} - \frac{\partial \xi_s}{\partial x} \right) - y'^2 \frac{\partial \xi_s}{\partial y} \right]$$

or

$$B_i \zeta_k - B_k \zeta_i = \sum_s c_{iks} \zeta_s ,$$

as claimed.

COMMENTS ON SECTION 10.2

In terms of the notation introduced in my comments to Section 10.1, this section proves that the prolongation differential operator

$$V(X) \rightarrow V(PT^d(X))$$

is a _*Lie algebra homomorphism of vector fields*_*, i.e., preserves Jacobi brackets. This property can, in fact, be deduced from the group-theoretic meaning of prolongation,* *and <u>without explicit calculation</u>, but it is typical of Lie's work that qualitative arguments (which one can guess that Lie understood perfectly well) are replaced by detailed (and tedious) calculations. Presumably, this was because mathematics in the 19-th century had to <u>appear</u> computational in order to be acceptable. It is amazing to note how the situation is now reversed!*

10.3 LINEARIZATION AT A FIXED POINT

Let us now suppose that both the quantities ξ_i, η_i vanish for $x = x_0$, $y = y_0$; geometrically, this means that all the infinitesimal transformations (of the given group) leave the point (x_0, y_0) fixed. Then the line elements through this point are transformed according to the infinitesimal transformation

$$\delta y' = \left[\frac{\partial \eta_i(x_0, y_0)}{\partial x_0} + y'\left(\frac{\partial \eta_i}{\partial y_0} - \frac{\partial \xi_i}{\partial x_0}\right) - y'^2 \frac{\partial \xi_i}{\partial y_0}\right] \delta t$$

$$= \zeta_i^{(0)} \delta t \quad , \tag{10.3.1}$$

which is a linear transformation of the quantity y'. And since the equation

$$B_i \zeta_k - B_k \zeta_i = \sum_s c_{iks} \zeta_s$$

becomes

$$\zeta_i^{(0)} \frac{d\zeta_k^{(0)}}{dy'} - \zeta_k^{(0)} \frac{d\zeta_i^{(0)}}{dy'} = \sum_s c_{iks}\zeta_s^{(0)}$$

when $x = x_0$, $y = y_0$, it follows that the linear infinitesimal transformations 10.3.1 always form a group. This proves the following general theorem:

Theorem 10.3.1. If the infinitesimal transformations of a group leave fixed a point of the plane, then they act as a group of linear transformations on the 1-parameter family of line-elements through that point.

COMMENTS ON SECTION 10.3

Here is the general background. Let G *be a group of diffeomorphisms of a manifold* X. *Let* $x \in X$ *be a fixed point of* X, *i.e.,*

$$gx = x$$

for all $g \in G$.

Then

$$g_* \text{ maps } X_x \to X_x \quad .$$

As g *varies over* G, *this defines a group of linear maps on the tangent space* X_x, *called the linearization of* G *at a fixed point.*

There is also a dual action

$$g \to g^{-1*}$$

of G *on* X_x^d, *called the* <u>*co-linearization of*</u> G *at the fixed point. This group passes to the quotient to act on*

$$PX_x^d \quad .$$

It is this action (in the case $\dim X = 2$*) that Lie describes in Theorem 10.3.1.*

As usual, there is also an infinitesimal version of these remarks. Let $\underset{\sim}{G}$ *be a Lie algebra of vector fields on* X. $\underset{\sim}{G}$ *has a fixed point at* x_0 *if the following condition is satisfied:*

$$\underset{\sim}{G}(x_0) = 0 \quad . \qquad (10.3.2)$$

With 10.3.2 satisfied, we can define a linear representation

$$\rho\colon \underset{\sim}{G} \to L(X_x)$$

by the following formula:

$$\rho(A)(B(x_0)) = [A,B](x_0) \qquad (10.3.3)$$

$$\text{for } B \in V(X), \ A \in \underset{\sim}{G} \quad .$$

(Since $A(x_0) = 0$, *the right hand side of 10.3.3 only depends on the value of* B *at* x_0.*)*

There is a dual action of $\underset{\sim}{G}$ *on* $X_{x_0}^d$*:*

$$\rho^d(A) = -\rho(A)^d \qquad (10.3.4)$$

This action, when translated into a vector field action on $X^d_{x_0}$, *is then projected via the map*

$$X^d_{x_0} - (0) \to PX^d_{x_0}$$

into a vector field action on the projective space. Again, this is the action referred to in Theorem 10.3.1.

10.4 LINEAR STABILITY SUBGROUP

This theorem can be extended to arbitrary transformation groups of the plane as follows.

Let us suppose that the infinitesimal transformations $A_1, \ldots, A_r$ of an arbitrary r-term group do not all leave fixed the point (x_o, y_o). Then in the expressions

$$\lambda_1 A_1 + \cdots + \lambda_r A_r = p \sum_k \lambda_k \xi_k + q \sum_k \lambda_k \eta_k$$

the constants λ_k can always be chosen so that $\sum \lambda_k \xi_k$ and $\sum \lambda_k \eta_k$ become zero when $x = x_0$, $y = y_0$. Then one finds that the group always contains $r-2$ (and in special circumstances may contain $r-1$) [independent] infinitesimal transformations $B_1, \ldots, B_\rho$ fixing the point (x_0, y_0). Now the B_k, being infinitesimal transformations of the group, satisfy relations of the form

$$B_i \circ B_k - B_k \circ B_i = \sum_s d_{iks} A_s \ .$$

And since the left, and therefore also the side, of this equation vanishes when $x = x_0$, $y = y_0$, it can be put in the form

$$B_i \circ B_k - B_k \circ B_i = \sum_s \alpha_{iks} B_s \quad .$$

But now we can conclude, exactly as in the preceding section (10.3), that the infinitesimal transformations B_k transform the line-elements through the fixed point according to a linear group. This gives the following theorem:

Theorem 10.4.1. Those infinitesimal transformations of a group which leave fixed a point of the plane, transform the line-elements through this point according to a linear group.

In this situation there are four essentially different cases, according as the linear group depends on 3, 2, 1 or 0 parameters. And correspondingly, there are four distinct types of transformation groups of a plane. We shall return to this principle of classification.

COMMENTS ON SECTION 10.4

Let $\underset{\sim}{G}$ *be a Lie algebra of vector fields on the manifold* X, *and let* x_0 *be a point of* X. *Set:*

$$\underset{\sim}{H} = \{A \varepsilon G: A(x_0) = 0\} \quad .$$

Now, if A,B *vanish at* x_0, *i.e., belong to* H, *so does their bracket and sum, i.e.,*

> $\underset{\sim}{H}$ *is a Lie subalgebra of* $\underset{\sim}{G}$.

The geometric property of $\underset{\sim}{H}$ *is that:*

> *Each one-parameter group generated by elements of* $\underset{\sim}{H}$ *leaves* x_0 *fixed.*

$\underset{\sim}{H}$ *is called the* ***stability subalgebra*** *at* x_0. *(It is also called the* ***isotropy subalgebra****.) Hence,* $\underset{\sim}{H}$ *acts on the tangent and cotangent vectors in a way described in the previous section.*

Chapter 11

INFINITESIMAL TRANSFORMATIONS OF VARIOUS ORDERS

11.1 ORDER OF AN INFINITESIMAL TRANSFORMATION AT A POINT

If $\delta x = \xi(x,y)\delta t$, $\delta y = \eta(x,y)\delta t$ or $A = \xi p + \eta q$ is the symbol of an infinitesimal transformation, then ξ and η can always be expanded in powers of $x-x_0$ and $y-y_0$:

$$\xi = a_0 + a_1(x-x_0) + b_1(y-y_0) + a_2(x-x_0)^2 + b_2(x-x_0)(y-y_0) + c_2(y-y_0)^2 + \ldots$$

$$\eta = \alpha_0 + \alpha_1(x-x_0) + \beta_1(y-y_0) + \alpha_2(x-x_0)^2 + \beta_2(x-x_0)(y-y_0) + \gamma_2(y-y_0)^2 + \ldots$$

If $a_0 \neq 0$ or $\alpha_0 \neq 0$, we say that our infinitesimal transformation is of <u>order zero</u> at the point (x_0,y_0). On the other hand, if both $a_0 = 0$ and $\alpha_0 = 0$ while at least one of $a_1,b_1,\alpha_1,\beta_1, \neq 0$, we say that the transformation is of <u>order one</u>.

More generally, we say that an infinitesimal transformation $\xi p + \eta q$ is of <u>order</u> s at (x_0,y_0) if, in the expansion of ξ and η in powers of $x-x_0$ and $\eta-\eta_0$, all the terms of order $0,1,\ldots,s-1$ are absent, while the term of order s is actually present.

In investigations of infinitesimal transformations it very often suffices to consider only the terms of lowest order in the power series expansions of ξ and η. When in the following I speak for example of an infinitesimal transformation

$$(x-x_0)q + \cdots ,$$

I understand by this an infinitesimal transformation $\xi p + \eta q$ of which ξ is of order ≥ 2 with respect to $x-x_0$ and $y-y_0$, while η contains a term of first order, namely $x-x_0$.

COMMENTS ON SECTION 11.1

Let $\underset{\sim}{G}$ *be a Lie algebra of vector fields on a manifold* X, *and let* x *be a point of* X. *In Chapter C I have explained how* x *determines a descending filtration*

$$\underset{\sim}{G} = \underset{\sim}{G}^0 = \underset{\sim}{G}^1 \supset \underset{\sim}{G}^2 \supset \cdots$$

of $\underset{\sim}{G}$. $\underset{\sim}{G}^1$ *consists of those vector fields which vanish at* x, *i.e., which vanish to "first order or higher".* $\underset{\sim}{G}^2$ *consists of those which vanish to* <u>*second or higher order*</u>*. According to Lie's definition, an* $A \in \underset{\sim}{G}$ <u>*is of order*</u> s *at* x *if:*

$$A \in \underset{\sim}{G}^s , \quad \textit{but} \quad A \notin \underset{\sim}{G}^{s+1}$$

The properties of these filtrations are fundamental to the techniques of this paper. (Indeed, an algebraist might be able to recast all of the results of this paper into a purely algebraic-filtered Lie algebra context.)

11.2 GEOMETRIC PROPERTIES

Under the transformation $\xi p + \eta q$ the coordinates x_0, y_0 of an arbitrary point acquire the increments $\xi(x_0,y_0)\delta t$ and $\eta(x_0,y_0)\delta t$, where $\xi(x_0,y_0)$, $\eta(x_0,y_0)$ are the terms of order zero in the expansion of ξ and η in powers of $x-x_0$ and $y-y_0$.

Hence if a given infinitesimal transformation is of order ≥ 1 at a point (x_0,y_0), then it doesn't change the position of this point.

If we want to investigate how such a transformation transforms the line-elements through (x_0,y_0), we use the formula 10.3.1:

$$\delta y' = \left[\frac{\partial\eta(x_0,y_0)}{\partial x_0} + y'\left(\frac{\partial\eta_0}{\partial y_0} - \frac{\partial\xi_0}{\partial x_0}\right) - y'^2\frac{\partial\xi_0}{\partial y_0}\right]\delta t \quad ,$$

which now takes the form

$$\delta y' = [\alpha_1 + (\beta_1 - a_1)y' - b_1 y'^2]\delta t \quad .$$

If the transformation is of order ≥ 2, *then obviously all the line-elements through* (x_0,y_0) *retain their position. The same happens if the transformation is of order one under the assumption that it has the form*

$$(x-x_0)p + (y-y_0)q + \cdots$$

On the other hand, if $\beta_1 \neq a_1$, then the line-elements through (x_0,y_0) are transformed linearly. In this case, there is always at least one, and in general two, elements satisfying the equation

$$0 = \alpha_1 + (\beta_1 - a_1)y' - b_1 y'^2$$

and which therefore retain their position unchanged.

11.3 DIMENSION OF INFINITESIMAL TRANSFORMATIONS OF VARIOUS ORDERS

If $A_1,\ldots,A_r$ are r independent infinitesimal transformations of an r-term group, whose general infinitesimal transformation is therefore of the form $\lambda_1 A_1 + \cdots + \lambda_r A_r$, then, if $r > 2$, it is always possible to choose such values for the λ_k that $\sum \lambda_k A_k$ is of order one at the point (x_0,y_0). Thus one can always find $r-2$ *transformations of order one*. If $r > 6$, then there are values of the λ_k for which $\sum \lambda_k A_k$ is of order 2 at (x_0,y_0), and so there are $r-6$ *transformations of order* 2.

Similarly, there are r-12 infinitesimal transformations of order 3, and so on.

We sometimes say that ρ first order transformations $B_1,\ldots,B_\rho$ are <u>independent first order</u> transformations if no linear combination of them is of order ≥ 2.

Hence, <u>a group contains at most four independent</u> [<u>infinitesimal</u>] <u>first order transformations</u>.

COMMENTS ON SECTION 11.3

Let $\underset{\sim}{G}$ *be a Lie algebra of vector fields on a manifold* X. *Let* x *be a point of* X. *Let:*

$\underset{\sim}{G}^r$ = *set of vector fields in* $\underset{\sim}{G}$ *which vanish to order* r *or greater at* x.

Then, $\underset{\sim}{G}^1$ *is the stability subalgebra at* x. *Let*

$$\rho: \underset{\sim}{G}^1 \to L(X_x)$$

be the linearization representation of $\underset{\sim}{G}^1$. *(See comments to Section 0.3.) It is readily seen that:*

$$\rho(\underset{\sim}{G}^2) = 0 .$$

Now,

$$[\underset{\sim}{G}^1,\underset{\sim}{G}^2] \subset \underset{\sim}{G}^{1+2-1} = \underset{\sim}{G}^2 .$$

Hence:

$\underset{\sim}{G}^2$ *is a Lie ideal of* $\underset{\sim}{G}^1$.

ρ *then passes to the quotient to define a linear representation of the Lie algebra* $\underset{\sim}{G}^1/\underset{\sim}{G}^2$.

Theorem 10.3.1. *This linearization representation*

$$\rho: \underset{\sim}{G}^1/\underset{\sim}{G}^2 \to L(X_x)$$

is one-one.

Proof. *Let* $A \in \underset{\sim}{G}^1$. *To say that*

$$\rho(A) = 0$$

is to say that:

$$[A,B](x) = 0 \quad .$$

$$\text{for \underline{all}} \quad B \in V(X) \quad .$$

Since $A(x) = 0$, *this condition means that* A *vanishes to the second (or higher) order at* x, *i.e.,* $A \in \underset{\sim}{G}^2$, *hence the image of* A *in* $\underset{\sim}{G}^1/\underset{\sim}{G}^2$ *is zero.*

Theorem 10.3.2.

$$\dim (\underset{\sim}{G}^1/\underset{\sim}{G}^2) \leq n^2 \qquad (10.3.1)$$

with: $n = \dim X$.

Proof. $L(X_x)$, *the space of linear maps* $X_x \to X_x$, *is identified with the space of* $n \times n$ *real matrices, which is of course of dimension* n^2.

Of course, if

$$n = 2 \quad ,$$

which is the case being considered, 10.3.1 specializes to Lie's statement.

11.4 PROPERTIES OF TERMS OF VARIOUS ORDER

It is easy to show that the independent first order transformations of a group must have certain definite forms according as the number of them is 4, 3, 2, 1, or 0.

It is at first thinkable that the group may contain <u>four</u> independent first order transformations. Then obviously these can be put in the form

$$(x-x_0)p + \cdots, \quad (y-y_0)p + \cdots, \quad (x-x_0)q + \cdots, \quad (y-y_0)q + \cdots \quad .$$

If the group has only <u>three</u> independent first order transformations, there are two essentially different cases according as there is a transformation of the form

$$(x-x_0)p + (y-y_0)q + \cdots \quad = \quad U + \cdots$$

or not. In the latter case the three infinitesimal transformations are of the form:

$$\begin{aligned} (x-x_0)q + \alpha U + \cdots &= B_1 + \cdots \quad , \\ (x-x_0)p - (y-y_0)q + \beta U + \cdots &= B_2 + \cdots \quad , \\ (y-y_0)p \qquad\qquad + \gamma U + \cdots &= B_3 + \cdots \quad . \end{aligned}$$

But one has

$$B_1 \circ B_3 - B_3 \circ B_1 = (x-x_0)p - (y-y_0)q \quad ,$$

and since $B_1 \circ B_3 - B_3 \circ B_1$ must be the symbol of an infinitesimal transformation of the group (Theorem 8.2.1), we see that we must have $\beta = 0$. Similarly, it follows from considering $B_1 \circ B_2 - B_2 \circ B_1$ and $B_3 \circ B_2 - B_2 \circ B_3$ that $\alpha = 0$ and $\gamma = 0$. Therefore:

Theorem 11.4.1. If a group contains three independent [infinitesimal] first order transformations, of which none has the form $(x-x_0)p + (y-y_0)q + \cdots$, then they are of the form

$$(x-x_0)q + \cdots, \quad (x-x_0)p - (y-y_0)q + \cdots, \quad (y-y_0)p + \cdots$$

Let us now suppose that the group contains three independent first order transformations, one of which is of the form

$$U = (x-x_0)p + (y-y_0)q + \cdots$$

Then the other two can be put in the form

$$\alpha_1(x-x_0)q + \beta_1[(x-x_0)p - (y-y_0)q] + \gamma_1(y-y_0)p + \cdots = C_1 + \cdots$$

$$\alpha_2(x-x_0)q + \beta_2[(x-x_0)p - (y-y_0)q] + \gamma(y-y_0)p + \cdots = C_2 + \cdots$$

where the first order expressions C_1 and C_2 are connected by a relation of the form

$$C_1 \circ C_2 - C_2 \circ C_1 = A_1 \cdot C_1 + A_2 \cdot C_2 .$$

In this, the constants A_1 and A_2 do not both vanish, since otherwise the three expressions

$$\alpha_1\beta_2 - \alpha_2\beta_1 , \quad \beta_1\gamma_2 - \beta_2\gamma_1 , \quad \gamma_1\alpha_2 - \gamma_2\alpha_1$$

would be zero, which is excluded by the independence of C_1 and C_2. Therefore it is no restriction to choose the infinitesimal transformations $C_1 + \cdots$, $C_2 + \cdots$ so that $A_1 = 1$, $A_2 = 0$. This gives a number of relations among the six constants $\alpha_1, \beta_1, \gamma_1, \alpha_2, \beta_2, \gamma_2$ by means of which they can be determined in the most general way. But I do not want to go any further into this.

COMMENTS

At this point Lie began the details of the classification of the Lie algebras of vector fields that can act on 2-dimensional manifolds. My plan is not to comment immediately on these details, but to go over the whole proof at the end of the paper. Of course, if there is a general fact involved, and if I believe it will help the reader, I will comment directly on it.

Chapter 12

GROUPS FIXING A FAMILY OF CURVES $\phi(x,y) = \text{const.}$

12.1 CONDITIONS THAT A FAMILY OF CURVES ADMIT AN INFINITESIMAL TRANSFORMATIONS

The infinitesimal transformation $A = \xi p + \eta q$ in general takes a given family of curves $\phi(x,y) = a = \text{const.}$ into a new family

$$\phi(x,y) + \left(\frac{\partial\phi}{\partial x}\xi + \frac{\partial\phi}{\partial y}\eta\right)\delta t = a \quad .$$

For the new family is to be identical with the given one, a necessary and sufficient condition is that

$$A(\phi) = \frac{\partial\phi}{\partial x}\xi + \frac{\partial\phi}{\partial y}\eta$$

be a function of ϕ. Now assume that ϕ is a solution of the differential equation

$$B(f) = X\frac{\partial f}{\partial x} + Y\frac{\partial f}{\partial y} = 0 \quad .$$

Therefore if one replaces f by ϕ in the identity

$$A(B(f)) - B(A(f)) = (AX - B\xi)\frac{\partial f}{\partial x} + (AY - B\eta)\frac{\partial f}{\partial y}$$

it follows from our assumption that both $B(\phi)$ and $B(A(\phi))$ are zero:

$$0 = (AX - B\xi)\frac{\partial\phi}{\partial x} + (AY - B\eta)\frac{\partial\phi}{\partial y} \quad .$$

Therefore ϕ is a solution of both the equations

$$(AX-B\xi)\frac{\partial f}{\partial x} + (AY-B\eta)\frac{\partial f}{\partial y} = 0 \ ,$$

$$X\frac{\partial f}{\partial x} + Y\frac{\partial f}{\partial y} = 0 \ ,$$

so that these equations are the same.

Therefore, a necessary and sufficient condition for the family of curves $\phi(x,y) = a$ determined by the equation $X(\partial f/\partial x) + Y(\partial f/\partial y) = 0$ to admit the infinitesimal transformation $A = \xi(\partial/\partial x) + \eta(\partial/\partial y)$ is that the equation

$$\frac{AX-B\xi}{X} = \frac{AY-B\eta}{Y} \tag{12.1.1}$$

hold identically.

Finally, we assume that the family of curves $\phi(x,y) = a$ is determined by a differential equation in implicit form

$$\psi(x,y,y') = 0 \ .$$

To decide whether the family $\phi = a$ admits the infinitesimal transformation $\xi p + \eta q = A$, we put (§10)

$$B = \xi\frac{\partial}{\partial x} + \eta\frac{\partial}{\partial y} + \left[\frac{\partial \eta}{\partial x} + y'\left(\frac{\partial \eta}{\partial y} - \frac{\partial \xi}{\partial x}\right) - y'^2\frac{\partial \xi}{\partial y}\right]\frac{\partial}{\partial y'}$$

and require that $B(\psi) = 0$ if and only if $\psi = 0$, i.e., that the equation which results by eliminating y' from the equations $\psi = 0$ and $B(\psi) = 0$ be an identity. This condition is equivalent to: the differential equation $\psi = 0$ admits the infinitesimal transformation A. Obviously, when ψ is of the form $Xy' - Y$ this condition reduces to the equation (12.1.1).

COMMENTS ON SECTION 12.1

Here are some general facts. Let X *be a manifold,* A *a vector field,* $\phi \in F(X)$ *a function, such that:*

$$d\phi \neq 0 \qquad \text{at each point of } X.$$

The subsets

$$\phi = \text{constant}$$

are then codimension one submanifolds, and determine a *hypersurface foliation* *of* X.

The condition that A *leave invariant this hypersurface foliation, in the sense that the group generated by* A *map a leaf into another leaf, is that:*

$$A(\phi) \text{ is a function of } \phi. \qquad (12.1.1)$$

Let:

$$V = \{B \in V(X) : B(\phi) = 0\}$$

V *is the* *vector field system* *determining (and determined by) the foliation. (In case* $\dim X = 2$, *there is but* *one* *independent vector field in* V; *the leaves are* *curves*.) *12.1.1 is then equivalent to the following condition*

$$[A,B] \in V \text{ for all } B \in V \ . \qquad (12.1.2)$$

In case

$$\dim X = 2 ,$$

(which is the case considered by Lie), 12.1.2 means that

$$[A,B] = fB \qquad (12.1.3)$$

for some function $f \in F(X)$

Now, let

$$PT^d(X)$$

be the <u>*contact manifold*</u> *defined by* X. *Let*

$$\psi \in F(PT^d(X))$$

be a function on this manifold. ψ *defines a family of submanifolds. Let* Y *be a manifold of dimension* $n-1$, *and let*

$$\alpha: Y \to X ,$$

be a submanifold map. Consider α *as a solution of the following first order, non-linear differential equations:*

$$(\partial\alpha)^*(\psi) = 0 . \qquad (12.1.4)$$

If A *is a vector field on* X, *let* $B \in V(PT^d(X))$ *be the prolongation to the contact manifold. Then,*

$$B(\psi) = f\psi , \qquad (12.1.5)$$

for some $f \in F(PT^d(X))$, *is the condition that* A *leave invariant the family of submanifolds of* X *determined by the differential equation 12.1.4.*

12.2 FAMILIES OF CURVES INVARIANT UNDER 2-TERM GROUPS

There are infinitely many families of curves $\phi = a$ which admit a given infinitesimal transformation. They are found by choosing the function Ω arbitrarily in the equation

$$\xi \frac{\partial \phi}{\partial x} + \eta \frac{\partial \phi}{\partial y} = \Omega(\phi)$$

and then setting an arbitrary solution ϕ of this equation equal to the constant a.

There are infinitely many invariant families of curves for any 2-term group A_1, A_2, where

$$A_1 \circ A_2 - A_2 \circ A_1 = c_1 A_1 + c_2 A_2 \quad . \qquad (12.2.1)$$

We shall give only one.

Let us first assume that c_1 and c_2 are not both zero. If ψ is any solution of $A_1(f) = 0$, then, by (12.2.1),

$$A_1(A_2(\psi)) = 0 \quad ,$$

which means, as before, that the family of curves $\psi = a$ admits both the infinitesimal transformations.

12.3 FAMILIES OF CURVES INVARIANT UNDER 3-TERM GROUPS

A 3-term group A_1, A_2, A_3 always has at least one, and in general only one, invariant family of curves, as will now be shown.

I put $A_i = \xi_i p + \eta_i q$ $(i=1,2,3)$ and for the determinant

$$\Delta = \begin{vmatrix} \xi_1 & \eta_1 & \dfrac{\partial\eta_1}{\partial x} + y' \dfrac{\partial\eta_1}{\partial y} - \dfrac{\partial\xi_1}{\partial x} - y'^2 \dfrac{\partial\xi_1}{\partial y} \\ \xi_2 & \eta_2 & \dfrac{\partial\eta_2}{\partial x} + y' \dfrac{\partial\eta_2}{\partial y} - \dfrac{\partial\xi_2}{\partial y} - y'^2 \dfrac{\partial\xi_2}{\partial y} \\ \xi_3 & \eta_3 & \dfrac{\partial\eta_3}{\partial x} + y' \dfrac{\partial\eta_3}{\partial y} - \dfrac{\partial\xi_3}{\partial y} - y'^2 \dfrac{\partial\xi_3}{\partial y} \end{vmatrix}$$

and claim that $\Delta = 0$ is a differential equation invariant under the group, assuming that it does not hold for all values x, y, y' identically.

I put

$$\xi_i \frac{\partial}{\partial x} + \eta_i \frac{\partial}{\partial y} + \left[\frac{\partial\eta_i}{\partial x} + y'\left(\frac{\partial\eta_i}{\partial y} - \frac{\partial\xi_i}{\partial x}\right) - y'^2 \frac{\partial\xi_i}{\partial y}\right] \frac{\partial}{\partial y'} = B_i$$

by the developments of the end of the preceding section (12.2), my assertion is equivalent to saying that the three equations $B_q(\Delta) = 0$ hold identically if $\Delta = 0$. For example, set $q = 2$ and

$$\frac{\partial\eta_i}{\partial x} + y'\left(\frac{\partial\eta_i}{\partial y} - \frac{\partial\xi_i}{\partial x}\right) - y'^2 \frac{\partial\xi_i}{\partial y} = \zeta_i \quad ;$$

then

$$B_2\Delta = \begin{vmatrix} B_2\xi_1 & B_2\eta_1 & B_2\zeta_1 \\ \xi_2 & \eta_2 & \zeta_2 \\ \xi_3 & \eta_3 & \zeta_3 \end{vmatrix} + \begin{vmatrix} \xi_1 & \eta_1 & \zeta_1 \\ B_2\xi_2 & B_2\eta_2 & B_2\zeta_2 \\ \xi_3 & \eta_3 & \zeta_3 \end{vmatrix}$$

$$+ \begin{vmatrix} \xi_1 & \eta_1 & \zeta_1 \\ \xi_2 & \eta_2 & \zeta_2 \\ B_2\xi_3 & B_2\eta_3 & B_2\zeta_3 \end{vmatrix}$$

Now, by (§ 10.2) there are relations of the form

$$B_2\xi_1 = B_1\xi_2 + c_1\xi_1 + c_2\xi_2 + c_3\xi_3 \ ,$$

$$B_2\eta_1 = B_1\eta_2 + c_1\eta_1 + c_2\eta_2 + c_3\eta_3 \ ,$$

$$B_2\zeta_1 = B_1\zeta_2 + c_1\zeta_1 + c_2\zeta_2 + c_3\zeta_3 \ ,$$

$$B_2\xi_3 = B_3\xi_2 + d_1\xi_1 + d_2\xi_2 + d_3\xi_3 \ ,$$

$$B_2\eta_3 = B_3\eta_2 + d_1\eta_1 + d_2\eta_2 + d_3\eta_3 \ ,$$

$$B_2\zeta_3 = B_3\zeta_2 + d_1\zeta_1 + d_2\zeta_2 + d_3\zeta_3 \ ,$$

where c and d are constants.

Using these values, one has

$$B_2(\Delta) = (c_1+d_3)\Delta + \begin{vmatrix} B_1\xi_2 & B_1\eta_2 & B_1\zeta_2 \\ \xi_2 & \eta_2 & \zeta_2 \\ \xi_3 & \eta_3 & \zeta_3 \end{vmatrix}$$

$$+ \begin{vmatrix} \xi_1 & \eta_1 & \zeta_1 \\ B_2\xi_2 & B_2\eta_2 & B_2\zeta_2 \\ \xi_3 & \eta_3 & \zeta_3 \end{vmatrix} + \begin{vmatrix} \xi_1 & \eta_1 & \zeta_1 \\ \xi_2 & \eta_2 & \zeta_2 \\ B_3\xi_2 & B_3\eta_2 & B_3\zeta_2 \end{vmatrix}$$

Now

$$B_i\xi_2 = A_i\xi_2 , \qquad B_i\eta_2 = A_i\eta_2 ,$$

$$B_i\zeta_2 = A_i\zeta_2 + \left(\frac{\partial\eta_2}{\partial y} - \frac{\partial\xi_2}{\partial x} - 2y' \frac{\partial\xi_2}{\partial y}\right)\zeta_i ,$$

so that

$$B_2(\Delta) = (c_1+d_3)\Delta + \frac{\partial\eta_2}{\partial y} - \frac{\partial\xi_2}{\partial x} - 2y' \frac{\partial\xi_2}{\partial y} \quad \Delta$$

$$+ \begin{vmatrix} A_1\xi_2 & A_1\eta_2 & A_1\zeta_2 \\ \xi_2 & \eta_2 & \zeta_2 \\ \xi_3 & \eta_3 & \zeta_3 \end{vmatrix} + \begin{vmatrix} \xi_1 & \eta_1 & \zeta_1 \\ A_2\xi_2 & A_2\eta_2 & A_2\zeta_2 \\ \xi_3 & \eta_3 & \zeta_3 \end{vmatrix}$$

$$+ \begin{vmatrix} \xi_1 & \eta_1 & \zeta_1 \\ \xi_2 & \eta_2 & \zeta_2 \\ A_3\xi_2 & A_3\eta_2 & A_3\zeta_2 \end{vmatrix}$$

To determine the sum of the three last determinants, we expand each determinant along the rows $(A_i\xi_2, A_i\eta_2, A_i\zeta_2)$, and then consider separately the coefficients of

$$\frac{\partial\xi_2}{\partial x}, \quad \frac{\partial\xi_2}{\partial y}, \quad \frac{\partial\eta_2}{\partial x}, \quad \frac{\partial\eta_2}{\partial y}, \quad \frac{\partial\zeta_2}{\partial x}, \quad \frac{\partial\zeta_2}{\partial y} \quad ;$$

both $\partial\xi_2/\partial x$ and $\partial\eta_2/\partial y$ have Δ as coefficient, while the coefficients of $\partial\xi_2/\partial y$, $\partial\eta_2/\partial x$, $\partial\zeta_2/\partial x$, $\partial\zeta_2/\partial y$ all vanish. Therefore

$$B_2(\Delta) = \left(c_1 + d_3 + 2\frac{\partial\eta_2}{\partial y} - 2y'\frac{\partial\xi_2}{\partial y}\right)\Delta ,$$

so that $B_2(\Delta) = 0$ if and only if $\Delta = 0$.

This proves that the differential equation $\Delta = 0$ admits the infinitesimal transformation B_2, and in a similar way one sees that it also admits the transformations B_1 and B_3.

Let us now assume that Δ does not vanish for all values of x, y, y'. Then there is one, or perhaps two, equations of the form

$$\sum_k \phi_k(x,y,y')\left[\xi_k \frac{\partial}{\partial x} + \eta_k \frac{\partial}{\partial y} + \zeta_k \frac{\partial}{\partial y'}\right] = 0 = \sum_k \phi_k B_k$$

But it is easy to see that two such equations cannot hold. Otherwise, elimination of B_3 would give an equation

$$\phi_1 B_1 + \phi_2 B_2 = 0 \quad ,$$

equivalent to the three equations

$$\phi_1\xi_1 + \phi_2\xi_2 = 0 \,, \quad \phi_1\eta_1 + \phi_2\eta_2 = 0 \,,$$

$$\phi_1\zeta_1 + \phi_2\zeta_2 = 0 \quad ;$$

and since ξ_1, η_1, as well as ξ_2, η_2, do not both vanish and depend only on x and y, we can assume that ϕ_1 and ϕ_2 depend only on x and y.

Therefore, it is no essential restriction to set ϕ_2 equal to -1. But then,

$$\xi_2 = \phi(x,y)\xi_1 \,, \quad \eta_2 = \phi(x,y)\eta_1 \,,$$

$$\frac{\partial\eta_2}{\partial x} + y'\frac{\partial\eta_2}{\partial y} - \frac{\partial\xi_2}{\partial x} - y'^2\frac{\partial\xi_2}{\partial y} = \phi\left(\frac{\partial\eta_1}{\partial x} + y'\frac{\partial\eta_1}{\partial y} - \frac{\partial\xi_1}{\partial x} - y'^2\frac{\partial\xi_1}{\partial y}\right) \,,$$

which implies

$$\eta_1\frac{\partial\phi}{\partial x} = 0 \,, \quad \eta_1\frac{\partial\phi}{\partial y} - \xi_1\frac{\partial\phi}{\partial x} = 0 \,, \quad \xi_1\frac{\partial\phi}{\partial y} = 0 \quad ;$$

and since ξ_1 and η_1 are not both zero, we see that ϕ must be a constant; but this contradicts the independence of B_1 and B_2. This proves that there cannot be more than one relation of the form $\sum \phi_k B_k = 0$.

If <u>one</u> such equation holds, then the two linear partial differential equations

$$B_1(f) = 0, \quad B_2(f) = 0$$

form a complete system, to which the equation $B_3(f) = 0$ also belongs. If $\psi(x,y,y')$ is a solution of this system, i.e., if $B_1(\psi) = B_2(\psi) = B_3(\psi) = 0$, then every differential equation of the form $\psi(x,y,y') = a = \text{const.}$ admits the three infinitesimal transformations B_1, B_2, B_3. In this case there is a 1-parameter family of families of curves invariant with respect to the group. Here it should be noted that ψ is independent of y'. In this exceptional case, $\psi = a$ is not a differential equation in the usual sense, and so we find here only the single family of invariant curves $\psi(x,y) = a$.

It can be shown that if Δ is not identically zero, then $\Delta = 0$ is the only first order differential equation which is invariant under our group. For if $\phi(x,y) = a$ is an arbitrary invariant family of curves, then there are three equations of the form

$$\xi_k \frac{\partial\phi}{\partial x} + \eta_k \frac{\partial\phi}{\partial y} = \Omega_k(\phi) \quad ,$$

from which follow, by differentiation,

$$\frac{\partial\xi_k}{\partial x}\frac{\partial\phi}{\partial x} + \frac{\partial\eta_k}{\partial x}\frac{\partial\phi}{\partial y} = \frac{d\Omega_k}{d\phi}\frac{\partial\phi}{\partial x} - \xi_k \frac{\partial^2\phi}{\partial x^2} - \eta_k \frac{\partial^2\phi}{\partial x\partial y} \quad ,$$

$$\frac{\partial\xi_k}{\partial y}\frac{\partial\phi}{\partial x} + \frac{\partial\eta_k}{\partial y}\frac{\partial\phi}{\partial y} = \frac{d\Omega_k}{d\phi}\frac{\partial\phi}{\partial y} - \xi_k \frac{\partial^2\phi}{\partial x\partial y} - \eta_k \frac{\partial^2\phi}{\partial y^2} \quad ;$$

eliminating $d\Omega_k/d\phi$ now gives:

$$- \frac{\partial\xi_k}{\partial y}\left(\frac{\partial\phi}{\partial x}\right)^2 + \left(\frac{\partial\xi_k}{\partial x} - \frac{\partial\eta_k}{\partial y}\right)\frac{\partial\phi}{\partial x}\frac{\partial\phi}{\partial y} + \frac{\partial\eta_k}{\partial x}\left(\frac{\partial\phi}{\partial y}\right)^2$$

$$= \xi_k \left(\frac{\partial^2\phi}{\partial x\partial y}\frac{\partial\phi}{\partial x} - \frac{\partial^2\phi}{\partial x^2}\frac{\partial\phi}{\partial y}\right) + \eta_k \left(\frac{\partial^2\phi}{\partial y^2}\frac{\partial\phi}{\partial x} - \frac{\partial^2\phi}{\partial x\partial y}\frac{\partial\phi}{\partial y}\right)$$

Therefore,

$$\begin{vmatrix} \xi_1 & \eta_1 & - \frac{\partial\xi_1}{\partial y}\left(\frac{\partial\phi}{\partial x}\right)^2 + \left(\frac{\partial\xi_1}{\partial x} - \frac{\partial\eta_1}{\partial y}\right)\frac{\partial\phi}{\partial x}\frac{\partial\phi}{\partial y} + \frac{\partial\eta_1}{\partial x}\left(\frac{\partial\phi}{\partial y}\right)^2 \\ \xi_2 & \eta_2 & - \frac{\partial\xi_2}{\partial y}\left(\frac{\partial\phi}{\partial x}\right)^2 + \left(\frac{\partial\xi_2}{\partial x} - \frac{\partial\eta_2}{\partial y}\right)\frac{\partial\phi}{\partial x}\frac{\partial\phi}{\partial y} + \frac{\partial\eta_2}{\partial x}\left(\frac{\partial\phi}{\partial y}\right)^2 \\ \xi_3 & \eta_3 & - \frac{\partial\xi_3}{\partial y}\left(\frac{\partial\phi}{\partial x}\right)^2 + \left(\frac{\partial\xi_3}{\partial x} - \frac{\partial\eta_3}{\partial y}\right)\frac{\partial\phi}{\partial x}\frac{\partial\phi}{\partial y} + \frac{\partial\eta_3}{\partial x}\left(\frac{\partial\phi}{\partial y}\right)^2 \end{vmatrix} = 0$$

which means that $\phi = a$ is an integral of the equation $\Delta = 0$. This proves the following theorem:

Theorem 12.3.1. Every 3-term group leaves invariant at least one family of curves $\phi(x,y) = a$, and in general only one.

COMMENTS ON SECTION 12.3

Let X *be a 2-dimensional manifold, with*

$$PT^d(X)$$

the space of first-order hypersurface contact elements to X. *We denote the space by* X'. *It is 3-dimensional.*

Let $\underset{\sim}{G}$ *be a 3-dimensional Lie algebra of vector fields on* X.

Each $A \in \underset{\sim}{G}$ *admits a prolonged vector field--which we now denote by* A' -- *to* X'. *The mapping*

$$A \to A'$$

is a Lie algebra isomorphism between $\underset{\sim}{G}$ *and a Lie algebra of vector fields on* $PT^d(X)$. *Set:*

$$Y = \{x' \varepsilon X' : \dim \underset{\sim}{G}'(x') \leq 2\} \qquad (12.2.1)$$

Notice that in terms of the local coordinate system (x,y,y') *for* X' *used constantly by Lie, there is a function* $\Delta(x,y,y')$ *such that*

$$\Delta = 0 \text{ defines } Y . \qquad (12.3.2)$$

Thus, if Δ *always has non-zero differential (which is the "generic" situation, and the one assumed by Lie)* Y *is a submanifold of dimension 2 of* X'.

Theorem 12.3.1. $\underset{\sim}{G}'$ *is tangent to the submanifold* Y.

Proof. *Let* $x' \in Y$. *Let*

$$t \to x'(t)$$

be a curve in X, *beginning at* x', *which is an orbit curve of a vector field* A' *in* $\underset{\sim}{G}'$. *Let* $t \to g'(t)$ *be the one-parameter group of diffeomorphisms of* X' *generated by* A'. *Then,*

$$x'(t) = g'(t)(x') \quad .$$

Hence,

$$\underset{\sim}{G}'(x'(t)) = g(t)_*(g(-t)_*(\underset{\sim}{G}')(x'))$$

But, since $g(t)$ *belongs to the group generated by* $\underset{\sim}{G}'$,

$$g(t)_*(\underset{\sim}{G}') = \underset{\sim}{G}' \quad .$$

Hence,

$$\dim \underset{\sim}{G}'(x'(t)) = \dim \underset{\sim}{G}'(x')$$

for all t ,

i.e.,

$$x'(t) \in Y$$

for all t .

This proves Theorem 12.3.1 using the "obvious" geometric fact that "$\underset{\sim}{G}'$ tangent to Y" is equivalent to the orbits of elements of $\underset{\sim}{G}'$, beginning at points of Y, will always be in Y.

Of course, Lie proves this <u>computationally</u> by showing that

$$A'(\Delta) \text{ is a multiple of } \Delta,$$
$$\text{for all } A \in \underset{\sim}{G} .$$

We can see also what Lie means by "in general" in Theorem 12.3.1. The "generic" points of X' are those where $d\Delta \neq 0$. If <u>all</u> points are "non-generic", then <u>either</u>:

$$\Delta \equiv 0 \tag{12.3.3}$$

or

$$\Delta = \text{non-zero constant} \tag{12.3.4}$$

12.3.3 means that

$$\dim \underset{\sim}{G}'(x') \leq 2 \tag{12.3.5}$$
$$\text{for all } x' \in X' ,$$

i.e., $\underset{\sim}{G}'$ acts <u>intransitively</u> on X'. This situation can be analyzed separately.

Similarly, if 12.3.4 is satisfied--or more generally if

$$\Delta \neq 0 \text{ at each point of } X',$$

then $\underset{\sim}{G}'$ acts transitively on X' <u>at each point</u>.

It is <u>*possible*</u> *that Lie's use of this term "in general" can be made precise in terms of statements about the "space" of "all" Lie algebra actions on* X. *This might be an interesting topic for research. The reader should keep in mind, however, that there is usually a common sense, intuitive and/or geometric meaning to the use of this term, which the 19-th century authors used so constantly!*

12.4 FAMILIES OF CURVES INVARIANT UNDER GENERAL GROUPS

Let there now be given an arbitrary group $A_1,\dots,A_r$. If a family of curves $\phi(x,y) = a$ admits three transformations of the group, say A_i, A_j, A_q, it follows in the same way as at the preceding section (§ 12.3), that ϕ is an integral of the equation

$$\left(\xi_i \eta_j \frac{\partial \eta_q}{\partial x} + y' \left(\frac{\partial \eta_q}{\partial y} - \frac{\partial \xi_q}{\partial x}\right) - y'^2 \frac{\partial \xi_q}{\partial y}\right) = 0 = \Delta_{ijq}$$

Thus, if the family $\phi = a$ admits all the transformations of the group, then ϕ is an integral of all the equations $\Delta_{ijq} = 0$.

It can be shown that, conversely, if all the equations $\Delta_{ijq} = 0$ have a common integral ϕ, then $\phi = a$ represents a family of curves invariant under the group.

In fact, proceeding exactly as at the beginning of the preceding section (§ 12.3), one can give the expression $B_\rho \Delta_{ijk}$ the form

$$\sum\sum\sum c_{uvw}\Delta_{uvw} + 2\left(\frac{\partial \eta_\rho}{\partial y} - y' \frac{\partial \xi_\rho}{\partial y}\right)\Delta_{ijq} ,$$

where c_{uvw} are constants. Thus we get the theorem:

<u>Theorem 12.4.1</u>. The group $A_1,\ldots,A_r$ leaves invariant a family of curves $\phi(x,y) = a$ if and only if ϕ is an integral of all the equations $\Delta_{ijq} = 0$.

If all the Δ_{ijq} vanish identically, the considerations of this section must be modified. Proceeding as in the preceding section (§ 12.3), one sees that two of the linear partial differential equations

$$\xi_k \frac{\partial f}{\partial x} + \eta_k \frac{\partial f}{\partial y} + \zeta_k \frac{\partial f}{\partial y'} = 0 = B_k(f)$$

form a complete system, to which all the remaining equations $B_k(f) = 0$ belong. If $\psi(x,y,y')$ is a solution of this system, then every differential equation of the form ψ = const. is invariant under the group. In particular, if ψ is independent of y', then $\psi = a$ is an invariant family of curves.

COMMENTS ON SECTION 12.4

Continue with the notation described in the comments to Section 12.3. Let $\underset{\sim}{G}$ *be a Lie algebra of vector fields on* X *of arbitrary dimension. For each triple*

$$(A_1, A_2, A_3)$$

of elements of $\underset{\sim}{G}$*, let*

$$\Delta(A_1, A_2, A_3)$$

be the function on X' *whose vanishing at a point* x' *is necessary and sufficient that*

$$A_1'(x'), A_2'(x'), A_3'(x')$$

be linearly <u>*dependent*</u>*.*

Let

$$\phi\colon X \to R$$

be a real valued function, with a non-zero differential at each point of X*.* $d\phi$ *is then a mapping*

$$d\phi\colon X \to T^d(X) - (0)$$

Follow this with the projection mapping

$$T^d(X) - (0) \to PT^d(X) \equiv X'$$

to obtain a map, which we denote <u>*also*</u> *by* $d\phi$*, i.e.,*

$$d\phi\colon X \to X' \quad .$$

Then, Lie asserts that $\underset{\sim}{G}$ *leaves invariant the foliation*

$$\phi = \textit{constant}$$

if and only if

$$d\phi^*(\Delta(A_1,A_2,A_3)) = 0$$

$$\textit{for each triple } (A_1,A_2,A_3)$$

$$\textit{of elements of } \underset{\sim}{G}'.$$

If the $\Delta(A_1,A_2,A_3)$ *vanish identically for* *__each__* *choice* (A_1,A_2,A_3) *then:*

$$\dim \underset{\sim}{G}'(x') \leq 2$$

$$\textit{for all } x' \in X' \quad .$$

This means that (at least "generically") there is a function

$$\psi: X' \to R$$

which is *__invariant__* *under* $\underset{\sim}{G}'$. *This determines a first order, non-linear partial differential equation, for hypersurfaces on* X, *which is invariant under the action of* $\underset{\sim}{G}$.

Chapter 13

THE FIRST ORDER INFINITESIMAL TRANSFORMATIONS DETERMINE WHETHER THERE IS AN INVARIANT FAMILY OF CURVES

It will now be shown that in order to decide whether a given group leaves invariant some family of curves, it suffices to consider its infinitesimal transformations of order zero or one at a general point (x_0, y_0).

13.1 A CONDITION FOR AN INVARIANT FAMILY

According to the preceding section (§ 12.4) one must investigate all the equations $\Delta_{ijq} = 0$ corresponding to the three infinitesimal transformations of the group and see whether these equations, for an arbitrary value $x = x_0$, $y = y_0$, are satisfied by one value of y'. Think of ξ and η as power series in $x-x_0$ and $y-y_0$; then Δ_{ijq} vanishes for $x = x_0$, $y = y_0$ whenever one of the transformations A_i, A_j, A_q is of order ≥ 2, or if two of these transformations are of order 1. Therefore, if the group contains fewer than two independent infinitesimal transformations of order 0, then all the Δ_{ijq} vanish identically.*

---

* Geometrically expressed, this is the case where there is a 1-parameter family of curves each of which is fixed by all transformations of the group.

Hence we can assume that our group contains two independent infinitesimal transformations of order zero at (x_0, y_0), hence of the form

$$A_1 = p + \cdots , \quad A_2 = q + \cdots$$

and we shall assume that there are ρ independent first-order infinitesimal transformations:

$$[a_k(x-x_0)+b_k(y-y_0)]p + [\alpha_k(x-x_0)+\beta_k(y-y_0)]q + \cdots$$

$$(k=1,2,\ldots,\rho)$$

where $\rho \leq 4$. We form the ρ expressions Δ_{12k}, which arise from a composition of A_1 and A_2 with a first-order transformation.

Setting $x = x_0$, $y = y_0$, then gives

$$\Delta_{12k}^{(0)} = \alpha_k + y'(\beta_k - a_k) - y'^2 b_k \ .$$

The group leaves a family of curves invariant if and only if the ρ equations

$$\alpha_k + y'(\beta_k - a_k) - y'^2 b_k = 0 \qquad (13.1.1)$$

are all satisfied by <u>one</u> value of y'.

13.2 SOME GENERAL RESULTS

If the group contains four first order infinitesimal transformations, which can be assumed of the form

$$(x-x_0)p+\cdots, \qquad (y-y_0)p+\cdots,$$

$$(x-x_0)q+\cdots, \qquad (y-y_0)q+\cdots,$$

then the four equations (13.1.1):

$$-y' = 0, \quad -y'^2 = 0, \quad 1 = 0, \quad y' = 0$$

must all hold, which is impossible. Hence a group with 4 independent infinitesimal transformations leaves invariant no family of curves $\phi(x,y) = a$.

If the group contains 3 independent first-order transformations none of which is of the form

$$(x-x_0)p + (y-y_0)q+\cdots,$$

then we know (Theorem 11.4.1) that these transformations can be put in the form

$$(x-x_0)q+\cdots, \qquad (x-x_0)p - (y-y_0)q+\cdots,$$

$$(y-y_0)p+\cdots$$

And therefore one obtains the three equations (13.1.1):

$$1 = 0, \quad -2y' = 0, \quad -y'^2 = 0,$$

which again are contradicting. Hence, a group with 3 infinitesimal transformations of order one, none of which is of the form

$$(x-x_0)p + (y-y_0)q+\cdots,$$

leaves no family of curves $\phi(x,y) = a$ invariant.

Let us now assume that one of the first order transformations is of the form $(x-x_0)p + (y-y_0)q + \cdots$. By the considerations at the end of Section 11.4, we can then assume that the other first order transformations are of the form

$$C_1 + \cdots = a_1[(x-x_0)p-(y-y_0)q] + b_1(y-y_0)p + \alpha_1(x-x_0)q + \cdots$$

$$C_2 + \cdots = a_2[(x-x_0)p-(y-y_0)q] + b_2(y-y_0)p + \alpha_2(x-x_0)q + \cdots$$

and satisfy the relation

$$C_1 \circ C_2 - C_2 \circ C_1 = C_1 \qquad (13.2.1)$$

Then the three equations of the form (13.1.1) reduce to the two

$$\alpha_1 - 2a_1y' - b_1y'^2 = 0 ,$$

$$\alpha_2 - 2a_2y' - b_2y'^2 = 0 ,$$

and the question is whether these equations are satisfied by <u>one</u> value of y', i.e., whether

$$4(a_1\alpha_2-\alpha_1a_2)(a_1b_2-a_2b_1) - (\alpha_1b_2-\alpha_2b_1)^2 = 0 \qquad (13.2.2)$$

In fact, this is always the case. For the equation 13.2.1 is equivalent to the three equations

$$\left.\begin{aligned} \alpha_1b_2 - \alpha_2b_1 &= a_1 \\ 2(b_1a_2-b_2a_1) &= b_1 \\ 2(a_1\alpha_2-a_2\alpha_1) &= \alpha_1 \end{aligned}\right\} \qquad (13.2.3)$$

and if one multiplies these by $2a_1$, α_1 and b_1, respectively, and then adds them, one gets

$$\alpha_1 b_1 + a_1^2 = 0 ,$$

and this equation gives the equation 13.2.2 by repeated use of the relations 13.2.3.

<u>Hence a group with three independent first-order infinitesimal transformations</u>, <u>one of which is of the form</u> $(x-x_0)p + (y-y_0)q + \cdots$ <u>always leaves invariant some family of curves</u> $\phi(x,y) = a$.

If a group contains only two independent first-order infinitesimal transformations, then two cases are possible.

If neither of these transformations is of the form $(x-x_0)p + (y-y_0)q + \cdots$, then they have the form

$$C_1 + \cdots = [a_1(x-x_0)+b_1(y-y_0)]p + [\alpha_1(x-x_0)+\beta_1(y-y_0)]q + \cdots$$

$$C_2 + \cdots = [a_2(x-x_0)+b_2(y-y_0)]p + [\alpha_2(x-x_0)+\beta_2(y-y_0)]q + \cdots$$

and it is no restriction to assume that

$$C_1 \circ C_2 - C_2 \circ C_1 = C_1 . \qquad (13.2.4)$$

Now we obtain two equations of the form 13.1.1:

$$\alpha_1 + (\beta_1 - a_1)y' - b_1 y'^2 = 0 ,$$

$$\alpha_2 + (\beta_2 - a_2)y' - b_2 y'^2 = 0 ,$$

and the question is whether these are satisfied by one value of y', i.e., whether

$$(a_1-\beta_1,\alpha_2)(a_1-\beta_1,b_2) - (\alpha_1,b_2)^2 = 0 \qquad (13.2.5)$$

This is always the case. For the equation 13.2.4 is equivalent to the three equations

$$\left.\begin{aligned} (\alpha_1,b_2) &= a_1 = -\beta_1 = \tfrac{1}{2}(a_1-\beta_1) \\ (b_1,a_2-\beta_2) &= b_1 \\ (a_1-\beta_1,\alpha_2) &= \alpha_1 \end{aligned}\right\} \qquad (13.2.6)$$

and if one multiplies these by $a_1-\beta_1$, α_1 and b_1, respectively, and then adds them, one gets

$$\alpha_1 b_1 + \tfrac{1}{4}(a_1-\beta_1)^2 = 0 ,$$

which gives 13.2.5 by using 13.2.6.

<u>Hence a group with two independent first-order infinitesimal transformations neither of which is of the form</u>

$$(x-x_0)p + (y-y_0)q + \cdots$$

<u>always leaves invariant some family of curves</u> $\phi(x,y) = a$.

On the other hand, if the given group does contain the transformation $(x-x_0)p + (y-y_0)q + \cdots$ and only one transformation independent of this, then one gets only one equations of the form 13.1.1. Hence such a group leaves invariant a family of curves. The same is obviously true of any group having at most one first-order transformation.

The considerations of this section prove the following fundamental theorem:

<u>Theorem 13.2.1</u>. <u>If a group of point-transformations of the plane leaves invariant no family of curves</u> $\phi(x,y) = a$, <u>then either the group contains four independent first-order [infinitesimal] transformations of the form</u>

$$(x-x_0)p + \cdots , \quad (y-y_0)p + \cdots ,$$

$$(x-x_0)q + \cdots , \quad (y-y_0)q + \cdots ,$$

<u>or else it has three such transformations of the form</u>

$$(x-x_0)p - (y-y_0)q + \cdots , \quad (x-x_0)q + \cdots , \quad (y-y_0)p + \cdots$$

Chapter 14

GROUPS LEAVING INVARIANT ALL CURVES OF A FAMILY $\phi(x,y) = a$

In this section we determine all groups which leave invariant all the curves of a family $\phi(x,y) = a$. But first we make some general remarks on groups which leave a family of curves $\phi(x,y) = a$ invariant only in the sense of interchanging the curves of the family.

14.1 A METHOD FOR SOLVING THE GENERAL PROBLEM

If $A_1,\ldots,A_r$ are the infinitesimal transformations of such a group, then each $A_k(\phi)$ can be expressed as a function of ϕ:

$$A_k(\phi) = \xi_k(\phi) \quad .$$

Hence if we introduce ϕ as our new x-variable, we obtain

$$A_k = \xi_k(x)p + \eta_k(x,y)q \quad .$$

The relations

$$A_i \circ A_k - A_k \circ A_i = \sum_s c_{iks} A_s$$

provide $\frac{1}{2}\, r(r-1)$ relations of the form

$$\xi_i \frac{d\xi_k}{dx} - \xi_k \frac{d\xi_i}{dx} = \sum_s c_{iks} \xi_s \quad .$$

Hence, if we put

$$\xi_k(x)p = B_k ,$$

and consider the B_k as infinitesimal transformations of the 1-dimensional manifold of the x's, we get

$$B_i \circ B_k - B_k \circ B_i = \sum_s c_{iks} B_s ,$$

from which it follows that the B_k form a group. By Part I it is therefore always possible to replace x by a function of x in such a way that the B_k's will assume the form

$$(a_0+a_1x+a_2x^2)p$$

and hence constitute a linear group.

There are now four possibilities, according as the group constituted by the B_k's has zero, one, two or three terms. And the problem of determining all groups of the plane leaving invariant a family of curves is correspondingly separated into four problems. And it is possible to treat these problems in such a sequence that the solution of each problem is essentially furthered by those of the preceding ones.

The infinitesimal transformations of these groups can be put in the form

$$A = (a_0+a_1x+a_2x^2)p + \eta q .$$

Now if the group has $r > 1$ transformations, then it contains $r-1$ transformations of the form

$$B = (b_0+b_1x)p + \eta q \quad ,$$

where the B_k satisfy relations of the form

$$B_i \circ B_k - B_k \circ B_i = \sum b_{iks}B_s \quad .$$

Moreover, there are always $r-2$ infinitesimal transformations of the form

$$C = c_0p + \eta q \quad ,$$

satisfying relations

$$C_i \circ C_k - C_k \circ C_i = \sum c_{iks}C_s \quad .$$

And finally there are always $r-3$ infinitesimal transformations of the form:

$$D = \phi(x,y)q \quad ,$$

with relations

$$D_i \circ D_k - D_k \circ D_i = \sum d_{iks}D_s \quad .$$

This provides the following method of solving our general problem. We first seek the most general ρ-term family of transformations of the form $D_k = \eta_k q$ which satisfy relations of the form

$$D_i \circ D_k - D_k \circ D_i = \sum d_{iks}D_s \quad .$$

Then we seek, in the most general way, one additional infinitesimal transformation of the form $C = p + \eta_0 q$ satisfying ρ relations of the form

$$C \circ D_k - D_k \circ C = \sum_s c_{ks} D_s \quad ;$$

we emphasize that the right side of these equations does not contain C. Then we seek, in the most general way, to determine an infinitesimal transformation of the form $B = xp + \eta_1 q$ satisfying $\rho+1$ relations of the form

$$B \circ D_k - D_k \circ B = \sum_s b_{ks} D_s \quad ,$$

$$B \circ C - C \circ B = -C + \sum_s \gamma_s D_s \quad .$$

And finally we seek, in the most general way, an infinitesimal transformation of the form $A = x^2 p + \eta_2 q$ satisfying $\rho+2$ relations of the form

$$A \circ D_k - D_k \circ A = \sum \alpha_{ks} D_s \quad ,$$

$$A \circ C - C \circ A = -2B + \sum \beta_s D_s \quad ,$$

$$A \circ B - B \circ A = -A + \sum \delta_s D_s \quad .$$

After all these determinations have been made, we verify that all the families of infinitesimal transformations we have obtained in this way actually do determine a group of finite transformations.

COMMENTS ON SECTION 14.1

Here is the general framework for the method Lie sketches in this section.

Let $\underset{\sim}{G}$ *be a Lie algebra of vector fields on a manifold* X. *Suppose there is a completely integrable vector field system*

$$V \subset V(X)$$

which is left invariant by $\underset{\sim}{G}$, *i.e.,*

$$[\underset{\sim}{G}, V] \subset V \quad .$$

Let

$$\pi : X \to X'$$

be a submersion mapping which is a *__quotient mapping__* *for* V, *i.e., the fibers of* π *are the leaves of* V. *Then, there is a Lie algebra* $\underset{\sim}{G}'$ *of vector fields on* X' *such that:*

$$\pi_*(\underset{\sim}{G}) = \underset{\sim}{G}'$$

Thus, $\underset{\sim}{G}$ *is a prolongation of the action of* $\underset{\sim}{G}'$ *on* X'.

Lie is now dealing with the problem of classifying the possibilities for $\underset{\sim}{G}$, *knowing the classification of the possibilities for* $\underset{\sim}{G}'$.

There are two aspects to this. One might call them the "algebraic" and the "geometric". First, consider π_*

purely as a Lie algebra homomorphism. Let $\underset{\sim}{H}$ *be the kernel. Then,*

$$[\underset{\sim}{G},\underset{\sim}{H}] \subset \underset{\sim}{H}$$

i.e., $\underset{\sim}{H}$ *is a Lie algebra ideal of* $\underset{\sim}{G}$, *and*

$$\underset{\sim}{G}' = \underset{\sim}{G}/\underset{\sim}{H} .$$

One says that $\underset{\sim}{G}$ *is an* *__extension__* *of* $\underset{\sim}{G}'$ *by* $\underset{\sim}{H}$.

In certain circumstances, it is possible to classify algebraically these extensions. For example, if $\underset{\sim}{H}$ *is abelian, the extensions are governed by the* *__second cohomology group of the Lie algebra__* $\underset{\sim}{G}'$, *__with coefficients in the representation deduced from__* Ad $\underset{\sim}{G}'$ *__acting on__* $\underset{\sim}{H}$. *(See VB, vol. II, Chapter 2 for a simple treatment of this.) In particular, if* $\underset{\sim}{G}'$ *is* *__semisimple__*, $\underset{\sim}{G}$ *is determined by this representation.*

Lie did not, of course, know any of these fancy algebraic techniques. (It would perhaps be interesting for an historian to analyze whether Lie, by the evidence of how he actually worked out his problems, developed special cases of these techniques.) His method is more "geometric"-- Find coordinate systems for X' *in which* $\underset{\sim}{G}'$ *takes its "canonical form", then analyze how these coordinates are put together with coordinates for the leaves of* V *to make up the vector fields in* $\underset{\sim}{G}$. *This method is probably only practical in case*

$$\dim X' = 1 ,$$

which is, of course, the case being treated here. (In case X' *is higher dimensional, there are probably too many canonical forms for* $\underset{\sim}{G}'$.*)*

Here is another general technique one can abstract from what Lie does. Choose a Lie subalgebra

$$\underset{\sim}{L}' \subset \underset{\sim}{G}'$$

and set:

$$\underset{\sim}{L} = \pi_*^{-1}(L')$$

$\underset{\sim}{L}$ *is now a subalgebra of* $\underset{\sim}{G}$, *and Lie deduces a classification for* $(\underset{\sim}{G},\underset{\sim}{G}')$ *based on a classification for* $(\underset{\sim}{L},\underset{\sim}{L}')$.

14.2 LIE ALGEBRAS OF INFINITESIMAL TRANSFORMATIONS IN ONE VARIABLE

Let there now be given an arbitrary family of transformations of the form $D_k = \eta_k q$ with relations

$$D_i \; D_k - D_k \; D_i = \sum c_{iks} D_s .$$

From among these we choose the r-1 independent transformations $D_1', \ldots, D_{r-1}'$ which are of order ≥ 1 at a general point (x_0, y_0). And since each expression $D_i' \circ D_k' - D_k' \circ D_i'$

is of order one, we see that such an expression is always a linear combination of the D_i'.

<u>Theorem 14.2.1.</u> <u>If</u> r <u>infinitesimal transformations of the form</u> $D_k = \eta_k q$ <u>satisfy relations of the form</u>

$$D_i \circ D_k - D_k \circ D_i = \sum c_{iks} D_s ,$$

<u>then among them there are</u> $r-1$ <u>independent transformations</u> D_k' <u>which satisfy the corresponding relations</u>

$$D_i' \circ D_k' - D_k' \circ D_i' = \sum d_{iks} D_s' .$$

By means of this theorem we can determine the most general group of the form $\eta_k q$ as follows. We first take the most general infinitesimal transformation of the form $D_1 = \eta_1(x,y)q$; we then determine, in the most general way, an infinitesimal transformation $D_2 = \eta_2(x,y)q$ satisfying a relation of the form

$$D_1 \circ D_2 - D_2 \circ D_1 = a_1 D_1 + a_2 D_2 .$$

Then we determine the most general transformation $D_3 = \eta_3 q$ satisfying two relations of the form

$$D_1 \circ D_3 - D_3 \circ D_1 = b_1 D_1 + b_2 D_2 + b_3 D_3$$

$$D_2 \circ D_3 - D_3 \circ D_2 = c_1 D_1 + c_2 D_2 + c_3 D_3 ,$$

and so on.

14.3 CONTINUATION

An infinitesimal transformation of the form $\eta_1 q$ can always be given the form $X_1(x)q$, where X_1 is an arbitrary function of x, by replacing y by a suitable function of x and y. To determine the most general 2-term family

$$D_1 = \eta_1 q , \quad D_2 = \eta_2 q ,$$

we remark that D_1 and D_2 can be so chosen that the relation between them assumes the form

$$D_1 \circ D_2 - D_2 \circ D_1 = \varepsilon D_1 ,$$

where $\varepsilon = 0$ or 1.

Now, by replacing y by a suitable function of x and y, we can give D_1 the form $D_1 = X_1(x)q$. We then obtain, for the determination of $D_2 = \eta q$, the equation

$$\frac{\partial \eta}{\partial y} = \varepsilon ,$$

whence

$$\eta = \varepsilon y + f(x) .$$

If $\varepsilon \neq 0$, then $\varepsilon = 1$ and we replace y by $y + f(x)$, and thereby our two transformations assume the form $X_1 q$, y_q.

Hence there are only two types of 2-term families of the form $\eta_k q$, namely

X_1q, X_2q

X_1q, yq

If $D_1 = X_1q$, $D_2 = X_2q$, $D_3 = \eta q$ is a family of transformations with relations of the form

$$[D_1, D_3] = a_1D_1 + a_2D_2 + a_3D_3 \quad ,$$

$$[D_2, D_3] = b_1D_1 + b_2D_2 + b_3D_3 \quad ,$$

then there is a transformation $D = \alpha_1D_1 + \alpha_2D_2$ such that $[D, D_3]$ is a linear combination of D_1 and D_2, and it is no real restriction to assume $D = D_1$. Then

$$X_1 \frac{\partial\eta}{\partial y} = a_1X_1 + a_2X_2 \quad ,$$

$$X_2 \frac{\partial\eta}{\partial y} = b_1X_1 + b_2X_2 + b_3\eta$$

and these equations are satisfied if η is an arbitrary function of x.

If η depends on y also, then, by the first equation, η_y is a function of x, and therefore in the last equation one has $b_3 = 0$. From this one concludes that η_y is a constant. For our equations show that each expression

$$(A_1X_1 + A_2X_2) \frac{\partial\eta}{\partial y}$$

is a linear combination of X_1 and X_2:

$$(A_1X_1+A_2X_2)\ \frac{\partial\eta}{\partial y} = B_1X_1 + B_2X_2$$

Similarly,

$$(B_1X_1+B_2X_2)\ \frac{\partial\eta}{\partial y} = C_1X_1 + C_2X_2$$

whence

$$(A_1X_1+A_2X_2)\left(\frac{\partial\eta}{\partial y}\right)^2 = C_1X_1 + C_2X_2$$

and in general

$$(A_1X_1+A_2X_2)\left(\frac{\partial\eta}{\partial y}\right)^2 = L_1X_1 + L_2X_2 \quad ,$$

where k is an arbitrary positive integer. For example, if one sets $A_1 = 1$, $A_2 = 0$, one obtains three equations of the form:

$$X_1\left(\frac{\partial\eta}{\partial y}\right)^0 = X_1 \quad ,$$

$$X_1\ \frac{\partial\eta}{\partial y} = \alpha_1X_1 + \alpha_2X_2 \quad ,$$

$$X_1\left(\frac{\partial\eta}{\partial y}\right)^2 = \beta_1X_1 + \beta_2X_2 \quad ,$$

where the right sides of these equations are linearly dependent.

$$\gamma_0X_1 + \gamma_1(\alpha_1X+\alpha_2X_2) + \gamma_2(\beta_1X_1 + \beta_2X_2) = 0$$

$\gamma_0, \gamma_1, \gamma_2$ not all 0. Therefore

$$X_1\left(\gamma_0 + \gamma_1 \frac{\partial\eta}{\partial y} + \gamma_2 \left(\frac{\partial\eta}{\partial y}\right)^2\right) = 0 \quad ,$$

from which it follows that ηy is a constant. Since this constant is $\neq 0$ by our earlier assumption, we can take

$$\eta = y + f(x) \quad ,$$

and in fact it is no real restriction to take $\eta = y$. <u>The 3-term family of transformations we have found has the form</u>

$$X_1 q, \; X_2 q, \; yq \quad .$$

Now let q, yq, ηq be three transformations satisfying the familiar relations. Then η is determined by two relations of the form:

$$\begin{aligned} \frac{\partial\eta}{\partial y} &= a_0 + 2a_1 y + a_2\eta \\ y\frac{\partial\eta}{\partial y} - \eta &= b_0 + 2b_1 y + b_2\eta \end{aligned} \qquad (14.3.1)$$

It is easy to see that $a_2 = 0$. Otherwise, elimination of ηy would show that η is a rational function of y, while integration of the first equation would show that η is a transcendental function of y. Hence we put a_2 = and then find, by integrating the first equation, that η is of the form

$$\eta = a_0 y + a_1 y^2 + f(x) \quad .$$

Since it is no real restriction to assume $a_0 = 0$, we get:

$$\eta = a_1 y^2 + f(x) \quad .$$

If we substitute this value into the second of the equations (14.3.1), we find that $a_1 = 0$ or $f(x) = 0$, so that

$$\eta = y^2 \qquad \text{or} \qquad \eta = f(x)$$

Combining this with our previous work we see that *the 3-term family of transformations we have sought has one of the forms*:

$$\boxed{X_1 q, X_2 q, X_3 q} \quad , \quad \boxed{X_1 q, X_2 q, yq} \quad , \quad \boxed{q, yq, y^2 q}$$

14.4 CONTINUATION

If four infinitesimal transformations of the form q, yq, $y^2 q$, ηq satisfy the familiar relations, then η is determined by three relations of the form:

$$\left.\begin{aligned} \frac{\partial \eta}{\partial y} &= a_0 + 2a_1 y + 3a_2 y^2 + a_3 \eta \quad , \\ y \frac{\partial \eta}{\partial y} - \eta &= b_0 + 2b_1 y + 3b_2 y + b_3 \eta \quad , \\ y^2 \frac{\partial \eta}{\partial y} - 2y\eta &= c_0 + 2c_1 y + 3c_2 y^2 + c_3 \eta \quad , \end{aligned}\right\} \quad (14.4.1)$$

and we see, as above that $a_3 = 0$. Therefore, integrating the first equation shows that η is of the form

$$\eta = a_0 y + a_1 y^2 + a_3 y^3 + f(x) \quad ,$$

and it is no real restriction to assume $a_0 = a_1 = 0$:

$$\eta = a_3 y^3 + f(x) \quad .$$

The second of the equations (14.4.1) shows that either $\eta = y^3$ or $\eta = f(x)$, both of which contradict the third of the equations (14.4.1). We conclude that <u>no 4-term group can have the form</u> $q, yq, y^2q, \eta q$.

We now seek, in the most general way, $r+1$ infinitesimal transformations of the form

$$X_1 q, X_2 q, \ldots, X_r q, \eta q$$

satisfying the familiar relations. Proceeding exactly as in the case $r = 2$ we obtain, for the determination of η, r equations of the form:

$$X_1 \frac{\partial \eta}{\partial y} = \sum \alpha_i X_i \quad ,$$

$$\cdot \qquad \cdot \quad \cdot$$

$$X_{r-1} \frac{\partial \eta}{\partial y} = \sum \lambda_i X_i \quad ,$$

$$X_r \frac{\partial \eta}{\partial y} = \sum \mu_i X_i + \mu \eta$$

where μ must in fact be $= 0$, as before. Finally, for any constants $A_1, \ldots, A_r$ there is a relation of the form

$$(A_1X_1+\cdots+A_rX_r)\frac{\partial\eta}{\partial y} = B_1X_1+\cdots+B_rX_r .$$

In particular,

$$(B_1X_1+\cdots+B_rX_r)\frac{\partial\eta}{\partial y} = C_1X_1+\cdots+C_rX_r ,$$

whence

$$(A_1X_1+\cdots+A_rX_r)\left(\frac{\partial\eta}{\partial y}\right)^2 = C_1X_1+\cdots+C_rX_r ,$$

and in general

$$(A_1X_1+\cdots+A_rX_r)\left(\frac{\partial\eta}{\partial y}\right)^k = L_1X_1+\cdots+L_rX_r .$$

In this equation let k take the values $0,1,2,\ldots,r$ successively and let $A_1,A_2,\ldots,A_r$ be fixed quantities. This gives $r+1$ equations, from which one obtains, by elimination, an equation of the form

$$\left(\sum A_iX_i\right)\left(k_0+k_1\frac{\partial\eta}{\partial y}+\cdots+k_r\left(\frac{\partial\eta}{\partial y}\right)^r\right) = 0 .$$

Therefore

$$k_0 + k_1\frac{\partial\eta}{\partial y}+\cdots+k_r\left(\frac{\partial\eta}{\partial y}\right)^r = 0 ,$$

in which all the constants k vanish only if $\eta_y = 0$. Hence η_y is a constant and $\eta = \alpha y + f(x)$, so that we can assume either $\eta = y$ or $\eta = f(x)$.

<u>Hence we find only the two families</u>

$$X_1q, X_2q, \ldots, X_rq, yq,$$

$$X_1q, X_2q, \ldots, X_rq, X_{r+1}q .$$

Finally, we seek, in the most general way, $r+2$ infinitesimal transformations of the form

$$X_1q, X_2q, \ldots, X_rq, yq, \eta q \qquad (r>1)$$

satisfying the familiar relations. Since $r > 1$ we get two relations of the form

$$\left.\begin{aligned} X_1\frac{\partial\eta}{\partial y} &= \sum \alpha_i X_i + \alpha_0 y + \alpha\eta \ , \\ X_2\frac{\partial\eta}{\partial y} &= \sum \beta_i X_i + \beta_0 y + \beta\eta \ , \end{aligned}\right\} \qquad (14.4.2)$$

in which it is no real restriction to assume $\alpha = 0$. Integrating the first equation gives

$$X_1\eta = y\sum \alpha_i X_i + \frac{1}{2}\alpha_0 y^2 + f(x) \ ,$$

and we now substitute the value of η this gives into the second of the equations (14.4.2).

Let us first assume $\beta \neq 0$; then η is of the form

$$\eta = \alpha_1 y + \phi(x) \ ,$$

and it is no restriction to take $\alpha_1 = 0$. The hypothesis $\beta \neq 0$ therefore leads to an $(r+2)$-term family having the same form as the given $(r+1)$-term family.

Now assume $\beta = 0$. Eliminating y from the equations (14.4.2) shows that η_y depends only on x. Therefore we obtain r equations of the form

$$X_k\frac{\partial\eta}{\partial y} = \sum c_{ks} X_s \ ,$$

from which it follows, as in the previous case, that η is of the form $\eta = \alpha y + f(x)$, and it is no real restriction to take $\alpha = 0$. Hence our family again has the same form as the given $(r+1)$-term family.

These considerations give the general theorem:

Theorem 14.4.1. If a family of infinitesimal transformations of the form $\eta_k q$ satisfies the familiar relations, then it is of one of the following forms

$$\boxed{\begin{matrix} X_1 q \\ X_2 q \\ \vdots \\ X_r q \end{matrix}} \qquad \boxed{\begin{matrix} X_1 q \\ X_2 q \\ \vdots \\ X_r q \\ yq \end{matrix}} \qquad \boxed{\begin{matrix} q \\ yq \\ y^2 q \end{matrix}}$$

COMMENTS ON SECTION 14.4

When Lie says that a set of vector fields "satisfies the familiar relations", he means that they are closed under Jacobi bracket, i.e., that their linear span forms a Lie algebra.

The classification problem of the previous three sections can be described as follows. Let $\underset{\sim}{G}$ *be a finite dimensional Lie algebra of vector fields on a manifold* Z, *and let* V *be a completely integrable vector field system on* Z *such that:*

$\underset{\sim}{G}$ *is tangent to* V

This means that each vector field $A \in \underset{\sim}{G}$ *is tangent to the leaves of* V. *Lie's problem is to:*

Classify the possibilities for $(Z,V,\underset{\sim}{G})$.

Lie is, of course, only classifying locally, which simplifies his task. (Perhaps only now is mathematics developing the differential-topological tools to tackle the job globally.)

We may then suppose that:

$$Z = X \times Y \quad ,$$

where X,Y *are manifolds, such that:*

The leaves of V *are the subsets*

$$\{(x,Y) \colon X \in X\} \quad .$$

Thus, each $A \in \underset{\sim}{G}$ *determines a vector field*

$$B_x$$

on the manifold $x \times Y$, *for each* $x \in X$.

Let

$$M(X,V(Y))$$

denote the space of C^∞ *mappings*

$$\phi: X \to V(Y)$$

(To say that the map is C^∞ *is to say that, for each* $f \in F(Y)$, *the map*

$$(x,y) \to \phi(x)(f)(y)$$

is C^∞*).*

M(X,V(Y)) *is a Lie algebra, which is infinite dimensional as a Lie algebra over the real numbers. (It is also a "Lie module" over* F(X), *and finite dimensional as such an object).* M(X,V(Y)) *is encountered quite often in physics (see LAQM, FA, VB, GPS, and volume VI and X) and I call it the gauge Lie algebra or, sometimes, the Lie algebra of currents.*

Return to $\underset{\sim}{G}$ *acting on* Z, *tangent to the fibers of* V. *Each* $A \in \underset{\sim}{G}$ *determines a map*

$$x \to B_x$$

of $X \to V(Y)$, *i.e., an element of* M(X,V(Y)). *Thus,* $\underset{\sim}{G}$ *may be identified with an element of* M(X,V(Y). *Then, the problem of these sections is really to:*

> *Classify the finite dimensional Lie subalgebras of* M(X,V(Y)).

By "classify", one means "up to an obvious equivalence". Namely, consider the diffeomorphisms on $X \times Y$ *of the form*

$$(x,y) \to (x'(x), y'(x,y))$$

i.e., which preserve the foliation determined by Y. *Two subalgebras which are equivalent under this group of diffeomorphisms are considered the same.*

There are relations to the *theory of deformations of Lie algebras and representations.* *(See Chapter D.) Namely, to each* $x \in X$, *set:*

$$\underset{\sim}{G}(x) = \phi(x)(\underset{\sim}{G}) .$$

$\{\underset{\sim}{G}(x)\}$ *is a family of Lie algebras, parameterized by* x. *This defines a Lie algebra deformation.* $\underset{\sim}{G}(x)$ *is also* *given* *as a Lie subalgebra of*

$$V(Y) .$$

As x *varies, this defines a Lie algebra* *homomorphism deformation*.

The formulas in Theorem 14.4.1 determine these deformations explicitly. Here,

$$X = Y = R .$$

There are, up to isomorphism, only *three* *finite dimensional Lie algebras which can appear as subalgebras of* V(R). *This explains why there are three boxes. For the third box, the Lie algebra is* *semisimple*. *In Chapter E I prove the vanishing of the relevant cohomology group for the semisimple case. This is why the formulas given in the third box do not depend on parameters.*

14.5 COMPLETION OF THE DETERMINATION OF GROUPS IN THE PLANE WHICH LEAVE INVARIANT A FAMILY OF CURVES

We now show that these families of infinitesimal transformations always determine a group. The equations

$$x' = x, \qquad y' = y + a_1X_1 + a_2X_2 + \cdots + a_rX_r$$

determine an r-term group of finite transformations, and the infinitesimal transformations of this group are the r quantities X_kq.

Moreover, it is clear that the equations

$$x' = x, \qquad y' = ay + a_1X_1 + \cdots + a_rX_r$$

determine an (r+1)-term group of finite transformations, and the infinitesimal transformations of this group are the $r+1$ quantities X_kq, yq.

Finally, it is well known that the equations

$$x' = x, \qquad y' = \frac{a_1y+a_2}{a_3y+1}$$

determine a 3-term group with the infinitesimal transformations q, yq, y^2q.

<u>We have now determined all groups of the plane which leave invariant all the curves of a family</u> $\phi(x,y) = a$.

COMMENTS ON SECTION 14.5

In this section, Lie gives the explicit formulas for the group actions corresponding to the Lie algebra deformations classified in Theorem 14.4.1. Thus, for each $x \in X$, *he defines a group* $G(x)$, *and a mapping*

$$G(x) \times Y \to Y$$

which defines a transformation group action of $G(x)$ *on* Y. *This example involves, first, a deformation* $\{G(x)\}$ *of group structures, then, second, a deformation of the transformation group actions. There is still much work needed to understand the general features of these objects!*

Chapter 15

SOME AUXILIARY THEORIES

In this section we develop a general theorem on arbitrary transformation groups. First some remarks on <u>linear</u> groups.

15.1 GENERAL REMARKS ON LINEAR GROUPS

An infinitesimal transformation of the form

$$\sum_{i=1}^{n} \sum_{k=1}^{n} c_{ik} x_i p_k$$

will be called a <u>linear</u> transformation. By means of this, the x_k receive the following increments

$$\delta x_k = \left(\sum_i c_{ik} x_i \right) \delta t \tag{15.1.1}$$

It follows from Cauchy's investigations of simultaneous systems of the form 15.1.1 that the x_i can be replaced by suitable linear combinations of themselves to give a system for which $c_{ik} = 0$ when $i > k$. <u>Thus every linear infinitesimal transformation can be put in the form</u>

$$c_{11}x_1p_1 + (c_{12}x_1 + c_{22}x_2)p_2 + (c_{13}x_1 + c_{23}x_2 + c_{33}x_3)p_3 + \cdots$$

$$+ (c_{1q}x_1 + c_{2q}x_2 + \cdots + c_{qq}x_q)p_q + \cdots \quad .$$

COMMENTS ON SECTION 15.1

The result stated here is that *<u>each linear transformation of a vector space into itself can</u>, <u>by appropriate choice of basis for the vector space</u>, <u>be put into triangular form</u>. This requires that the scalar field of the vector space be <u>algebraically closed</u> (e.g., the complex numbers), a hypothesis that Lie does not seem to state explicitly.*

15.2 CONTINUATION

We now suppose given two linear infinitesimal transformations $A = \sum \eta_i p_i$ and $B = \sum \xi_i p_i$ such that

$$[A,B] = A \tag{15.2.1}$$

We shall show that it is possible to replace $x_1,\ldots,x_n$ by new independent variables so that A and B are simultaneously put into a remarkable form.

We first put A in the form discussed in the preceding section (15.1), so that the coefficient of p_1 is of the form εx_1. In doing this it can happen that more of the η_k, say $\eta_2,\eta_3,\ldots,\eta_q$, also assume the simple form εx_k, so that

$$A = \varepsilon(x_1p_1+x_2p_2+\cdots+x_qp_q) + \eta_{q+1}p_{q+1} +\cdots+ \eta_np_n \quad .$$

It is no real restriction to assume that $c_1x_1 + \cdots + c_qx_q$ is the most general linear function of x satisfying a relation

$$A\left(\sum c_i x_i\right) = \varepsilon \sum c_i x_i \quad .$$

I first claim that $\varepsilon = 0$. For if $\varepsilon \neq 0$, it is no real restriction to assume $\varepsilon = 1$, and we will show that the integer n cannot then be finite.

The equation (15.2.1) is equivalent to the n equations

$$A(\xi_i) - B(\eta_i) = \eta_i \qquad (15.2.2)$$

and hence

$$A(\xi_i) - \eta_i = x_i \qquad \text{for } i=1,2,\ldots,q$$

From this we conclude that $\xi_1,\ldots,\xi_q$, $x_1,\ldots,x_q$ are independent. For if there were a relation

$$\nu_1\xi_1 + \cdots + \nu_q\xi_q + \mu_1x_1 + \cdots + \mu_qx_q = 0 \quad ,$$

applying A to it would yield

$$\nu_1(\xi_1+x_1) + \cdots + \nu_q(\xi_q+x_q) + \mu_1x_1 + \cdots + \mu_qx_q = 0 \quad ,$$

from which there would follow the impossible equation

$$\nu_1x_1 + \cdots + \nu_qx_q = 0 \quad .$$

Therefore, $n \geq 2q$, and it is permissible to set

$$x_{q+1} = \xi_1,\ldots,x_{2q} = \xi_q \quad .$$

Applying A gives

$$A(x_{q+i}) = A(\xi_i) = \xi_i + x_i \qquad (i=1,\ldots,q)$$

or, since $A(x_{q+i}) = \eta_{q+i}$,

$$\eta_{q+i} = x_{q+i} + x_i \qquad (i=1,\ldots,q)\ .$$

This determines the q quantities $\eta_{q+1},\ldots,\eta_{2q}$. But

$$A(\xi_j) - B(\eta_j) = \eta_j\ ,$$

and therefore the q quantities $\xi_{q+1},\ldots,\xi_{2q}$ satisfy relations of the form

$$A(\xi_{q+i}) - \xi_{q+i} - \xi_i = \eta_{q+i} = x_{q+i} + x_i \qquad (i=1,\ldots,q)$$

or

$$A(\xi_{q+i}) = \xi_{q+i} + 2x_{q+i} + x_i \qquad (i=1,\ldots,q)$$

from which it easily follows that $\xi_{q+1},\ldots,\xi_{2q}, x_1,\ldots,x_{2q}$ are linearly independent. Hence $n \geq 3q$ and it is no real restriction to put

$$x_{2q+1} = \xi_{q+1},\ldots,x_{3q} = \xi_{2q}\ .$$

Applying A gives

$$A(x_{2q+i}) = A(\xi_{q+i}) = \xi_{q+i} + 2x_{q+i} + x_i \qquad (i=1,\ldots,q)$$

or, since $A(x_{2q+i}) = \eta_{2q+i}$,

$$\eta_{2q+i} = x_{2q+i} + 2x_{q+i} + x_i \qquad (i=1,\ldots,q)$$

Continuing in the same way as before, one sees that $\xi_{2q+1},\ldots,\xi_{3q},x_1,\ldots,x_q,\ldots,x_{2q},\ldots,x_{3q}$ are independent and hence $n \geq 4q$, and so on. Since n is a finite integer, we see that $\epsilon = 0$.

15.3 CONTINUATION

Hence we can put

$$A = 0.p_1 + \cdots + 0.p_q + \eta_{q+1}p_{q+1} + \cdots ,$$

$$B = \xi_1 p_1 + \cdots + \xi_q p_q + \xi_{q+1}p_{q+1} + \cdots ,$$

and from 15.2.1 it follows that

$$A(\xi_1) = 0,\ A(\xi_2) = 0,\ldots,A(\xi_q) = 0 ,$$

from which it follows, by our previous hypotheses, that $\xi_1,\ldots,\xi_q$ depend only on $x_1,\ldots,x_q$. Now I can choose x_{q+1} so that there is a relation of the form

$$A(x_{q+1}) = c_1x_1 + \cdots + c_qx_q + cx_{q+1} ;$$

then

$$\eta_{q+1} = c_1x_1 + \cdots + c_qx_q + cx_{q+1} .$$

It is easy to see that $c = 0$. One proves this by setting $x_1 = \cdots = x_q = 0$ in A and B and then applying the considerations of the preceding section (15.2) to the resulting expressions $A^{(0)}$ and $B^{(0)}$, which are $\neq 0$ if $c \neq 0$.

Hence we can put

$$A(x_{q+1}) = c_1x_1 +\cdots+ c_qx_q = \eta_{q+1} ;$$

it is thinkable that $\eta_{q+2},\ldots,\eta_{q'}$, for example are still linear functions of $x_1,\ldots,x_q$. In this it is no real restriction to assume that $\alpha_1x_1+\cdots+\alpha_qx_q+\cdots+\alpha_{q'}x_{q'}$ is the most general quantity satisfying a relation of the form

$$A\left(\sum\alpha_ix_i\right) = \beta_1x_1 +\cdots+ \beta_qx_q .$$

By 15.2.1 there are then $q'-q$ relations of the form

$$A(\xi_{q+1}) = \delta_1x_1 +\cdots+ \delta_qx_q \qquad (i=1,\ldots,q'-q)$$

and therefore $\xi_{q+1},\ldots,\xi_{q'}$ are functions of $x_1,\ldots,$ $x_q,\ldots,x_{q'}$.

We continue in this way. Choose $x_{q'+1}$ so that there is a relation of the form

$$A(X_{q'+1}) = c_1x_1 +\cdots+ c_qx_q +\cdots+ c_{q'}x_{q'} + cx_{q'+1} ,$$

and we see, as before, that $c = 0$. Therefore,

$$\eta_{q'+1} = c_1x_1 +\cdots+ c_{q'}x_{q'} ,$$

and it is thinkable that still more of the η's, for example, $\eta_{q'+2},\ldots,\eta_{q''}$, have this form. It is no real restriction to assume that the general solution of the equation

$$A(\phi) = c_1x_1 + \cdots + c_{q'}x_{q'}$$

is

$$\phi = \delta_1x_1 + \cdots + \delta_{q''}x_{q''} .$$

But by 15.2.1 there are relations of the form

$$A(\xi_{q'+i}) = c_1x_1 + \cdots + c_{q'}x_{q'} ,$$

and therefore $\xi_{q'+1},\ldots,\xi_{q''}$ are functions of $x_1,\ldots,x_{q''}$, and so on.

Hence there is an increasing sequence of integers $q,q',q'',q''',\ldots,n$ *such that* $\eta_1,\ldots,\eta_q$ *are zero*; $\eta_{q+1},\ldots,\eta_{q'}$ *depend only on* $x_1,\ldots,x_q$; $\eta_{q'+1},\ldots,\eta_{q''}$ *depend only on* $x_1,\ldots,x_{q'}$, *and so on*; *and that at the same time* $\xi_1,\ldots,\xi_q$ *depend only on* $x_1,\ldots,x_q$; $\xi_{q+1},\ldots,\xi_{q'}$ *depend only on* $x_1,\ldots,x_{q'}$; $\xi_{q'+1},\ldots,\xi_{q''}$ *depend only on* $x_1,\ldots,x_{q''}$, *and so on*.

15.4 A NORMAL FORM FOR A 2-TERM LINEAR GROUP

It only remains to make certain simple transformations. A linear transformation of $x_1,\ldots,x_q$ leaves A and B in the form we have found for them. In particular, we can arrange that $\xi_1p_1+\cdots+\xi_qp_q$, which depends only on $x_1,\ldots,x_q$, $p_1,\ldots,p_q$, assumes the form mentioned in Section 15.1 by such a transformation. Then make a linear transformation of $x_{q+1},\ldots,x_{q'}$ so that

$$\xi_1 p_1 + \cdots + \xi_q p_q + \cdots + \xi_{q'} p_{q'}$$

assumes the canonical form of 15.1, and so on. Continuing in this way, we obtain the theorem

Theorem 15.4.1. If two linear [infinitesimal] transformations $A = \sum \eta_i p_i$, $B = \sum \xi_i p_i$ satisfy

$$[A,B] = A ,$$

then it is possible to choose the independent variables $x_1,\ldots,x_n$ in such a way that η_i depends only on $x_1,\ldots,x_{i-1}$, while each ξ_i is a function of $x_1,\ldots, x_{i-1},x_i$.*

COMMENTS

This is a special case of what is now known as Lie's theorem in Lie algebra theory; namely, the matrices of a linear representation of a solvable Lie algebra may always be put into triangular form. See Samelson (1), VB, vol. II; and vol. VII of IM.

---

* Without doubt this theorem has been known for a long time, although perhaps in a different form. It is a special case of a much more general theorem, which I have established on a different occasion.

15.5 THE STRUCTURE CONSTANTS OF A LIE ALGEBRA

We now turn to quite arbitrary groups of point-transformations of an n-dimensional manifold $x_1,\ldots,x_n$. Let $A_1,\ldots,A_r$ be independent [infinitesimal] transformations with relations

$$[A_i,A_k] = \sum_s c_{iks}A_s \quad ,$$

we will now establish certain relations among the c_{iks}. In the well-known Jacobi identity

$$[[A_i,A_k],A_s] + [[A_k,A_s],A_i] + [[A_s,A_i],A_k] = 0 \quad ,$$

substitute the above values of the quantities $[A_\mu,A_\nu]$; and do this again. This produces an equation of the form

$$C_1A_1 + C_2A_2 + \cdots + C_rA_r = 0 \quad ,$$

where the C_k are functions of the $c_{\mu\nu\rho}$. Since the A_k are independent infinitesimal transformations, it follows that

$$C_1 = C_2 = \cdots = C_r = 0 \quad ,$$

i.e.,

$$\sum_\rho (c_{ik\rho}c_{\rho s\sigma} + c_{ks\rho}c_{\rho i\sigma} + c_{si\rho}c_{\rho k\sigma}) = 0 \quad ,$$

in which i,k,s,ρ vary over $1,2,\ldots,r$.

COMMENTS

The (c_{ijk}) *are called the* *structure constants* *of the Lie algebra with respect to the bases* $A_1,\ldots,A_r$. *The quadratic relations given in the text are the* *Lie relations*, *and are the necessary and sufficient conditions that a given set* (c_{ijk}) *of constants be the structure constants of a Lie algebra. In the older literature on Lie groups and algebras the role of these structure constants was emphasized, but this has been replaced by basis-free methods.*

15.6 THE LINEAR ADJOINT GROUP

Again, let $A_1,\ldots,A_r$ be arbitrary (i.e., not necessarily linear) infinitesimal transformations which are independent and satisfy relations

$$[A_i,A_k] = \sum_s c_{iks}A_s$$

I claim that the linear infinitesimal transformations

$$B_i = \sum_s \frac{\partial f}{\partial x_s} \sum_k c_{isk}x_k$$

satisfy

$$[B_i,B_j] = \sum_s c_{ijs}B_s \quad .$$

Indeed, substitution gives

$$[B_i, B_j] = \sum_\sigma \frac{\partial f}{\partial x_\sigma} \sum_s \sum_k (c_{j\sigma s} c_{isk} - c_{i\sigma s} c_{jsk}) x_k \quad ,$$

and our claim is that this expression can be brought to the form

$$\sum_s c_{ijs} \sum_\sigma \frac{\partial f}{\partial x_\sigma} \sum_k c_{s\sigma k} x_k \quad ,$$

or, equivalently, that the sum

$$\sum_s (c_{j\sigma s} c_{isk} - c_{i\sigma s} c_{jsk} - c_{ijs} c_{s\sigma k}) = -\sum_s (c_{j\sigma s} c_{sik} + c_{\sigma is} c_{sjk} + c_{ijs} c_{s\sigma k})$$

vanishes identically. In the preceding section (15.5) we saw that this is, in fact, the case. Thus:

<u>Theorem 15.6.1</u>. <u>If</u> $A_1, \ldots, A_r$ <u>are independent</u> [<u>infinitesimal</u>] <u>transformations with relations</u>

$$[A_i, A_j] = \sum_s c_{ijs} A_s \quad ,$$

<u>then the linear expressions</u>

$$B_i = \sum_s \frac{\partial f}{\partial x_s} \sum_k c_{isk} x_k$$

<u>satisfy</u>

$$[B_i, B_j] = \sum_s c_{ijs} B_s$$

Theorems 15.4.1 and 15.6.1, as well as the well-known theorem of Section 15.1 have an important application in the next two sections.

COMMENTS

Let $\underset{\sim}{G}$ *be a Lie algebra,* V *a vector space,* L(V) *the space of linear maps* $V \to V$. *A* <u>*linear representation of*</u> $\underset{\sim}{G}$ *is defined as a linear mapping*

$$\rho: G \to L(V)$$

such that:

$$\rho([A,B]) = \rho(A)\rho(B) - \rho(B)\rho(A) \qquad (15.6.1)$$

for all $A,B \in \underset{\sim}{G}$.

The <u>*adjoint representation*</u> *(which is the modern name for the representation defined by the* B_i *in the text) is such a representation obtained by setting:*

V = *underlying vector space of* $\underset{\sim}{G}$.

$\rho(A)(B) = [A,B] \equiv \mathrm{Ad}\ A(B)$

for $A \in \underset{\sim}{G}$, $B \in \underset{\sim}{G} \equiv V$.

The Jacobi identity guarantees that 15.6.1 is satisfied, i.e., that ρ really defines a Lie algebra representation.

Chapter 16

DETERMINATION OF ALL GROUPS WHICH TRANSFORM THE CURVES OF A FAMILY $\phi(x,y) = a$ IN A 1-TERM WAY

We now determine all families of transformations of the form

$$\eta_1 q, \eta_2 q, \ldots, \eta_r q, p+rq$$

satisfying the familiar relations. Recall that the η_k have one of the forms determined in Section 14. We also remark that each $[\eta_k q, p+\eta q]$ is a linear combination of the $\eta_k q$ alone.

COMMENTS

Let X,Y *be manifolds,*

$$Z = X \times Y$$

$$\pi: Z \to X$$

the Cartesian projection map. Let $\underset{\sim}{G}$ *be a finite dimensional Lie algebra of vector fields on* Z. *Suppose that each* $A \in \underset{\sim}{G}$ *is projectable under* π, *i.e., there is an* $A' \in V(Y)$ *such that:*

$$\pi^*(A'(f')) = A(\pi^*(f'))$$

$$\textit{for all } f' \in F(Y) \quad .$$

Let $\underset{\sim}{G}'$ *be the collection of such* A''*s. Then,*

$$\pi_*: \underset{\sim}{G} \to \underset{\sim}{G}'$$

is an onto *__Lie algebra homomorphism__.*

Let

$$\underset{\sim}{H} = \{A\varepsilon G: \pi_*(A) = 0\} .$$

Then, $\underset{\sim}{H}$ *is an ideal of* $\underset{\sim}{G}$, *and*

$$\underset{\sim}{G}' = \underset{\sim}{G}/\underset{\sim}{H} .$$

$\underset{\sim}{G}$ *is an* *__extension of__* $\underset{\sim}{G}'$ *__by__* $\underset{\sim}{H}$. *However, the "geometric", prolongation, feature of the situation is also important. (This combination of algebraic and geometric features is, of course, typical of Lie's work.)*

Here,

$$X = Y = R .$$

In this chapter Lie deals with the case:

$$\dim \underset{\sim}{G}' = 1 .$$

16.1 CONTINUATION

__We first determine all families of the form__ q, yq, y^2q, $p+\eta q$.

The quantity η is determined by equations of the form

$$\left.\begin{aligned} \frac{\partial\eta}{\partial y} &= a_0 + 2a_1y + 3a_2y^2 \\ y\frac{\partial\eta}{\partial y} - \eta &= b_0 + 2b_1y + 3b_2y^2 \\ y^2\frac{\partial\eta}{\partial y} - 2y\eta &= c_0 + 2c_1y + 3c_2y^2 \end{aligned}\right\} \qquad (16.1.1)$$

The first of these shows that

$$\eta = a_0 y + a_1 y^2 + a_2 y^3 + f(x) \quad ,$$

or, since it is no real restriction to set $a_0 = a_1 = 0$,

$$\eta = a_2 y^3 + f(x) \quad .$$

The last two of the equations (16.1.1) show that $a_2 = f(x) = 0$, so that $\eta = 0$. Hence, the family sought is of the form

$$q, \; yq, \; y^2q, \; p \quad .$$

16.2 CONTINUATION

Next we seek all families of the form

$$X_1q, X_2q, \ldots, X_rq, p+\eta q \quad .$$

We can assume $r > 0$, since when $r = 0$ we are dealing with the single transformation $p+\eta q$, and this takes the simple form p if y is replaced by a suitable function of x and y.

There are then equations of the form

$$[X_kq, p+\eta q] = c_{k1}X_1q + \cdots + c_{kr}X_rq \quad .$$

It is possible to replace $X_1, \ldots, X_r$ by linear combinations of them, $X_i' = \alpha_{i1}X_1 + \cdots + \alpha_{ir}X_r$, so that in the transformed equation

$$[X_k'q, p+\eta q] = c_{k1}'X_1'q + \cdots + c_{kr}'X_r'q \qquad (16.2.1)$$

the coefficients c_{ki}' have very simple values.

Let $\Delta = \det(\alpha_{ik})$ and let β_{ik} be the subdeterminant with respect to α_{ik}, so that

$$\Delta \cdot X_k = \beta_{ik}X_1' + \cdots + \beta_{rk}X_r' \quad ;$$

computation then shows

$$c_{k\sigma}' = \frac{1}{\Delta} \sum_j \sum_\rho \alpha_{kj}c_{j\rho}\beta_{\sigma\rho} \qquad (16.2.2)$$

We now make some remarks to orient ourselves in the discussion of these complicated expressions. In the linear infinitesimal transformation

$$B = \sum_j \sum_\rho \frac{\partial}{\partial x_j} c_{j\rho}x_\rho$$

we make the substitution

$$x_i' = \sum_j \alpha_{ij}x_j \quad ,$$

so that $\partial/\partial x$ is transformed according to

$$\frac{\partial}{\partial x_j} = \sum_k \alpha_{kj} \frac{\partial}{\partial x_k'}$$

Therefore,

$$B = \frac{1}{\Delta} \sum_j \sum_\rho \sum_k \sum_\sigma \frac{\partial}{\partial x_k'} \alpha_{kj}x_{j\rho}\beta_{\sigma\rho}x_\sigma' \quad ,$$

so that setting

$$B = \sum_k \sum_\sigma \frac{\partial}{\partial x'_k} c'_{k\sigma} x'_\sigma$$

gives

$$c'_k = \frac{1}{\Delta} \sum_j \sum_\rho \alpha_{kj} c_{j\rho} \beta_{\sigma\rho} \quad ,$$

which is the same as formula 16.2.2.

We have seen in Section 15.1 that any infinitesimal transformation B can be put in the form

$$c'_{11}x'_1p'_1 + (c'_{21}x'_1+c'_{22}x'_2)p'_2 + (c'_{31}x'_1+c'_{32}x'_2+c'_{33}x'_3)p'_3$$

$$+ (c'_{q1}x'_1+\cdots+c'_{qq}x'_q)p'_q + \cdots$$

by introducing appropriate new variables. Hence it is possible to choose the constants α_{ik} so that $c'_{ks} = 0$ for $s > k$. Thus we have the following important theorem:

Theorem 16.2.1. If $r+1$ infinitesimal transformations $X_1q,\ldots,X_rq$, $p+\eta q$ satisfy the familiar relations, then it is no essential restriction to assume that these relations have the following form:

$$[X_1q.p+\eta q] = c_{11}X_1q$$

$$[X_2q,p+\eta q] = c_{21}X_1q + c_{22}X_2q$$

$$\cdot \quad \cdot \quad \cdot \qquad \cdot \qquad \cdot$$

$$[X_kq,p+\eta q] = c_{k1}X_1q + \cdots + c_{kk}X_kq$$

$$\cdot \quad \cdot \quad \cdot \qquad \cdot \qquad \cdot$$

The first of these equations gives

$$X_1 \frac{\partial \eta}{\partial y} - \frac{dX_1}{dx} = c_{11}X_1 \quad ,$$

or, since it is no restriction to set $X_1 = 1$,

$$\frac{\partial \eta}{\partial y} = c_{11}, \qquad \eta = cy + f(x) \quad .$$

To simplify this value of η, put

$$y' = y + \phi(x) \quad .$$

Then differentiation gives

$$\delta y' = \delta y + \frac{\partial \phi}{\partial x} \delta x \quad ,$$

whence

$$p + \eta q = p + \left[c(y'-\phi)+f+ \frac{\partial \phi}{\partial x} \right] q' \quad .$$

Now choose ϕ so that

$$\frac{d\phi}{dx} + f - c\phi = 0 \quad ;$$

then

$$p + \eta q = p + cy'q' \quad .$$

This change of variables leaves fixed the form of the transformations X_kq. <u>Hence this family has the form</u>

$$q, X_2q, X_3q, \ldots, X_rq, p+cyq \quad ,$$

<u>where</u> X_k <u>is determined as a function of</u> x <u>by the integrable differential equation</u>

$$cX_k - \frac{dX_k}{dx} = c_{k1}X_1 + \cdots + c_{kk}X_k \qquad (16.2.3)$$

16.3 CONTINUATION

Finally we seek all families of the form

$$X_1q,\ X_2q, \ldots, X_rq,\ yq,\ p+\eta q \quad .$$

Each $[X_kq, p+\eta q]$ is a linear combination of the X_{kq} and yq. Thus there are r relations of the form

$$X_k \frac{\partial\eta}{\partial y} - \frac{dX_k}{dx} = \sum_i c_{ki}X_i + c_k y \quad ,$$

whence

$$Y_k\eta = y\left(\frac{dX_k}{dx} + \sum_i c_{ki}X_i\right) + \frac{1}{2} c_k y^2 + f_k(x) \quad .$$

On the other hand, considering $[yq, p+\eta q]$ gives an equation of the form

$$y\frac{\partial\eta}{\partial y} - \eta = \sum_i \delta_i X_i + \delta y \quad , \qquad (16.3.1)$$

into which we introduce the value of η just found. This shows that all the c_k are 0, and hence that there are r equations of the form

$$[X_kq+p+\eta q] = c_{k1}X_1q + \cdots + c_{kr}X_rq \quad .$$

By a linear transformation of the X_i, as in Section 16.2, we can bring these equations to the simpler form

$$[X_kq, p+\eta q] = c_{k1}X_1q + \cdots + c_{kk}X_kq$$

Setting $X_1 = 1$ as in the preceding section, gives

$$\frac{\partial\eta}{\partial y} = c_{11}, \qquad \eta = c_{11}y + f(x) \quad ,$$

in which it is no real restriction to set the constant c_{11} equal to zero. If we substitute the value $\eta = f(x)$ into the equation (16.3.1), we find that η can be set equal to zero since it is a linear function of the X_k.

Hence our infinitesimal transformations have the form

$$q, X_2q, \ldots, X_rq, yq, p \quad ,$$

where the X_k are determined as functions of x by the integrable differential equations

$$-\frac{dX_k}{dx} = c_{k1}X_1 + \cdots + c_{kk}X_k \qquad (16.3.2)$$

The developments of this Section 16 give us the following theorem:

Theorem 16.3.1. A family of infinitesimal transformations of the form $\eta_1 q, \ldots, \eta_r q$, $p+\eta q$ satisfying the familiar relations has one of the following forms

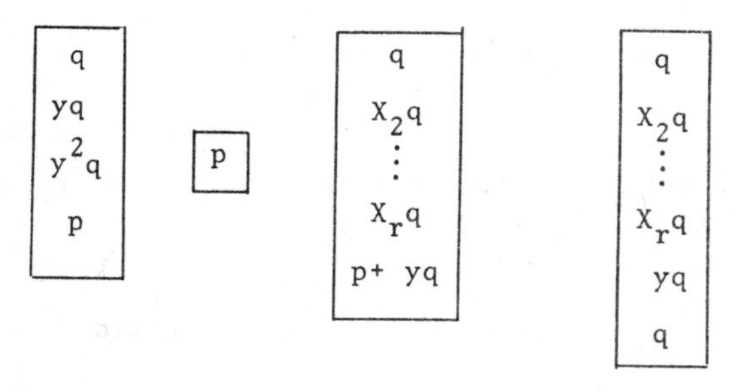

COMMENTS ON SECTION 16.3

Let $\underset{\sim}{G}$ *be a Lie algebra, with*

$$\underset{\sim}{H} \subset \underset{\sim}{G}$$

an _ideal_ *such that:*

$$\dim (\underset{\sim}{G}/\underset{\sim}{H}) = 1 .$$

Suppose that the field of scalars of $\underset{\sim}{G}$ _is the complex numbers._

Let B *be any element of* $\underset{\sim}{G}$ *which is not in* $\underset{\sim}{H}$. *Then,*

$$\text{Ad } B(\underset{\sim}{H}) \subset \underset{\sim}{H} .$$

Then, as recalled in Chapter 15, Ad B *can, by choice of bases of* $\underset{\sim}{H}$, *be put into triangular form. This means that there is a basis of* $\underset{\sim}{H}$ *of the form:*

$$A_1, \ldots, A_r$$

and relations of the following form:

$$\begin{aligned} [B,A_1] &= c_{11}A_1 \\ [B,A_2] &= c_{22}A_2 + c_{12}A_1 \\ &\vdots \end{aligned} \qquad (16.3.3)$$

Return to the geometric situation:

$$Z = X \times Y$$

$$X = C = Y = \textit{The complex numbers}$$

Now, assume (as Lie does implicitly) that the vector fields $\underset{\sim}{G}$ *are complex-holomorphic vector fields on* Z, *that* x *is a complex variable on* X, y *is a complex variable on* Y. *(The results obtained by working over the complexes can readily be interpreted over the reals, as well.)*

Suppose:

$$\pi_*(\underset{\sim}{G}) = \underset{\sim}{G}'$$

$$\pi_*(\underset{\sim}{H}) = 0 .$$

Then, $A_1,\ldots,A_n$ *have the form indicated in the text, i.e.,*

$$A_1 = \eta_1(x,y)\frac{\partial}{\partial y}$$

and so forth.

B *is of the form*

$$B = \eta\frac{\partial}{\partial y} + \frac{\partial}{\partial x}$$

The $A_1,\ldots,A_r$ *further have three forms determined by Theorem 14.4.1. Lie now uses relation 16.3.1 to determine the remaining quantities. Theorem 16.3.1 is the consequence.*

16.4 DETERMINATION OF THE GROUP TRANSFORMATIONS

It is now easy to verify that each of these families actually does provide a group of finite transformations:

1. The first family gives the 4-term group

$$y' = \frac{a_1 y + a_2}{a_3 y + 1}, \quad x' = x + a_4 .$$

2. The infinitesimal transformation p gives the 1-term group

$$y' = y, \quad x' = x + a$$

3. The third family gives the $(r+1)$-term group

$$y' = ay + a_1 + a_2 X_2 + \cdots + a_r X_r ,$$

$$x' = x + \frac{1}{\varepsilon} \log a .$$

4. The fourth family gives the $(r+2)$-term group

$$y' = ay + a_1 + a_2 X_2 + \cdots + a_r X_r ,$$

$$x' = x + a_0 .$$

These formulas are the group-relations corresponding to the Lie algebra relations in Theorem 16.3.1. They are readily derived by finding the orbit curves of the vector fields listed in Theorem 16.3.1.

Chapter 17

GROUPS FOR WHICH THE CURVES OF A FAMILY $\phi(x,y) = a$ ARE TRANSFORMED IN A 2-TERM WAY

We now determine all families of infinitesimal transformations of the form

$$\eta_1 q, \ldots, \eta_r q, \ p+\eta_0 q, \ xp+\eta q \quad ,$$

satisfying the familiar relations.

17.1. CONTINUATION

<u>There are two essentially different families of the form</u>

$$p, \ xp+\eta q$$

For, in any case one has

$$\frac{\partial \eta}{\partial x} = 0 \, , \qquad \eta = f(y) \quad .$$

If $f = 0$, one obtains the family p, xp. If $f \neq 0$, then one can replace y by a function of y so that f will become equal to 1. This gives the two forms:

$$p, xp \qquad \text{and} \qquad p, \ xp+q \quad .$$

17.2 CONTINUATION

In determining all families of the form

$$q,\ yq,\ y^2q,\ p,\ xp+\eta q \quad ,$$

one obtains the equations

$$\frac{\partial\eta}{\partial y} = a_0 + a_1y + a_2y^2 \quad ,$$

$$y\frac{\partial\eta}{\partial y} - \eta = b_0 + b_1y + b_2y^2 \quad ,$$

$$y^2\frac{\partial\eta}{\partial y} - 2y\eta = c_0 + c_1y + c_2y^2 \quad ,$$

which show that $\eta = 0$. This gives only the family:

$$q,\ yq,\ y^2q,\ p,\ xp \quad .$$

17.3 CONTINUATION

We now seek all families of the form

$$X_1q, \ldots, X_rq,\ p+\eta_1q,\ xp+\eta_2q \quad .$$

There are relations of the form

$$[X_kq,\ p+\eta_1q] = \sum_s c_{ks}X_sq \quad ,$$

$$[X_kq,\ xp+\eta_2q] = \sum_s d_{ks}X_sq \quad .$$

Again, it is possible to replace $X_1, \ldots, X_r$ by linear combinations of them

$$X_i' = \alpha_{i1}X_1 + \cdots + \alpha_{ir}X_r$$

so that in the transformed equations

$$[X_k'q, p+\eta_1 q] = \sum_s c_{ks}' X_s' q ,$$

$$[X_k'q, xp+\eta_2 q] = \sum_s d_{ks}' X_s' q ,$$

the new coefficients c_{ks}', d_{ks}' have very simple values.

As in §16.2, let $\Delta = \det(\alpha_{ik})$ and let β_{ik} be the subdeterminant corresponding to α_{ik}, so that

$$\Delta \cdot X_k = \beta_{ik}X_1' + \cdots + \beta_{rk}X_r'$$

and computation gives

$$\left.\begin{aligned} c_{k\sigma}' &= \frac{1}{\Delta} \sum_j \sum_\rho \alpha_{kj} c_{j\rho} \beta_{\sigma\rho} \\ d_{k\sigma}' &= \frac{1}{\Delta} \sum_j \sum_\rho \alpha_{kj} d_{j\rho} \beta_{\sigma\rho} \end{aligned}\right\} \qquad (17.3.1)$$

Now in the linear infinitesimal transformation

$$A = \sum_j \sum_\rho \frac{\partial}{\partial x_j} c_{j\rho} x_\rho ,$$

$$B = \sum_j \sum_\rho \frac{\partial}{\partial x_j} d_{j\rho} x_\rho$$

for which

$$[A,B] = A$$

by Theorem 15.6.1, make the substitution

$$x_\rho = \frac{1}{\Delta} \sum_\sigma \beta_{\sigma\rho} x'_\sigma$$

Then

$$x'_i = \sum_j \alpha_{ij} x_j$$

and

$$\frac{\partial}{\partial x_j} = \sum_k \alpha_{kj} \frac{\partial}{\partial x'_k}$$

so that

$$A = \frac{1}{\Delta} \sum_j \sum_\rho \sum_k \sum_\sigma \frac{\partial}{\partial x'_k} \alpha_{kj} c_{j\rho} \beta_{\sigma\rho} x'_\sigma \quad ,$$

$$B = \frac{1}{\Delta} \sum_j \sum_\rho \sum_k \sum_\sigma \frac{\partial}{\partial x'_k} \alpha_{kj} d_{j\rho} \beta_{\sigma\rho} x'_\sigma$$

Putting

$$A = \sum_k \sum_\sigma \frac{\partial}{\partial x'_k} c'_{k\sigma} x'_\sigma$$

$$B = \sum_k \sum_\sigma \frac{\partial}{\partial x'_k} d'_{k\sigma} x'_\sigma$$

gives

$$c'_{k\sigma} = \frac{1}{\Delta} \sum_j \sum_\rho \alpha_{kj} c_{j\rho} \beta_{\sigma\rho} \quad ,$$

$$d'_{k\sigma} = \frac{1}{\Delta} \quad {}_{j} \quad {}_{\rho} \quad \alpha_{kj} d_{j\rho} \beta_{\sigma\rho} \quad ,$$

these formulas being the same as 17.3.1. But we have seen (Theorem 15.4.1) that two linear infinitesimal transformations A and B for which $[A,B] = A$ can be simultaneously brought to canonical forms by introducing suitable new variables. In this way we see that the constants α_{ik} can be chosen so that $c'_{k\sigma} = 0$ when $\sigma > k-1$, and $d'_{k\sigma} = 0$ when $\sigma > k$. Thus we have the theorem:

<u>Theorem 17.3.1</u>. <u>If</u> $r+2$ <u>infinitesimal transformations of the form</u> $X_1q, \ldots, X_rq$, $p+\eta_1q$, $xp+\eta_2q$ <u>satisfy the familiar relations</u>, <u>then it is no essential restriction to assume that these relations have the following form</u>:

$$\left.\begin{aligned} [X_kq,\ p+{}_1q] &= c_{k1}X_1q + \cdots + c_{k,k-1}X_{k-1}q \\ [X_kq,\ xp+{}_2q] &= d_{k1}X_1q + \cdots + d_{kk}X_kq \end{aligned}\right\} \quad (17.3.2)$$

Since X_1 can be set equal to 1, one has first

$$\frac{\partial\eta_1}{\partial y} = 0, \qquad \eta_1 = f(x) \quad .$$

Replacing y by $y+\phi(x)$ shows that we can set $\eta_1 = 0$.

Hence setting $k = 2$ in the first of the equations (17.3.2) gives

$$-\frac{dX_2}{dx} = c_{21} \quad ,$$

which implies that one can take $X_2 = x$. Similarly, one obtains the values

$$X_3 = x^2, \ldots, X_r = x^{r-1} .$$

On the other hand, if one sets $k = 1$ in the second of the equations (17.3.2), one finds:

(No formula is given in the text (p.66).) *(*)*

Hence the sought family of transformations has the form:

$$q, xq, \ldots, x^r q, p, xp+[Ky+f(x)]q .$$

But one has

$$[p, xp+\eta_2 q] = p + (\nu_0+\nu_1 x+\cdots+\nu_r x^r)q ,$$

so that

$$\frac{df(x)}{dx} = \nu_0 + \nu_1 x + \cdots + \nu_r x^r ,$$

and so one can take $f(x) = Rx^{r+1}$. We shall show that, in general, it is no restriction to take $R = 0$.

Put

$$y' = y + Lx^{r+1}, \qquad x' = x,$$

so that

$$\delta y' = \delta y + L(r+1)x^r \delta x, \qquad \delta x' = \delta x ;$$

therefore

$$q' = q, xq = x'q', \ldots, x^r q = x'^r q' ,$$

$$p = p' + L(r+1)x^r q' ,$$

$$xp + (Ky+Rx^{r+1})q = x'p' + \{Ky'+[R+L(r+1-K)]x^{r+1}\}q'$$

If $K \neq r+1$, then L can be chosen so that

$$R + L(r+1-K) = 0 .$$

Therefore the family sought has one of the following two forms:

q
xq
$\vdots$
$x^r q$
p
$xp+Kyq$

q
xq
$\vdots$
$x^r q$
p
$xp + [(r+1)y+Rx^{r+1}]q$

If $R \neq 0$ in the second case, then it is no restriction to take $R = 1$.

17.4 CONTINUATION

<u>It remains to determine all families of the form</u>

$$X_1 q, \ldots, X_r q,\ yq,\ p+\eta_1 q,\ xp+\eta_2 q$$

There are $2r$ relations of the form

$$X_k \frac{\partial \eta_1}{\partial y} - \frac{dX_k}{dx} = \sum_i c_{ki} X_i + c_k y ,$$

$$X_k \frac{\partial \eta_2}{\partial y} - x \frac{dX_k}{dx} = \sum_i d_{ki} X_i + d_k y \quad ,$$

whence

$$X_k \eta_1 = y\left(\frac{dX_k}{dx} + \sum_i c_{ki} X_i\right) + \frac{1}{2} c_k y^2 + f_1(x) \quad ,$$

$$X_k \eta_2 = y\left(x \frac{dX_k}{dx} + \sum_i d_{ki} X_i\right) + \frac{1}{2} d_k y^2 + f_2(x) \quad ,$$

Forming the expressions $[yq,\ p+\eta_1 q]$, $[yq,\ xp+\eta_2 q]$ now gives two equations of the form

$$y \frac{\partial \eta_1}{\partial y} - \eta_1 = \sum_i \gamma_i X_i + \gamma y \quad ,$$

$$y \frac{\partial \eta_2}{\partial y} - \eta_2 = \sum_i \delta_i X_i + \delta y \quad .$$

Substituting into these equations the values just found for η_1 and η_2 shows that all the c_k and d_k are zero, and likewise that γ and δ are zero.

Hence there are $2r+3$ equations of the form

$$[X_k q,\ p+\eta_1 q] = \sum_i c_{ki} X_i q \quad ,$$

$$[X_k q,\ xp+\eta_2 q] = \sum_i d_{ki} X_i q \quad ,$$

$$[yq,\ p+\eta_1 q] = \sum_i \gamma_i X_i q \quad ,$$

$$[yq,\ xp+\eta_2 q] = \sum_i \delta_i X_i q \quad ,$$

$$[p+\eta_1 q,\ xp+\eta_2 q] = p + \eta_1 q + \lambda yq + \sum_i \mu_i X_i q \quad .$$

In the last equation one can attain $\lambda = 0$ by replacing η_1 by $\eta_1+\lambda y$. Then the infinitesimal transformations $X_k q$, $p+\eta_1 q$, $xp+\eta_2 q$ are in the same relation as in Section 17.3. And we see, as before, that it is no restriction to set

$$[X_k q,\ p+\eta_1 q] = c_{k1} X_1 q + \cdots + c_{k,k-1} X_{k-1} q \quad ,$$

$$[X_k q,\ xp+\eta_2 q] = d_{k1} X_1 q + \cdots + d_{k,k} X_k q \quad .$$

Setting $X_1 = 1$ gives

$$\frac{\partial \eta_1}{\partial y} = 0, \qquad \eta_1 = f(x) \quad ,$$

and forming $[yq,\ p+\eta_1 q]$ shows that it is no restriction to set $f(x) = 0$. Similarly, one finds that

$$\frac{\partial \eta_2}{\partial y} = K, \qquad \eta_2 = Ky + F(x) \quad ,$$

where it is no restriction to set $K = 0$ and $F(x) = 0$. Finally, one sees, as in Section 17.3, that one can set

$$X_2 = x,\ X_3 = x^2, \dots, X_r = x^{r-1} \quad .$$

Hence the sought family has the form:

$$q,\ xq, \dots, x^r q,\ yq,\ p,\ xp \quad .$$

The developments of this section give the theorem:

Theorem 17.4.1. *If the infinitesimal transformations*

$$\eta_1 q, \dots, \eta_r q,\ p+\dot{\eta}_0 q,\ xp+\eta q$$

satisfy the familiar relations, then the transformations can be put in the form

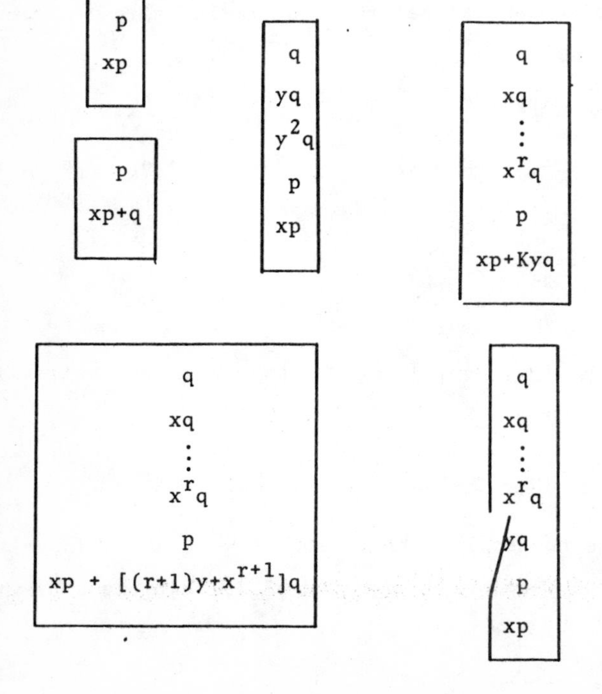

COMMENTS ON SECTION 17.4

The method used in this chapter may be regarded as a generalization of the method used in Chapter 16. Our problem is now to determine prolongation homomorphisms

$$\pi_*: \underset{\sim}{G} \to \underset{\sim}{G}' \quad ,$$

such that:

$$\dim \underset{\sim}{G}' = 2 \quad .$$

Logically, there are two parts to the problem, one ***algebraic****, the other* ***geometric****.*

The first is to classify all homomorphisms of this type in a purely algebraic way, the second is to find which of these possibilities is realizable with:

$$\dim X = \dim Y = 1 \quad ,$$

and with the kernel

$$\underset{\sim}{H} = \pi_*^{-1}(0) \subset \underset{\sim}{G}$$

determined by Theorem 14.4.1.

There are then two possibilities for $\underset{\sim}{G}'$*:*

a) $\underset{\sim}{G}'$ *is abelian*

b) $\underset{\sim}{G}'$ *is solvable, but not abelian*

There are the three ***families*** *of possibilities of* $\underset{\sim}{H}$*, listed in Theorem 14.4.1.*

Case 1: $\underset{\sim}{H}$ *is semi-simple*

Then, using standard properties of semi-simple Lie algebras (e.g., the complete reducibility property and the Levi-Malcev theorem), one sees that:

$\underset{\sim}{G}$ *is a direct sum of the ideal* $\underset{\sim}{H}$ *and an ideal isomorphic to* $\underset{\sim}{G}'$.

We denote this as follows:

$$\underset{\sim}{G} = \underset{\sim}{H} \oplus \underset{\sim}{G}'$$

We have then:

$$[\underset{\sim}{G}', \underset{\sim}{H}] = 0 \quad .$$

Case 2: $\underset{\sim}{H}$ *is not semi-simple.*

We see from Theorem 14.4.1 that in this case $\underset{\sim}{H}$ *does not even contain a semi-simple subalgebra.*

Now, apply the Levi-Malcev theorem. (See Samelson [1], *VB, vol. II, Chapter 2.) Write:*

$\underset{\sim}{G}$ = *vector space direct sum of a solvable ideal* $\underset{\sim}{R}$ *and a semi-simple subalgebra* $\underset{\sim}{S}$.

Now, the homomorphic image of a semi-simple Lie algebra is semi-simple. Hence,

$\pi_*(\underset{\sim}{S})$ *is semi-simple*

Since we are assuming $\underset{\sim}{G}'$ *is solvable, we have:*

$$\pi_*(\underset{\sim}{S}) = 0 \quad ,$$

i.e.,

$$\underset{\sim}{S} \subset \underset{\sim}{H} \quad ,$$

hence:

$$\underset{\sim}{S} = 0 \quad .$$

In particular,

$\underset{\sim}{G}$ *is* <u>*solvable*</u>

Lie's theorem on the triangular form of Ad $\underset{\sim}{G}$ *is now available.*

<u>*Case 3:*</u> $\underset{\sim}{G}$ <u>*is the semi-direct sum of*</u> $\underset{\sim}{H}$ <u>*and*</u> $\underset{\sim}{G}'$.

This means that there is a Lie <u>*subalgebra*</u> $\underset{\sim}{G}'' \subset \underset{\sim}{G}$ *such that,* <u>*as a vector space,*</u>

$$\underset{\sim}{G} = \underset{\sim}{H} \oplus \underset{\sim}{G}'' \quad .$$

π_* *maps* $\underset{\sim}{G}''$ *isomorphically onto* $\underset{\sim}{G}'$, *so that the Lie algebra structure of* $\underset{\sim}{G}''$ *is determined. In this case, the triangular form of*

Ad $\underset{\sim}{G}''$ *acting on* $\underset{\sim}{H}$

is the tool to determine $\underset{\sim}{G}$.

Case 4: $\underset{\sim}{G}$ is not a semi-direct sum of $\underset{\sim}{H}$ and $\underset{\sim}{G}'$.

In this case, choose

$$\underset{\sim}{G}'' \subset \underset{\sim}{G}'$$

as a one-dimensional ideal of $\underset{\sim}{G}'$. Set:

$$\underset{\sim}{H}' = \pi_*^{-1}(\underset{\sim}{G}'')$$

Then, $(\underset{\sim}{H}',\underset{\sim}{H},\pi,\underset{\sim}{G}'')$ is determined by the work of Chapter 16. One must then see how an element of $\underset{\sim}{G}'$ which is not in $\underset{\sim}{G}'$ may be fitted into the canonical forms found in Theorem 16.3.1.

17.5 GROUPS GENERATED BY THE LIE ALGEBRAS

It is now easy to verify that each of these families actually does provide a group of finite transformations.

For example, the fourth family gives the group

$$y' = a^k y + a_1 + a_2 x + \cdots + a_{r+1} x^r ,$$

$$x' = ax + a_{r+2} .$$

The fifth family gives the group

$$y' = e^{(r+1)a} y + e^{(r+1)a} a x^{r+1} + a_1 + \cdots + a_{r+1} x^r$$

$$x' = e^a x + a_0$$

with the parameters $a_0, a, a_1, \ldots, a_{r+1}$. Finally, the last family gives the group

$$y' = ay + a_1 + a_2x + \cdots + a_{r+1}x^r \quad ,$$

$$x' = bx + b_1 \quad ,$$

with the parameters $a, a_1, \ldots, a_{r+1}, b, b_1$.

Chapter 18

GROUPS FOR WHICH THE CURVES OF A FAMILY $\phi(x,y) = a$ ARE TRANSFORMED IN A 3-TERM WAY

We now determine all groups which transform the curves of a family $\phi(x,y) = a$ in a 3-term way. By Chapter 17 there are six different cases, which are quickly dealt with as follows.

18.1 FIRST CASES

All groups of the form $p,\ xp,\ x^2p+\eta q$ are determined by the equations

$$\frac{\partial\eta}{\partial x} = 0, \qquad x\,\frac{\partial\eta}{\partial x} = \eta ,$$

so that we obtain only the group $p,\ xp,\ x^2p$.

All groups of the form

$$p,\ xp+yq,\ x^2p+\eta q$$

are determined by the equations

$$\frac{\partial\eta}{\partial x} = 2y, \qquad x\,\frac{\partial\eta}{\partial x} + y\,\frac{\partial\eta}{\partial y} - \eta = \eta ,$$

from which it follows that η is of the form $\eta = 2xy + B^2$. Hence one finds the three transformations

$$p,\ xp+yq,\ x^2p+(2xy+By^2)q\ .\ ^*$$

---

* If $B \neq 0$, then it is no essential restriction to take $B = 1$.

All groups of the form

$$q,\ yq,\ y^2q,\ p,\ xp,\ x^2p + \eta q$$

are determined by the equations

$$\frac{\partial\eta}{\partial y} = a_0 + a_1y + a_2y^2 \quad ,$$

$$y\frac{\partial\eta}{\partial y} - \eta = b_0 + b_1y + b_2y^2 \quad ,$$

which show that one can set $\eta = 0$.

All groups of the form

$$q,\ xq,\ x^2q,\ldots,x^rq,\ p,\ xp+Kyq,\ x^2p+\eta q$$

satisfy the equations

$$\frac{\partial\eta}{\mu y} = \sum\nu_i x^i, \qquad \frac{\partial\eta}{\partial x} = \sum\mu_i x^i + 2Ky$$

from which it follows that one can set

$$\eta = Ly + Mx^{r+1} + 2Kxy \quad .$$

Moreover,

$$[xp+Kyq,\ x^2p+\eta q] = x^2p + \eta q + \sum\lambda_i x^i \quad ,$$

whence

$$M(r-K) = 0, \qquad L = 0 \quad .$$

Moreover,

$$[x^rq,\ x^2p+\eta q] = \sum\rho_i x^i q \quad ,$$

whence

$$2K = r \quad .$$

Hence if $r \neq 0$ one obtains the family

$$q, xq, \ldots, x^r q, p, 2xp+ryq, x^2p+rxyq \quad ;$$

while if $r = 0$ one obtains the family

$$q, p, xp, x^2p+Mxq \quad ,$$

and, since the hypothesis $M = 0$ gives nothing new, we can set $M = 1$. Now replace y by e^y, to obtain the infinitesimal transformations

$$yq, p, xp, x^2p+xyq \quad ,$$

which are all <u>linear</u> transformations.

18.2 NEXT CASES

To determine all families of the form

$$q, xq, \ldots, x^r q, p, xp+[(r+1)y+x^{r+1}]q, x^2p+\eta q$$

we first have the equations

$$\frac{\partial \eta}{\partial y} = \nu_0 + \nu_1 x + \cdots + \nu_r x^r \quad ,$$

$$\frac{\partial \eta}{\partial x} = 2(r+1)y + 2x^{r+1} + \sum \mu_i x^i \quad ,$$

whence

$$\eta = 2(r+1)xy + \frac{2}{r+2} x^{r+2} + Mx^{r+1} + Ny \quad .$$

Substituting this into the equation

$$[xp+\eta_0 q,\ x^2p+\eta q] = x^2p + \eta q + \sum \lambda_i x^i q \quad ,$$

shows that one must have $r = -1$; but this is absurd.

18.3 FINAL CASES

Finally, we seek the most general family of the form

$$q,\ xq, \ldots, x^r q,\ yq,\ p,\ xp,\ x^2p+\eta q \quad .$$

Since η satisfies three equations of the form

$$\frac{\partial \eta}{\partial y} = \sum \alpha_i x^i + \alpha y \quad ,$$

$$\frac{\partial \eta}{\partial x} = \sum \beta_i x^i + \beta y \quad ,$$

$$y\frac{\partial \eta}{\partial y} - \eta = \sum \gamma_i x^i + \gamma y \quad ,$$

one has $\eta = Byx$. Applying the operation $x^r q$ shows that $B = r$. Therefore, the family sought is

$$q,\ xq, \ldots, x^r q,\ yq,\ p,\ xp,\ x^2p+rxyq \quad .$$

The developments of this Section 18 give the theorem:

Theorem 18.3.1. *If the infinitesimal transformations* $\eta_1 q, \ldots, \eta_r q,\ p+\phi_0 q,\ xp+\phi_1 q,\ x^2p+\phi_2 q$ *satisfy the familiar relations, then they can always be put into one of the following forms*

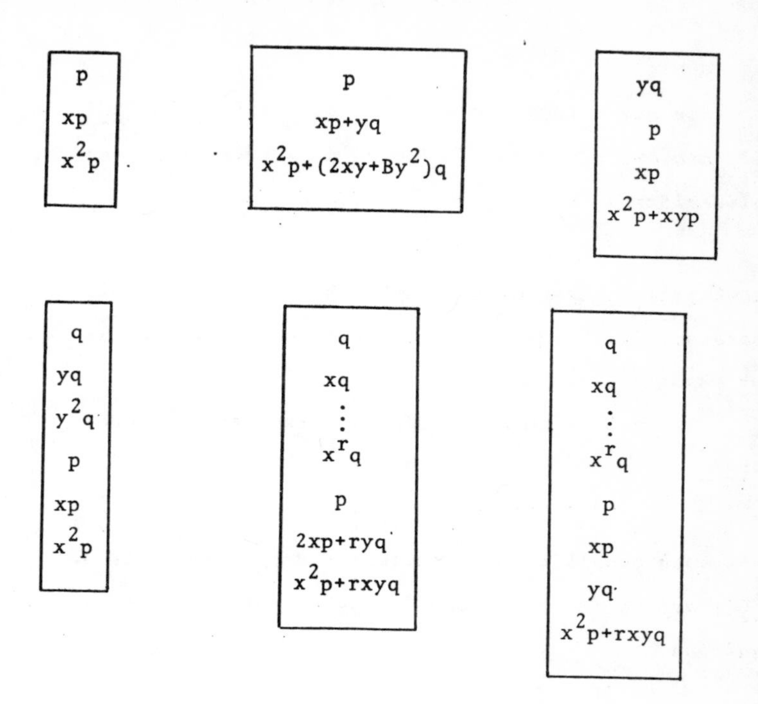

COMMENTS ON SECTION 18.3

Again, the Levi-Malcev theorem and J.H.C. Whitehead theorems on "the splitting of extension of semi-simple algebras" provide general insights into the results of this chapter. Let $\underset{\sim}{G}$ *be a Lie algebra*

$$\pi_*: \underset{\sim}{G} \to \underset{\sim}{G}'$$

a Lie algebra homomorphism. This chapter is concerned with the case where:

$$\dim \underset{\sim}{G}' = 3 .$$

Now, this case, together with the assumption that $\underset{\sim}{G}'$ *is a Lie algebra of vector fields on a 1-dimensional manifold, implies that:*

$$\underset{\sim}{G}' \textit{ is semi-simple.}$$

The Whitehead theorems now imply that π_* *is a "split" extension, i.e., there is a Lie subalgebra* $\underset{\sim}{G}'$ *of* $\underset{\sim}{G}$ *such that:*

$$\underset{\sim}{G} = \underset{\sim}{H} \oplus \underset{\sim}{G}' \quad \textit{(direct sum vector space)}$$

where

$$\underset{\sim}{H} = \textit{kernel } \pi_*$$

Each possible $\underset{\sim}{H}$ *is determined by Theorem 14.4.1.* Ad $\underset{\sim}{G}'$ *acting in* $\underset{\sim}{H}$ *is completely reducible. Again, the condition*

$$[\underset{\sim}{G}', \underset{\sim}{H}] \subset \underset{\sim}{H}$$

serves to determine $\underset{\sim}{G}'$.

Here is one general pattern to this. First, classify one sort of Lie algebra action, then determine in how many ways this can be extended to a larger Lie algebra action. I will develop some general features of this problem at the end of the paper.

Chapter 19

SOME GROUPS WHICH LEAVE INVARIANT NO FAMILY OF CURVES $\phi(x,y) = a$

By Chapter 13 the groups of the plane which leave invariant no family of curves $\phi(x,y) = a$ are characterized by the property that their first-order infinitesimal transformations are either of the form

$$(x-x_0)p+\cdots,\ (y-y_0)p+\cdots,\ (x-x_0)q+\cdots,\ (y-y_0)q+\cdots$$

or

$$(x-x_0)p - (y-y_0)q+\cdots,\ (x-x_0)q+\cdots,\ (y-y_0)p+\cdots$$

In this section we determine all those groups which, at an arbitrary point (x_0,y_0), possess not only first- but also higher-order transformations.

19.1 TRANSFORMATIONS OF MAXIMAL ORDER

The groups we seek have, in any case, three first-order transformations of the form

$$(x-x_0)p - (y-y_0)q+\cdots,\ (x-x_0)q+\cdots,\ (y-y_0)p+\cdots\ ;$$

moreover, they contain transformations of higher order. Let s be the maximal order of such a transformation. We shall show that $s = 2$.

If $\xi p+\eta q$ is an infinitesimal transformation of order s, then ξ or η is of order s; assume ξ is. Then we can put

$$\xi p + \eta q = p \sum_i \alpha_i (x-x_0)^i (y-y_0)^{s-i} + \eta q + \cdots ,$$

where the coefficients $\alpha_s, \ldots, \alpha_1, \alpha_0$ are not all zero. By Theorem 8.2.1,

$$[(x-x_0)q+\cdots, \xi p+\eta q] = p \sum \alpha_i (s-i)(x-x_0)^{i+1}(y-y_0)^{s-i+1}$$

$$+ \eta_1 q + \cdots$$

is an infinitesimal transformation of our group of order s, call it $\xi_1 p + \eta_1 q$. Then

$$[(x-x_0)q+\cdots, \xi_1 p+\eta_1 q]$$

is again an infinitesimal transformation of order s, and so on. Continuing in this way one finally obtains a transformation of order s

$$((x-x_0)^s + \alpha\rho (x-x_0)^{s-\rho}(y-y_0)^{\rho} + \cdots)p + \eta q + \cdots = G$$

whose ξ-term contains $(x-x_0)^s$.

To continue we first simplify the notation by setting $x_0 = 0$, $y_0 = 0$, which amounts to moving our arbitrary point to the origin. We then construct the infinitesimal transformation

$$[xp-yq+\cdots, (x^s+\alpha\rho x^{s-\rho}y^{\rho}+\cdots)p + \eta q + \cdots] ,$$

which is of the form

$$((s-1)x^s + (s-2\rho-1)\alpha\rho x^{s-\rho}y^{\rho} + \cdots)p + \eta q + \cdots = H .$$

Hence there is always a transformation of order s, namely $H - (s-2\rho-1)G$, which is of the form

$$(x^s+\alpha_{\rho+1}x^{s-\rho-1}y^{\rho+1}+\cdots)p + \eta q \quad .$$

In the same way we see that there is a transformation of the form

$$(x^s+\alpha_{\rho+2}x^{s-\rho-2}y^{\rho+2}+\cdots)p + \eta q \quad ,$$

and so on; finally we find an infinitesimal transformation of order s in the group of the form

$$x^sp + \eta q + \cdots \quad ,$$

i.e., whose ξ-term contains only one term of order s, namely x^s.

In Section 13.1 we saw that the group we seek contains two independent transformations of order zero:

$$p+\cdots, \ q+\cdots \quad .$$

Therefore it also contains the transformation of order $s-1$

$$[p+\cdots, \ x^sp+\eta q+\cdots] \quad ,$$

which is of the form

$$x^{s-1}p + \eta_1 q + \cdots \quad .$$

Moreover, it contains the transformation

$$[x^{s-1}p+{}_1q, \ x^sp+\eta q] + \cdots \ = \ x^{2s-2}p + \eta_2 q + \cdots \quad ,$$

whose order is $2s-2$. Thus we see that $2s-2 \leq s$, i.e., $s \leq 2$; this gives the theorem:

Theorem 19.1.1. A group which leaves invariant no family of curves $\phi(x,y) = a$ possesses no infinitesimal transformation whose order is >2 at a general point.

19.2 FORM OF TRANSFORMATIONS OF ORDER 2

It is now not difficult to determine the number and form of all infinitesimal transformations of order 2, assuming, as we do, their existence.

We have found that there is always a transformation of the form

$$x^2p + (\alpha x^2+\beta xy+\gamma y^2)q + \cdots = G ,$$

and therefore we also have the transformation

$$[xp-yq+\cdots, G] = x^2p + (3\alpha x^2+\beta xy-\gamma y^2)q + \cdots$$

$$= K ,$$

as well as

$$(\alpha x^2-\gamma y^2)q + \cdots = \frac{1}{2}(K-G) = L .$$

One has

$$[xq+\cdots, L] = -2\gamma xyq + \cdots ,$$

$$[xq+\cdots, -2\gamma xyq+\cdots] = -2\gamma x^2q + \cdots$$

Moreover,

$$[\gamma xyq+\cdots, \gamma x^2q+\cdots] = -\gamma^2x^3q +\cdots$$

which shows that $\gamma = 0$. Thus, $L = \alpha x^2q+\cdots$.

Now

$$[\alpha x^2q+\cdots, yp+\cdots] = \alpha(x^2p-2xyq) +\cdots ,$$

$$[\alpha x^2q+\cdots, \alpha(x^2p-2xyq)+\cdots] = -4\alpha^2x^3q +\cdots ,$$

so that $\alpha = 0$. Hence the infinitesimal transformation G has the form

$$G = x^2p + \beta xyq +\cdots .$$

To determine β we form the transformations

$$[yp+\cdots, G] = (2-\beta)xyp + \beta y^2q +\cdots = H ,$$

and

$$[G,H] = (\beta-1)(2-\beta)x^2yp + (\beta-1)2\beta xy^2q +\cdots ,$$

which shows that

$$(\beta-1)(2-\beta) = 0, \qquad (\beta-1)2\beta = 0 ,$$

so that $\beta = 1$.

This shows that the group sought contains two second-order infinitesimal transformations of the forms

$$x^2p + xyq +\cdots, \quad xyp + y^2q +\cdots .$$

If there are other second-order transformations

$$(ax^2+bxy+cy^2)p + (\alpha x^2+\beta xy+\gamma y^2)q +\cdots = U +\cdots ,$$

then

$$[x^2p+xyq,\ U] = 0, \qquad [xyp+y^2q,\ U] = 0 .$$

But

$$xyp + y^2q = \frac{y}{x}(x^2p+xyq) ,$$

so that

$$[x^2p+xyq,\ U] = 0, \qquad \left[\frac{y}{x},\ U\right] = 0 ,$$

showing that U is a function of $x^2p + xyq$ and y/x. Since U is a polynomial of degree 2 in x and y, it is of the form

$$A(x^2p+xyq) + B(xyp+y^2q) ,$$

which means that the group contains only the two second-order transformations found before.

Therefore, the group contains the first-order transformation

$$[p+\cdots,\ x^2p+xyq+\cdots] = 2xp + yq + \cdots ,$$

and also the transformations

$$xp - yq + \cdots,\ xq + \cdots,\ yp + \cdots ,$$

so that it contains <u>four</u> independent first-order transformations. Thus:

<u>Theorem 19.2.1</u>. <u>If the infinitesimal transformations of a group which leaves invariant no family of curves</u> $\phi(x,y) =$ <u>are not all of order</u> 0 <u>or</u> 1, <u>then they are of the form</u>

$$p+\cdots,\ q+\cdots,\ xp+\cdots,\ yp+\cdots,\ xq+\cdots,\ yq+\cdots,$$
$$x^2p+xyq+\cdots,\ xyp+y^2q+\cdots \quad .$$

19.3 DETERMINATION OF THE PRIMITIVE ALGEBRAS

We now determine the relations among the eight infinitesimal transformations we have found.

First, the following nine relations clearly hold:

$$[x^2p+xyq+\cdots,\ xyp+y^2q+\cdots] = 0 \ ,$$
$$[xp+\cdots,\ x^2p+xyq+\cdots] = x^2p + xyq + \cdots \ ,$$
$$[xp+\cdots,\ xyp+y^2q+\cdots] = 0 \ ,$$
$$[yq+\cdots,\ x^2p+xyq+\cdots] = 0 \ ,$$
$$[yq+\cdots,\ xyp+y^2q+\cdots] = xyp + y^2q + \cdots \ ,$$
$$[xq+\cdots,\ x^2p+xyq+\cdots] = 0 \ ,$$
$$[xq+\cdots,\ xyp+y^2q+\cdots] = x^2p+xyq + \cdots \ ,$$
$$[yp+\cdots,\ x^2p+xyq+\cdots] = xyp + y^2q + \cdots$$
$$[yp+\cdots,\ xyp+y^2q+\cdots] = 0 \ .$$

Moreover, there are relations of the form

$$[xp+\cdots,\ yq+\cdots] = A_1(x^2p+xyq+\cdots) + B_1(xyp+y^2q+\cdots) \ ,$$

$$[xp+\cdots,\ xq+\cdots] = (xq+\cdots) + A_2(\ ) + B_2(\qquad) ,$$

$$[yq+\cdots,\ xq+\cdots] = -(xq+\cdots) + A_3(\ ) + B_3(\qquad) ,$$

$$[xp+\cdots,\ yp+\cdots] = -(yp+\cdots) + A_4(\ ) + B_4(\qquad) ,$$

$$[yq+\cdots,\ yp+\cdots] = (yp+\cdots) + A_5(\ ) + B_5(\qquad) ,$$

$$[xq+\cdots,\ yp+\cdots] = (xp-yq+\cdots) + A_6(\ ) + B_6(\qquad) ,$$

where the A_k and B_k are unknown constants.

To simplify these equations we put

$$x'p'+\cdots = (xp+\cdots) + \alpha_1(x^2p+xyq+\cdots) + \beta_1(xyp+y^2q+\cdots)$$

$$y'q'+\cdots = (yq+\cdots) + \alpha_2(\qquad) + \beta_2(\qquad)$$

$$x'q'+\cdots = (xq+\cdots) + \alpha_3(\qquad) + \beta_3(\qquad)$$

$$y'p'+\cdots = (yp+\cdots) + \alpha_4(\qquad) + \beta_4(\qquad)$$

and introduce these quantities as first-order infinitesimal transformations. We make the abbreviations

$$x^2p + xyq+\cdots = S_1 ,$$

$$xyp + y^2q+\cdots = S_2 ,$$

so that

$$[x'p'+\cdots,\ y'q'+\cdots] = (A_1+\alpha_2)S_1 + (B_1-\beta_1)S_2 ,$$

$$[x'p'+\cdots,\ x'q'+\cdots] = (x'q'+\cdots) + (A_2-\beta_1)S_1 + (B_2-\beta_3)S_2$$

$$[y'q'+\cdots,\ x'q'+\cdots] = -(x'q'+\cdots) + (A_3+\alpha_3-\beta_2)S_1 + (B_3+2\beta_3)S_2 \quad ,$$

$$[x'p'+\cdots,\ y'p'+\cdots] = -(y'p'+\cdots) + (A_2+2\alpha_4)S_1 + (B_4-\alpha_1+\beta_4)S_2 \quad ,$$

$$[y'q'+\cdots,\ y'p'+\cdots] = (y'p'+\cdots) + (A_s-\alpha_4)S_1 + (B_5-\alpha_2)S_2$$

$$[x'q'+\cdots,\ y'p'+\cdots] = (x'p'-y'q'+\cdots) + (A_6+\beta_4-\alpha_1+\alpha_2)S_1 + (B_6-\alpha_3-\beta_1+\beta_2)S_2$$

In this, we can always assume that the α_k and β_k have been chosen so that

$$A_1 + \alpha_2 = 0, \qquad B_1 - \beta_1 = 0,$$

$$A_3 + \alpha_3 - \beta_2 = 0, \qquad B_2 - \beta_3 = 0,$$

$$A_5 - \alpha_4 = 0, \qquad B_4 - \alpha_1 + \beta_4 = 0 \ .$$

Hence it is no restriction to assume that we have equations of the form

$$[xp+\cdots,\ yq+\cdots] = 0 \quad ,$$

$$[xp+\cdots,\ xq+\cdots] = (xq+\cdots) + A_2S_1 \quad ,$$

$$[yq+\cdots,\ xq+\cdots] = -(xq+\cdots) + B_3S_2 \ ,$$

$$[xp+\cdots,\ yp+\cdots] = -(yp+\cdots) + A_4S_1 ,$$

$$[yq+\cdots,\ yp+\cdots] = (yp+\cdots) + B_5S_2 ,$$

$$[xq+\cdots,\ yp+\cdots] = (xp-yq+\cdots) + A_6S_1 + B_6S_2 .$$

We shall show that the remaining A_k, B_k must be 0. For this, we form the <u>Jacobi</u> identity

$$\begin{aligned} 0 = & [[xp+\cdots,\ yq+\cdots],\ xq+\cdots] \\ & + [[yq+\cdots,\ xq+\cdots],\ xp+\cdots] \\ & + [[xq+\cdots,\ xp+\cdots],\ yq+\cdots] , \end{aligned}$$

which gives, by substituting the values above,

$$-[xq+\cdots,\ xp+\cdots] - [xq+\cdots,\ yq+\cdots] = 0$$

and finally

$$(xq+\cdots) + A_2S_1 - (xq+\cdots) + B_3S_2 = 0 ,$$

so that $A_2 = B_3 = 0$. Similarly we find that $A_4 = B_5 = 0$ by applying the Jacobi identity to $xp+\cdots,\ yq+\cdots,\ yp+\cdots$.

We form the identity

$$\begin{aligned} 0 = & [[xq+\cdots,\ yp+\cdots],\ xp+\cdots] \\ & + [[yp+\cdots,\ xp+\cdots],\ xq+\cdots] \\ & + [[xp+\cdots,\ xq+\cdots],\ yp+\cdots] , \end{aligned}$$

whence

$$0 = -A_6S_1 + [yp+\cdots,\ xq+\cdots] + [xq+\cdots,\ yp+\cdots] ,$$

so that $A_6 = 0$. Similarly, applying this identity to $xq+\cdots$, $yp+\cdots$ and $yq+\cdots$, we see that $B_6 = 0$. <u>We have thus proved that all the</u> A_k <u>and</u> B_k <u>are</u> 0.

It remains to put the relations among $p+\cdots$, $q+\cdots$ and the other transformations into their simplest form.

First, there are equations of the form

$$[p+\cdots, x^2p+xyq+\cdots] = 2(xp+\cdots) + (yq+\cdots) + A_1S_1 + A_2S_2$$

$$[p+\cdots, xyp+y^2q+\cdots] = (yp+\cdots) + B_1S_1 + B_2S_2$$

We replace p by a transformation of the form

$$p + \alpha(xp+\cdots) + \beta(xq+\cdots) + \gamma(yp+\cdots) + \delta(yq+\cdots) \quad ,$$

noting that we can choose the constants $\alpha,\beta,\gamma,\delta$ so that the two equations above assume the simple form

$$[p+\cdots, x^2p+xyq+\cdots] = 2(xp+\cdots) + (yq+\cdots)$$

$$[p+\cdots, xyp+y^2q+\cdots] = (yp+\cdots) \quad .$$

Similarly, it is possible to choose the transformation $q+\cdots$ so that

$$[q+\cdots, x^2p+xyq+\cdots] = xq+\cdots \quad ,$$

$$[q+\cdots, xyp+y^2q+\cdots] = (xp+\cdots) + 2(yq+\cdots) \quad .$$

Moreover, there is a relation of the form

$$[p+\cdots, xp+\cdots] = (p+\cdots) + \alpha(xp+\cdots) + \beta(yp+\cdots)$$
$$+ \gamma(xq+\cdots) + \delta(yq+\cdots) + \mu S_1 + \nu S_2 \quad ;$$

we have

$$0 = [[p+\cdots, xp+\cdots], S_1] + [[xp+\cdots, S_1], p+\cdots] + [[S_1, p+\cdots], xp+\cdots] ,$$

whence

$$[[p+\cdots, xp+\cdots], S_1] + [S_1, p+\cdots] = 0$$

or

$$\alpha S_1 + \beta S_2 = 0 ,$$

so that $\alpha = \beta = 0$. On the other hand, applying the Jacobi identity to $p+\cdots$, $xp+\cdots$, and S_2 shows that $\gamma = \delta = 0$. Therefore,

$$[p+\cdots, xp+\cdots] = (p+\cdots) + \mu_1 S_1 + \mu_2 S_2$$

By completely analogous computations one finds equations of the form

$$[p+\cdots, yp+\cdots] = \nu_1 S_1 + \nu_2 S_2 ,$$

$$[p+\cdots, yq+\cdots] = \alpha_1 S_1 + \alpha_2 S_2 ,$$

$$[p+\cdots, xq+\cdots] = (q+\cdots) + \beta_1 S_1 + \beta_2 S_2$$

Now replacing $p+\cdots$ by a suitable quantity of the form $(p+\cdots) + c_1 S_1 + c_2 S_2$ shows that $\mu_1 = \mu_2 = 0$. Now applying the Jacobi identity to $p+\cdots$, $xp+\cdots$, $yp+\cdots$, and to $p+\cdots$, $yq+\cdots$, $xp+\cdots$ shows that $\nu_1 = \nu_2 = \alpha_1 = \alpha_2 = 0$. Therefore:

$$[p+\cdots, xp+\cdots] = p+\cdots, \quad [p+\cdots, yp+\cdots] = 0,$$

$$[p+\cdots, yq+\cdots] = 0 .$$

Completely analogous considerations show that the infinitesimal transformation $q+\cdots$ can be chosen so that

$$[q+\cdots, yq+\cdots] = q+\cdots ,$$

$$[q+\cdots, xq+\cdots] = 0 ,$$

$$[q+\cdots, xp+\cdots] = 0 ,$$

$$[q+\cdots, yp+\cdots] = (p+\cdots) + \gamma_1 S_1 + \gamma_2 S_2$$

Finally, applying the Jacobi identity to $p+\cdots$, $xq+\cdots$, $yq+\cdots$ shows that $\beta_1 = \beta_2 = 0$ and

$$[p+\cdots, xq+\cdots] = q+\cdots ;$$

an analogous computation shows that

$$[q+\cdots, yp+\cdots] = p+\cdots .$$

Now we must compute $[p+\cdots, q+\cdots]$. There is an equation of the form

$$[p+\cdots, q+\cdots] = A(p+\cdots) + B(q+\cdots) + C(xp+\cdots) + D(xq+\cdots) + E(yp+\cdots) + F(yq+\cdots) + GS_1 + HS_2 .$$

Applying the Jacobi identity first to $p+\cdots$, $q+\cdots$, S_1, and then to $p+\cdots$, $q+\cdots$, S_2 and finally to $p+\cdots$, $q+\cdots$, $xp+\cdots$, shows that

$$[p+\cdots, q+\cdots] = 0 .$$

The results of this section (19.3) are collected in the following theorem:

Theorem 19.3.1. The relations among the eight infinitesimal transformations

$$p+\cdots, q+\cdots, xp+\cdots, yp+\cdots, xq+\cdots, yq+\cdots,$$

$$x^2p+xyq+\cdots, xyp+y^2q+\cdots$$

have exactly the same form as the relations among the eight linear infinitesimal transformations.

$$p, q, xp, yp, xq, yq, x^2p+xyq, xyp+y^2q .$$

19.4 CANONICAL FORM

It is now extremely easy to bring the eight infinitesimal transformations we have found into a simple canonical form by introducing suitable new independent variables x', y'. For this we need the following lemma:

Theorem 19.4.1. If three infinitesimal transformations A_1, A_2, A_3 where $A_i = \xi_i p + \eta_i q$, satisfy the relations

$[A_i, A_k] = 0 \quad (1 \leq i,k \leq 3)$ <u>and if</u> $A_1(f) = 0$ <u>and</u> $A_2(f) = 0$ <u>are independent linear partial differential equations</u>, <u>then</u> A_3 <u>is a linear combination of</u> A_1 <u>and</u> A_2.

To prove this theorem introduce suitable new variables x', y' so that the A_k assume the form

$$A_1 = p', \quad A_2 = q', \quad A_3 = \xi' p' + \eta' q' \quad ;$$

then

$$\frac{\partial \xi'}{\partial x'} = \frac{\partial \xi'}{\partial y'} = 0, \quad \frac{\partial \eta'}{\partial x'} = \frac{\partial \eta'}{\partial y'} = 0 \quad ,$$

so that actually

$$A_3 = c_1 p' + c_2 q' = c_1 A_1 + c_2 A_2 \quad ,$$

for some constants c_1, c_2.

The infinitesimal transformations $p+\cdots$, $q+\cdots$, of Section 19.3 satisfy $[p+\cdots, q+\cdots] = 0$, and the linear partial differential equations

$$p+\cdots = 0, \quad q+\cdots = 0$$

are obviously independent since when $x = 0$ and $y = 0$ they become $p = 0$ and $q = 0$, respectively. Hence we can choose variables x' and y' so that $p+\cdots$ and $q+\cdots$ assume the form

$$p+\cdots = p', \quad q+\cdots = q' \quad .$$

To determine the quantity

$$xq+\cdots = \xi p' + \eta q'$$

in terms of the new variables we use the relations

$$[q+\cdots, xq+\cdots] = 0, \quad [p+\cdots, xq+\cdots] = q+\cdots ,$$

or, equivalently,

$$[q', \xi p'+\eta q'] = 0, \quad [p', \xi p'+\eta q'] = q' .$$

Putting

$$\xi p' + \eta q' = x'q' + \xi'p' + \eta'q'$$

gives

$$[q', \xi'p'+\eta'q'] = 0, \quad [p', \xi'p'+\eta'q'] = 0 ,$$

from which it follows by Theorem 19.4.1, that $\xi'p' + \eta'q'$ is of the form $c_1p' + c_2q'$. Therefore,

$$xq+\cdots = x'q' + c_1p' + c_2q' .$$

By completely analogous computations one finds that

$$xp+\cdots = x'p' + d_1p' + d_2q' ,$$

$$yp+\cdots = y'p' + e_1p' + e_2q' ,$$

$$yq+\cdots = y'q' + f_1p' + f_2q' ,$$

$$x^2p + xyq+\cdots = x'^2p' + x'y'q' + g_1p' + g_2q' ,$$

$$xyp + y^2q+\cdots = x'y'p' + y'^2q' + k_1p' + k_2q' .$$

In this it is not necessary to determine the constants c, d, e, f, g, k more closely. For it is well-known that the infinitesimal transformations $q, p, xq, xp, yp, yq, x^2p+xyq, xyp+y^2q$ are the infinitesimal transformations of the general linear fractional group.

Thus we have the following fundamental theorem:

Theorem 19.4.2. If a group leaves invariant no family of curves $\phi(x,y) = a$ and if the group contains infinitesimal transformations of order >1 at a general point (as well as ones of orders 0 and 1), then a suitable change of variables transforms this group to the general linear fractional group of the plane.

COMMENTS ON CHAPTER 19

This chapter is a tour-de-force of calculations. Lie's approach is powerful and straightforward. It is possible to give a shortened treatment (but one involving more algebraic machinery) using ideas developed by Cartan in his extension of Lie's work in his "Infinite Lie Group" papers.

Chapter 20

DETERMINATION OF ALL GROUPS WHICH LEAVE INVARIANT NO FAMILY OF CURVES $\phi(x,y) = a$

It remains to determine all groups whose infinitesimal transformations are either of the form

$$p+\cdots, \; q+\cdots, \; xp+\cdots, \; yp+\cdots, \; xq+\cdots, \; yq+\cdots$$

or of the form

$$p+\cdots, \; q+\cdots, \; xq+\cdots, \; yp+\cdots, \; xp-yq+\cdots \quad .$$

<u>These groups have either six or five parameters</u>.

20.1 THE SIX-PARAMETER CASE

To abbreviate the formulas we put

$$p+\cdots = P, \qquad q+\cdots = Q, \qquad xp+\cdots = XP,$$

$$yq+\cdots = YQ, \qquad xq+\cdots, = XQ, \qquad yp+\cdots, = YP \quad ,$$

noting that, for example, the symbol XP doesn't denote the product of two quantities X and P but is an <u>irreducible</u> symbol. Our first concern is to bring the relations among these six infinitesimal transformations into their simplest form.

Among the four first-order infinitesimal transformations it is obvious that there are the following relations:

$$\begin{aligned}
[XQ,YP] &= XP - YQ \;, \\
[XQ,XP-YQ] &= -2XQ \;, \\
[YP,XP-YQ] &= 2YP \;, \\
[XQ,XP+YQ] &= [YP,XP+YQ] = [XP-YQ,XP+YQ] = 0 \;.
\end{aligned}$$

To simplify these formulas denote $XP + YQ$ by U and let T denote the three quantities XQ, YP and $XP-YQ$. Then there are relations of the form:

$$\begin{aligned}
[P,XQ] &= Q + \sum \lambda_k T_k + \lambda U \;, \\
[P,XP-YQ] &= P + \sum \mu_k T_k + \mu U \;, \\
[P,YP] &= \sum \nu_k T_k + \nu U \;, \\
[Q,XQ] &= \sum \alpha_k T_k + \alpha U \;, \\
[Q,XP-YQ] &= -Q + \sum \beta_k T_k + \beta U \;, \\
[Q,YP] &= P + \sum \gamma_k T_k + \gamma U \;.
\end{aligned}$$

Now replace $q+\cdots$ and $p+\cdots$ by transformations of the form $Q+\varepsilon U$, $P+\delta U$. In this way we see that it is no restriction to take $\lambda = \gamma = 0$. Now the equation

$$[[P,XQ],XP-YQ] - 2[XQ,P] - [P,XQ] + \sum T = 0$$

implies

$$[Q,XP-YQ] + [P,XQ] + \sum T = 0 \;,$$

so that $\beta = 0$. Similarly, $\mu = 0$. Finally, the equation

$$[[P,XP-YQ],YP] - 2[YP,P] + \sum T = 0$$

gives an equation of the form

$$3[P,YP] + \sum T = 0 ,$$

so that $\nu = 0$. Similarly, $\alpha = 0$.

We shall now show that in the two equations

$$[Q,XQ] = \alpha_1 XQ + \alpha_2(XP-YQ) + \alpha_3 YP ,$$

$$[Q,XP-YQ] = -Q + \beta_1 XQ + \beta_2(XP-YQ) + \beta_3 YP$$

it is no restriction to take the α_i and β_i to be zero. For this purpose we introduce the quantity

$$Q + AXQ + B(XP-YQ) + CYP ,$$

in which A,B,C denote unknown constants, in place of $q+\cdots$. In this way we can make

$$\alpha_1 = \alpha_2 = \beta_1 = 0 .$$

Now the identity

$$[[Q,XQ],XP-YQ] - 2[XQ,Q] + [[XP-YQ,Q],XQ] = 0$$

implies

$$5\alpha_3 YP - 2\beta_2 XQ + \beta_3(XP-YQ) = 0 ,$$

which shows that $\alpha_3 = \beta_2 = \beta_3 = 0$. Therefore

$$[Q,XQ] = 0, \quad [Q,XP-YQ] = -Q .$$

We determine more of the constants as follows. There is a relation of the form

$$[Q,YP] = P + a_1XQ + a_2(XP-YQ) + a_3YP \quad ,$$

and it is permissible to replace P by the right side of this equation, which gives

$$[Q,YP] = P \quad .$$

Now the identity

$$[[Q,YP],XQ] + [YQ-XP,Q] = 0$$

implies

$$[P,XQ] = Q \quad ;$$

and the identity

$$[[Q,YP],XP-YQ] + 2[YP,Q] + [Q,YP] = 0$$

implies

$$[P,XP-YQ] = P \quad .$$

To determine the constants in

$$[P,YP] = b_1XQ + b_2(XP-YQ) + b_3YP$$

we use the identity

$$[[P,YP],XP-YQ] + 2[YP,P] - [P,YP] = 0$$

which gives

$$[P,YP] = 0 \quad .$$

Finally, there is an equation of the form

$$[P,XP+YQ] = P + c_1XQ + c_2(XP-YQ) + c_3YP + c_4(XP+YQ) \quad ;$$

using now the identity

$$[[P,XP+YQ],XP-YQ] - [P,XP+YQ] = 0$$

we see that

$$[P,XP+YQ] = P .$$

Similarly,

$$[Q,XP+YQ] = Q .$$

It now remains only to compute

$$[P,Q] = \alpha P + \beta Q + \gamma XQ + \delta YP + \varepsilon(XP-YQ) + \phi(XP+YQ) .$$

The identities

$$[[P,Q],XP-YQ] = 0 ,$$

$$[[P,Q],XP] - [P,Q] = 0$$

show that

$$[P,Q] = 0 .$$

Thus we have shown that <u>the six infinitesimal transformations</u> P, Q, XQ, YP, XP, YQ <u>satisfy the same relations as the linear infinitesimal transformations</u> p, q, xq, yp, xp, yq.

20.2 CANONICAL FORM FOR THE SIX-PARAMETER CASE

Since P and Q are independent infinitesimal transformations of order zero at the origin and $[P,Q] = 0$, we can always choose the variables x and y so that $P = p$, $Q = q$. Then the remaining transformations assume the form

$$XP = xp + \alpha_1 p + \beta_1 q \quad ,$$

$$XQ = xq + \alpha_2 p + \beta_2 q \quad ,$$

$$YP = yp + \alpha_3 p + \beta_3 q \quad ,$$

$$YQ = yq + \alpha_4 p + \beta_4 q \quad ,$$

and this proves the following theorem:

Theorem 20.2.1. _If a 6-parameter group leaves invariant no family of curves_ $\phi(x,y) = a$, _then a suitable change of variables brings it to the linear form_

$$x' = ax + by + c, \qquad y' = \alpha x + \beta y + \gamma \quad .$$

20.3 CONTINUATION

It is now easy to determine all groups whose infinitesimal transformations are of the form

$$p+\cdots = P, \qquad q+\cdots = Q, \qquad xq+\cdots = XQ,$$

$$xp-yq+\cdots = XP-YQ, \qquad yp+\cdots = YP \quad .$$

The two first-order transformations satisfy the equations

$$[XQ,YP] = XP - YQ \quad ,$$

$$[XQ,XP-YQ] = -2XQ \quad ,$$

$$[YP,XP-YQ] = 2YP \quad .$$

Denoting XQ, $XP-YQ$, YP by the common symbol T, one has six relations of the form

$$[P,XQ] = Q + \sum \lambda_k T_k ,$$

$$[P,XP-YQ] = P + \sum \mu_k T_k ,$$

$$[P,YP] = \sum \nu_k T_k ,$$

$$[Q,XQ] = \sum \alpha_k T_k ,$$

$$[Q,XP-YQ] = -Q + \sum \beta_k T_k ,$$

$$[Q,YP] = P + \sum \gamma_k T_k$$

and by Section 20.2 it is no restriction to assume that these equations have the simpler form

$$[P,XQ] = Q, \qquad [P,XP-YQ] = P, \qquad [P,YP] = 0,$$

$$[Q,XQ] = 0, \qquad [Q,XP-YQ] = -Q, \qquad [Q,YP] = P .$$

One has

$$[P,Q] = \alpha P + \beta Q + \gamma XQ + \delta(XP-YQ) + \varepsilon YP ,$$

and now the identities

$$[[P,Q],XP-YQ] = 0, \qquad [[P,Q],XQ] = 0$$

show that all the constants are zero, and hence that

$$[P,Q] = 0 .$$

Therefore:

<u>Theorem 20.3.1</u>. <u>If a 5-parameter group leaves invariant no family of curves</u> $\phi(x,y) = a$, <u>a suitable change of variables</u>

brings it to the form

$$x' = ax + by + \alpha, \qquad y' = cy + dx + \beta \quad ,$$

with $ac-bd = 1$.

Putting together the results of this section and the last, we obtain the following theorem:

Theorem 20.3.2. If a group of the plane leaves invariant no family of curves $\phi(x,y) = a$, *then it contains* 8, 6, *or* 5 *parameters. A suitable change of variables transforms it to a linear group, in fact to the general linear group or to a* 6- *or* 5-*term subgroup of the general linear group.*[*]

[*] This theorem extends to n dimensions. In my next paper on transformation groups I hope to be able to give a rigorous proof of this generalization.

Chapter 21

ENUMERATION OF ALL GROUPS OF THE PLANE

There is a natural division of all the groups of point transformations of a plane into five classes:

1. those leaving invariant no family of curves $\phi(x,y) = a$,
2. those leaving invariant exactly one family $\phi = a$,
3. those leaving invariant exactly two families $\phi = a$,
4. those leaving invariant an ∞^1 of families,
5. those leaving invariant an ∞^∞ of families.

21.1 FOLIATIONS ADMITTING VECTOR FIELDS

If a family $\phi = a$ admits an infinitesimal transformation, say q, then there are two possible cases. Either each curve of the family admits the transformation q, or q permutes the curves of the family. In the first case the family consists of the curves $x = \text{const.}$; in the second case its defining equation is of the form

$$y + f(x) = \lambda = \text{const.} \quad ,$$

where λ denotes an arbitrary constant.

If a family admits the two transformations q and $X(x)q$, then its defining equation is either of the form

$$y + f(x) = \lambda$$

or

$$x = \text{const.} \quad ,$$

and since

$$X(x) \frac{\partial}{\partial y} (y+f(x)) = X(x)$$

is not a function of $y+f(x)$, we see *that* $x = \text{const.}$ *is the only family which admits both the transformations* q *and* Xq.

If a family admits both the transformations q and yq, then it likewise is of one of the forms

$$y + f(x) = \lambda \qquad \text{or} \qquad x = \lambda \quad ,$$

and since y is a function of $y+f(x)$ only if $f(x) = \text{cons}$ we see *that* $x = \text{const.}$ *and* $y = \text{const.}$ *are the only families which admit both the transformations* q *and* yq.

If a family admits the transformations q and p, then again it is either of the form $x = \text{const.}$ or of the form $y+f(x) = \lambda$. And since in the latter case $f'(x)$ is not a function of $y+f(x)$ unless it is a constant, we conclude that *the family* $ay+bx = \lambda$, *depending on the (arbitrary) parameter* a/b, *is the most general family which admits the transformations* p *and* q.

21.2 CLASSIFICATION OF GROUPS IN THE PLANE

Applying these considerations to all the previously determined groups, we obtain the following exhaustive classification of all groups of the plane.

A) <u>Groups leaving invariant no family</u> $\phi(x,y) = a$:

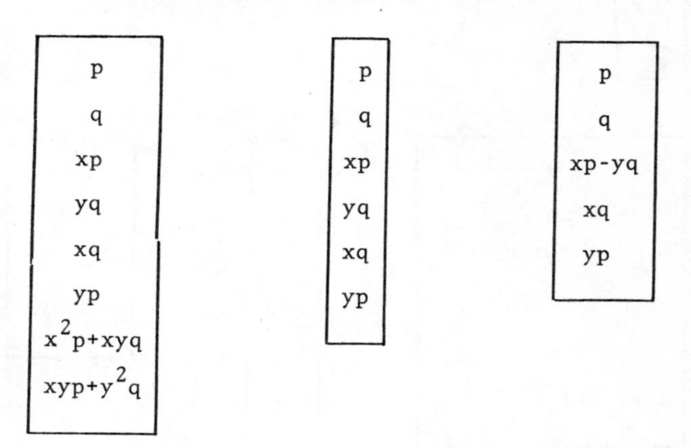

These groups are all linear groups.

B) <u>Groups leaving invariant exactly one family</u>:

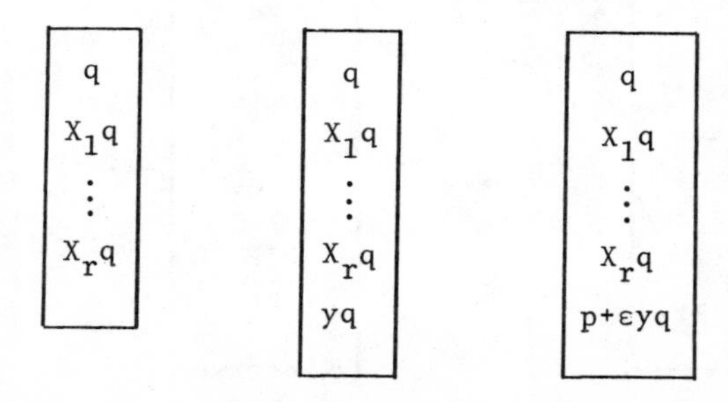

$$\begin{array}{c} q \\ X_1 q \\ \vdots \\ X_r q \\ yq \\ p \end{array}$$

$$\begin{array}{c} q \\ xq \\ \vdots \\ x^r q \\ p \\ xp+Kyq \end{array}$$

$$\begin{array}{c} p \\ xp+yq \\ x^2p+2xyq \end{array}$$

$$\begin{array}{c} q \\ xq \\ \vdots \\ x^r q \\ p \\ xp+[(r+1)y+x^{r+1}]q \end{array}$$

$$\begin{array}{c} q \\ xq \\ \vdots \\ x^r q \\ yq \\ p \\ xp \end{array}$$

$$\begin{array}{c} yq \\ p \\ xp \\ x^2p+xyq \end{array}$$

$$\begin{array}{c} q \\ xq \\ \vdots \\ x^r q \\ p \\ 2xp+ryq \\ x^2p+rxyq \end{array}$$

$$\begin{array}{c} q \\ xq \\ \vdots \\ x^r q \\ p \\ xp \\ yq \\ x^2p+rxyq \end{array}$$

It is assumed that $r > 0$; only for the seventh group can one have $r = 0$.

C) <u>Groups leaving invariant exactly two families</u>:

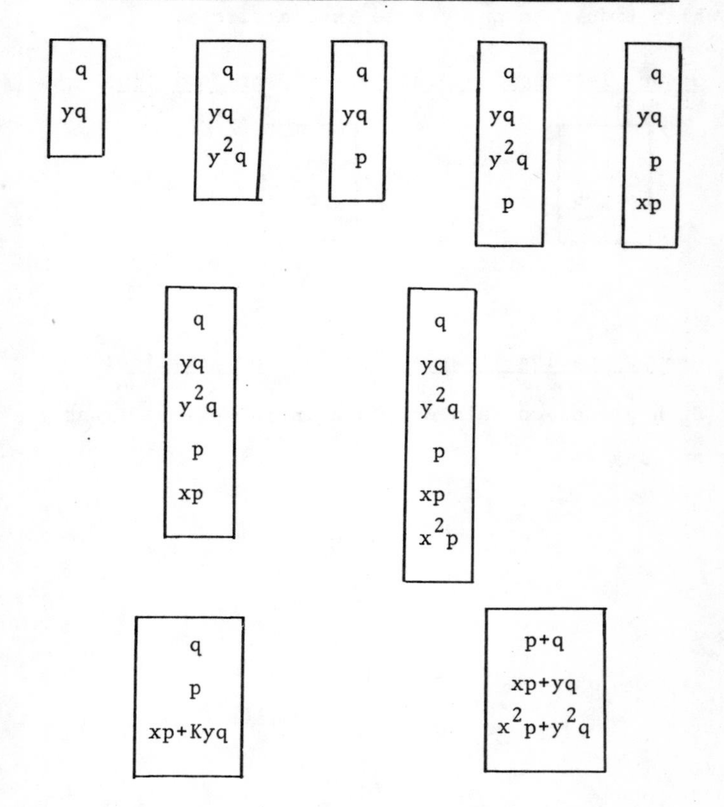

The last group is a new form of a group found earlier: p, $xp+yq$, $x^2p + (2xy+y^2)q$. In the penultimate group one must have $K \neq 1$. All the groups occurring here are subgroups of the 6-term group q, yq, y^2q, p, xp, x^2p whose simplest geometric form is the group of all point transformations which transform any circle into a circle.

D) <u>Groups leaving invariant an ∞^1 of families</u>:

q
$p+\varepsilon yq$

q
p
$xp+yq$

E) <u>Groups leaving invariant an ∞^∞ of families</u>:

Such groups contain only one infinitesimal transformation, say

p

.

Chapter 22

GENERAL OBSERVATIONS

22.1 TRANSFORMATION GROUPS AS A NEW CONCEPT

As already stated in the introduction, I believe that my theory of transformation groups, whose first elements have been developed in this paper, is to be considered a new theory, even though it has many points of contact with several mathematical disciplines, especially with the Galois theory of substitutions, with geometry and the modern theory of manifolds, with the theory of differential equations and, finally, also with invariant theory.

I shall permit myself to make my conception more precise. For this purpose I shall discuss all the older investigations (published before 1874) I am aware of which are more or less related to mine.* At the same time I shall make some general remarks on the new thoughts underlying my investigations.

22.2 ABEL'S RESULT

In his first paper in Crelle's Journal, Abel determined the most general symmetric function $F(x,y)$ satisfying

* Since my knowledge of the mathematical literature is incomplete, I must fear that the citations of the text are incomplete. I will receive with thanks any correction that I can use in my later publications on transformation groups.

a relation of the form

$$F(F(x,y),z) = F(x,F(y,z)) \quad .$$

This problem is a special case of the simplest problem of my theory. For the task of determining the most general 1-term group of a 1-dimensional manifold amounts to finding the most general functions F and ϕ satisfying a functional equation

$$F(F(x,a),b) = F(x,\phi(a,b)) \quad .$$

22.3 GALOIS THEORY OF ALGEBRAIC EQUATIONS

My transformation theory is closely related to the theory of substitutions.* Both the analogy and the distinction between these two disciplines are based on the fact that the substitution theory deals with discrete manifolds, while the transformation theory deals with continuous manifolds, and, on the other hand, that any two operations of a substitution group are finitely distinct, while the operations of a transformation group depend on continuous parameters.**

---

* Cf. Camille Jordan's "Traité des substitutions".

** Here I must mention C. Jordan's determination of all groups of motions. He considers two groups to be equivalent if one can be transformed to the other by an orthogonal transformation. In my investigations on the other hand, two groups are considered equivalent if one can be transformed into the other by an arbitrary analytic transformation.

22.4 TOWARDS A GALOIS THEORY OF DIFFERENTIAL EQUATIONS

It is well known that in the theory of algebraic equations the theory of substitutions plays a fundamental role. Similarly, the theory of transformation groups will play a not unimportant role in the theory of differential equations. And indeed, my theory has significance not only for differential equations which admit a transformation group, but also for arbitrary differential equations. This is based essentially on the following remarks.

The question of whether a given differential equation $G = 0$ (or a system of such equations) can be brought into a certain form $F = 0$ by a suitable point transformation or contact transformation in each case requires for its resolution only those operations which one is accustomed to consider permissible in the integral calculus. Indeed, if both $G = 0$ and $F = 0$ admit a transformation group, it is first of all necessary that one group can be transformed into the other. If one has found that neither $F = 0$ nor $G = 0$ admits a transformation group, while $G = 0$ can be brought into the form $F = 0$, then <u>this transformation can be accomplished by permissible operations</u>.

If a differential equation or a system of differential equations admits a transformation group, then, as I have already shown, or at least indicated, this situation lets us determine the general integral or at least certain distinguished classes of integrals.*

22.5 RELATIONS TO INVARIANT THEORY

The relation between my transformation theory and invariant theory is based on the fact that the former deals with differential equations which are invariant under arbitrary point transformations or contact transformations. Here I am thinking not only of the invariant theory of Cayley and Sylvester, but also of the (unfortunately almost unknown to me) investigations of Lipschitz and Christoffel on the transformation of differential expressions.

22.6 THE RIEMANN-HELMHOLTZ SPACE-FORM PROBLEM

The Riemann-Helmholtz investigations of the facts which are at the basis of geometry stand in a direct connection with the theory of transformation groups. Helmholtz' well-known note (Göttinger Nachrichten 1868, No. 9) deals, in my terminology, with the determination of a certain 6-term group of a 3-dimensional space.** The Riemann-Helmholtz

---

* The transformation group of a given differential equation may contain infinitely many parameters. On the other hand, it is important to note that differential equations may admit groups of infinitely-many-valued transformations.

** I thank Klein for pointing this out to me and for the reference to Helmholtz' note.

investigations are restricted to metrical geometry. The theory of transformation groups gives, among other things, a similarly penetrating discussion of the projective geometry of an n-dimensional space, as I shall demonstrate on another occasion.

22.7 CONTACT TRANSFORMATIONS

It is well-known that geometry often deals with transformation groups, for example, with the general linear group, the orthogonal group, the group of all conformal transformations, and so on. In my first geometric works I considered some new groups. In the note "Über die Reziprozitätsverhaltnisse des Reyeschen Komplexes"* ("On the reciprocity of the Reye complex") I dealt with the group of all contact transformations which leave invariant a certain second order partial differential equation closely connected with the tetrahedral line complex. I then investigated, together with Klein, who had already been occupied with applications of the theory of substitutions to geometry, the surfaces invariant under a doubly-infinite commutative family of linear transformations.** Further, in a paper on

* Göttinger Nachrichten, January 1870 [Collected Papers, vol. I, paper V.]

** Comptes Rendus 1870, Math. Ann. IV [Collected Papers, vol. I, papers VI, XIV.]

complexes (Math. Ann., vol. V [Collected Papers, vol. II, paper I]) I determined, among other things, an important new group, namely the group of all contact transformations whose curvature lines are invariant curves; I showed that this group can be transformed into the general projective group of space by means of a remarkable contact transformation. Finally, in his program "Vergleichende Betrachtungen über neuere geometrische Forschungen"* ("Comparative Observations on modern geometric researches"), Klein developed the conception that the methods of mathematics, especially of geometry, in many respects can be characterized by the transformation group which they adjoin, i.e., by the group of those changes which, in the sense of the given method, are considered inessential.

22.8 INFINITESIMAL TRANSFORMATIONS

In investigations of first-order partial differential equations I remarked that the formulas which occur in this discipline can be given a remarkable intuitive interpretation through the use of the concept of an infinitesimal transformation. In particular, the so-called Poisson-Jacobi theorem and the well-known Jacobi identity are in the closest relation

*Erlangen, 1872.

with the theory of the composition of infinitesimal transformations.* Pursuing this remark, I reached the astounding result that all transformation groups of a 1-dimensional manifold can be reduced to linear form by introducing suitable variables, and also that all groups of an n-dimensional manifold can be determined by integrating ordinary differential equations. This discovery, whose first traces go back to Abel and Helmholtz, was the starting point of my many years of investigations of transformation groups.

22.9 MY PREVIOUS WORK

My first publication on this subject (Göttinger Nachrichten 1874, No. 22 [Collected Papers, vol. V, paper I] contains not only a resume of all the results of the present paper but also the determination of all groups of contact transformations of a plane, as well as indications of the applications of my theory to differential equations. I then gave a full edition of my most important results in five papers,** which were printed in the Norwegian "Archiv for

---

* One sometimes encounters the idea that the famous Jacobi identity has only a subordinate value. I should like to remark here that this identity is the analytic foundation of my transformation theory.

** Several inaccuracies in the proofs crept into these preliminary works. Some of the mistakes in the first two works printed in Christiania were pointed out to me by Mayer. It is difficult to avoid inaccuracies in the proofs when one is editing extensive theories, found by mixed methods, in the language of pure analysis.

Mathematik og Naturvidenskab", vol. I, III and IV [Collected Papers, vol. V, papers II-VI]. I intend to develop, in the same place the theory of an n-dimensional space, in particular, of ordinary space, and to make some applications of my theory to differential equations. I have already given a first such application in the paper "Klassifikation der Flächen nach der Transformationsgruppe ihrer geodätischen Kurven" ("Classification of surfaces by the transformation groups of their geodesics"), Universitätsprogramm, Christiania 1879 [Collected Papers, vol. I, paper XXIV].

In closing, I cannot hold back the following remark. In my definition of the concept of a transformation group I explicitly require that the group contain the inverse of each of its transformations. For the group of all transformations leaving invariant a differential equation, this requirement is unnecessary.

Christiania, December 1879.

Chapter E

SOME GENERAL TECHNIQUES FOR THE CLASSIFICATION OF LIE ALGEBRAS OF VECTOR FIELDS

1. INTRODUCTION

Lie's local classification (in this paper) of the finite dimensional Lie algebras of vector fields in the plane is a masterpiece of direct and forceful computation. For better or worse, the tendency in modern mathematics is to replace (if possible) such brute force calculations with more subtle general arguments.

In my comments in the text, I have suggested certain general features that appear when reading Lie's work. In this chapter, I want to develop some of Lie's results in greater generality, and with modern tools.

As I have already indicated, there are (at least) two general topics intermingled in Lie's proof of the classification of Lie algebras in the plane--the general structure theory of filtered Lie algebras and ideas of the theory of deformations of Lie algebra structures and their representations. Here is another general problem of great interest:

> Given a finite dimensional Lie algebra $\underset{\sim}{H}$ of vector fields on a manifold X,

how many finite dimensional subalgebras
of V(X) are there containing $\underset{\sim}{H}$?

2. ABELIAN LIE ALGEBRAS OF VECTOR FIELDS

Lie's constant objective was to derive canonical forms for classes of Lie algebras of vector fields. His motivation comes from his interest in applying transformation group theory to the theory of differential equations in the 19-th century sense. For, often, one finds differential equations carrying along certain Lie algebras of vector fields, and it turns out that the coordinate systems in which these vector fields take their canonical form is a useful coordinate system in which to solve the differential equations. (Note the 19-th century emphasis on the practical aspects of a problem--some of the greatest theoretical work was done with such a motivation!)

In particular, abelian Lie algebras of vector fields play a key role in Lie's work. Basically, this is because they often determine the relevant canonical coordinate systems. We have seen a good illustration of this in Part I, in the classification of Lie algebras of vector fields on one-dimensional manifolds. The technique used there was to choose a vector field A in the algebra which did not vanish at a point, and to choose the coordinate x so that A took its canonical form, i.e., $A = \partial/\partial x$. It turns out that the whole Lie algebra takes its canonical form in this coordinate system.

Cartan too made initial use of certain types of abelian Lie algebras in his _algebraic structure theory of semi-simple Lie algebras_. (We now call them _Cartan subalgebras_.)

Here is one result which Lie uses often.

Theorem 2.1. Let $\underset{\sim}{A}$ be an abelian Lie algebra of vector fields on a manifold X, and let $x \in X$ be a point such that:

$$\underset{\sim}{A}(x) = X_x \quad , \tag{2.1}$$

i.e., $\underset{\sim}{A}$ acts _locally transitively on_ X _at_ x. Then,

$$\dim \underset{\sim}{A} = \dim X \quad . \tag{2.2}$$

Proof. Let $A_1,\ldots,A_n$ be elements of $\underset{\sim}{A}$, whose values at x form a basis for X_x. Then, by continuity,

$$A_1(y),\ldots,A_n(y)$$

forms a basis for X_y, for all y sufficiently close to x. In particular, there is an open subset O of X containing x such that $A_1,\ldots,A_n$ form an $F(O)$-module basis of $V(O)$.

Let A be another element of $\underset{\sim}{A}$. Then, there are functions $f_1,\ldots,f_n$ such that:

$$A = f_1A_1+\cdots+f_nA_n \quad .$$

Now, use the fact that $\underset{\sim}{A}$ is abelian.

$$0 = [A_1,A] = A_1(f_1)A_1 +\cdots+ A_1(f_n)A_n \quad ,$$

hence:

$$A_1(f_1) = 0 = \cdots = A_1(f_n)$$

Repeating the process for $A_2,\ldots,A_n$, we see (because of formula 2.1) that

$$\begin{aligned} f_1 &= \text{constant} \\ &\vdots \\ f_n &= \text{constant} \quad , \end{aligned}$$

i.e., A is linearly dependent on $A_1,\ldots,A_n$. This proves that:

$$\dim \underset{\sim}{A} = n = \dim S \quad .$$

<u>Remark</u>. Recall that Lie used this in the case "$\dim X = 1$" in a key way. If A,B are two vector fields, with $[A,B] = 0$ then they are linearly dependent.

Here is a more direct relation between abelian Lie algebras.

Theorem 2.2. Let $\underset{\sim}{A}$ be an abelian Lie algebra of vector fields on X. Let x be a point of X such that:

$$m = \dim \underset{\sim}{A} = \dim (\underset{\sim}{A}(x)) \quad ,$$

i.e., no non-zero element of $\underset{\sim}{A}$ leaves x fixed. Then, there is a coordinate system for X,

$$x^1,\ldots,x^n \quad ,$$

valid in a neighborhood of x, such that: The vector fields

$$\frac{\partial}{\partial x^1},\ldots,\frac{\partial}{\partial x^m}$$

are a basis for $\underset{\sim}{A}$ in that neighborhood.

Proof. For $m = 1$, the result is well-known. (See DGCV.)

Proceed to prove the theorem by induction on n. Let

$$(y^1,\ldots,y^n)$$

be a coordinate system about x such that:

$$A_1 \equiv \frac{\partial}{\partial y^1} \in \underset{\sim}{A} \quad . \tag{2.3}$$

Suppose that the vector fields

$$\begin{aligned} A_2 &= a_1 \frac{\partial}{\partial y^1} + a_2 \frac{\partial}{\partial y^2} + \cdots \\ A_3 &= b_1 \frac{\partial}{\partial y^1} + b_2 \frac{\partial}{\partial y^2} + \cdots \\ &\vdots \end{aligned} \tag{2.4}$$

together with A_1, form an R-basis for $\underset{\sim}{A}$, and also are linearly independent at x. Use the relation:

$$[A_1, A_2] = 0 = [A_1, A_3] = \cdots \quad (2.5)$$

Since $A_1 = \partial/\partial y^1$, we see that the coefficients $a_1, \ldots$ in 2.4 are <u>independent of</u> y^1.

Consider the vector fields

$$A_2' = a_2 \frac{\partial}{\partial y^2} + \cdots$$

$$A_3' = b_3 \frac{\partial}{\partial y^2} + \cdots$$

The relation $[\underset{\sim}{A}, \underset{\sim}{A}] = 0$ implies that

$$[A_2', A_3'] = 0 = \cdots ,$$

i.e., the $A_2', \ldots, A_3'$ <u>define an abelian Lie algebra of vector fields in</u> R^{n-1}. The induction hypothesis may be applied--there is a coordinate system for R^{n-1} in which the vector fields are the partial derivatives with respect to the coordinates. Putting this coordinate system for R^{n-1} together with y^1 defines a coordinate system

$$(z^1, \ldots, z^n)$$

in which the A's take the following form:

$$A_1 = \frac{\partial}{\partial z^1}$$

$$A_2 = a_1 \frac{\partial}{\partial z^1} + \frac{\partial}{\partial z^2}$$

$$A_3 = b_1 \frac{\partial}{\partial z^1} + \frac{\partial}{\partial z^3} ,$$

and so forth.

The coefficients $a_1, b_1, \ldots$ of $\partial/\partial z^1$ are independent of z^1. Thus, the relation

$$[A_2, A_3]$$

forces

$$\frac{\partial a_1}{\partial z^2} = 0$$

Continuing in this way, we see that a_1 is independent of $z^3, \ldots,$ i.e., a_1 _is a constant_. Similarly, $b_1, \ldots$ are constants. Now, a simple linear coordinate change determines a coordinate system $(x^1, \ldots, x^n)$ such that

$$A_1 = \frac{\partial}{\partial x^1}, \ldots, A_m = \frac{\partial}{\partial x^m}$$

Remark. It is important to keep in mind that this _canonical coordinate system_ for the abelian Lie algebra $\underset{\sim}{A}$ may be found by solving a succession of _ordinary_ differential equations. Lie always had in mind developing a vast generalization of the _Galois theory of equations_ to cover _differential equations_. Such a generalization was developed, partially following Lie's ideas, by Picard, Vessiot and Drach, but the whole circle of ideas is even today very incomplete. What is now known as the "Galois theory of differential equations", involves "differential algebra" (see Kaplansky [1], Kolchin [1]), and is considerably different in spirit, outlook, and results from what Lie had in mind.

3. SEMI-SIMPLE LIE ALGEBRAS OF VECTOR FIELDS

The chief technical advance in Lie algebra theory since Lie's time has been the theory of <u>semi-simple Lie algebras</u>. We shall see that the way such algebras can act in manifolds is relatively restricted. The chief tools will be the <u>Killing form</u> (and <u>Cartan's theorem</u> that it is non-degenerate if and only if the algebra is semi-simple).

Let $\underset{\sim}{G}$ be a finite dimensional Lie algebra. (The scalar field may be the real or complex numbers, or, in fact, any field K of characteristic zero). Let

$$\beta: \underset{\sim}{G} \times \underset{\sim}{G} \to K$$

be the (symmetric, bilinear) <u>Killing form</u>. Then, for $A, B \in G$,

$$\beta(A,B) = \text{trace}\,(\text{Ad}\,A\ \text{Ad}\,B) \quad .$$

For the properties of the Killing form that we need, see Chapter D.

Now suppose:

$$\underset{\sim}{G} = \underset{\sim}{G}^0 \supset \underset{\sim}{G}^1 \supset \underset{\sim}{G}^2 \supset \cdots$$

is a filtration of $\underset{\sim}{G}$. Recall that this requires that:

$$[\underset{\sim}{G}^j, \underset{\sim}{G}^k] \subset \underset{\sim}{G}^{j+k-1} \qquad (3.1)$$

$$\text{for } j \geq 1, k \geq 0 \quad .$$

Theorem 3.1.

$$\beta(\underset{\sim}{G}^3, \underset{\sim}{G}) = 0 \qquad (3.2)$$

Proof. Suppose $A \in \underset{\sim}{G}$, $B \in \underset{\sim}{G}^3$. Now, using 2.1,

$$(\text{Ad } B)(\underset{\sim}{G}^j) \subset \underset{\sim}{G}^{j+2}$$

Hence,

$$(\text{Ad } A \text{ Ad } B)(\underset{\sim}{G}^j) \subset \underset{\sim}{G}^{j+1}$$

Thus, if we compute the trace of (Ad A Ad B) with respect to a basis of increasing filtration, we see that:

$$\text{trace } (\text{Ad } A \text{ Ad } B) = 0 \quad ,$$

hence $\beta(A,B) = 0$, which forces 3.2, since A is an arbitrary element of $\underset{\sim}{G}$, B an arbitrary element of $\underset{\sim}{G}^2$.

Recall that the radical, denoted by $\underset{\sim}{R}$, of the Lie algebra $\underset{\sim}{G}$ is the largest solvable Lie ideal of $\underset{\sim}{G}$.

<u>Theorem 3.2</u>. $\underset{\sim}{G}^3$ is contained in the radical $\underset{\sim}{R}$ of $\underset{\sim}{G}$.

<u>Proof</u>. Formula 3.2 implies that $\underset{\sim}{G}^3$ is contained in the subset:

$$\underset{\sim}{H} = \{A \varepsilon G\colon \beta(A, \underset{\sim}{G}) = 0\}$$

($\underset{\sim}{H}$ so defined is called the <u>radical of</u> β.) Now it is readily seen that:

$\underset{\sim}{H}$ is a Lie ideal of $\underset{\sim}{G}$.

The Killing form of $\underset{\sim}{H}$ (as a Lie algebra) is the restriction to $\underset{\sim}{H}$ of the Killing form of $\underset{\sim}{G}$.

Hence, the Killing form of $\underset{\sim}{H}$ is zero. A theorem by Cartan (see Chapter D) now implies that:

$\underset{\sim}{H}$ is a solvable Lie algebra.

Hence,

$$\underset{\sim}{G}^3 \subset \underset{\sim}{H} \subset \underset{\sim}{R} \; .$$

<u>First Corollary to Theorem 3.2</u>. If $\underset{\sim}{G}$ is semi-simple and finite dimensional, with a filtration

$$\underset{\sim}{G} = \underset{\sim}{G}^0 \supset \underset{\sim}{G}^1 \supset \underset{\sim}{G}^2 \supset \cdots$$

then:

$$\underset{\sim}{G}^3 = 0 \; .$$

Second Corollary to Theorem 3.2. If $\underset{\sim}{G}$ is a finite dimensional semi-simple Lie algebra of vector fields on a manifold X, and if $A \in \underset{\sim}{G}$, $x \in X$ is such that:

A vanishes to the third order at x,

then $A = 0$.

This is a remarkable geometric property of semi-simple Lie algebras.

Theorem 3.3. Suppose that $\underset{\sim}{G}$ is a finite dimensional Lie algebra, with a filtration $\underset{\sim}{G} = \underset{\sim}{G}^0 \supset \underset{\sim}{G}^1 \supset \underset{\sim}{G}^2 \supset \cdots$. Then,

$$\underset{\sim}{G}^2 \text{ is contained in the radical of } \underset{\sim}{G}^1. \tag{3.3}$$

Proof. Again, this follows formula 3.2 and the use of the Killing form. For $A \in \underset{\sim}{G}^1$, $B \in \underset{\sim}{G}^2$,

$$(\text{Ad } A \text{ Ad } B)(\underset{\sim}{G}^j) \subset \underset{\sim}{G}^{j+1} \quad ,$$

hence:

$$\beta'(A,B) = 0 \quad ,$$

where "β'" is now the Killing form of $\underset{\sim}{G}^1$. We deduce that:

$$\beta'(\underset{\sim}{G}^2,\underset{\sim}{G}^1) = 0 \quad ,$$

which implies statement 3.3.

Theorem 3.4. Let $\underset{\sim}{G}$ be a finite dimensional filtered Lie algebra, and let β be the Killing form of $\underset{\sim}{G}$. Then,

$$\beta(\underset{\sim}{G}^j, \underset{\sim}{G}^k) = 0 \tag{3.4}$$

for $k \geq 2,\ j \geq 1$.

Proof. The argument is the same as led to Theorem 3.3.

Formula 3.4 leads, in the case that $\underset{\sim}{G}$ is real and semi-simple, to considerable limitations on $\underset{\sim}{G}^2$. For, it says that $\underset{\sim}{G}^2$ is an isotropic subspace of the Killing form. From the theory of real orthogonal forms, we see that, if β has p plus signs and q minus signs in a canonical form, then:

$$\dim \underset{\sim}{G}^2 \leq \min (p,q) \tag{3.5}$$

For example, suppose:

$$\underset{\sim}{G} = SL(2,R) \quad .$$

Then,

$$p = 2, \quad q = 1 \quad .$$

Hence,

$$\underset{\sim}{G}^2 \leq 1 \quad ,$$

which is a result found by Lie by direct calculation. Later on in this section we shall find a more powerful method for finding the dimensions of $\underset{\sim}{G}^2$.

Of course, if $\underset{\sim}{G}$ is a compact Lie algebra, i.e., its Killing form is negative definite, then:

$$\underset{\sim}{G}^2 = 0 \quad .$$

Theorem 3.5. Let $\underset{\sim}{G}$ be a finite dimensional, semi-simple, filtered Lie algebra. Then

$$[\underset{\sim}{G}^2,\underset{\sim}{G}^2] = 0 .$$

Proof. This is really a corollary to Theorem 3.3. For, $\underset{\sim}{G}$ semi-simple (i.e., no radical) implies that

$$\underset{\sim}{G}^3 = 0 .$$

But,

$$[\underset{\sim}{G}^2,\underset{\sim}{G}^2] \subset \underset{\sim}{G}^3 .$$

Continue with

$$\underset{\sim}{G} = \underset{\sim}{G}^0 \supset \underset{\sim}{G}^1 \supset \cdots$$

as a filtered Lie algebra such that:

$$\underset{\sim}{G}^3 = 0 .$$

Set

$$(\underset{\sim}{G}^2)^{\perp} = \{A\varepsilon\underset{\sim}{G}: \beta(A,\underset{\sim}{G}^2) = 0\}$$

$(\underset{\sim}{G}^2)^{\perp}$ is (as the notation indicates) the orthogonal complement of $\underset{\sim}{G}^2$ in $\underset{\sim}{G}$.

The Killing form β is invariant under Ad $\underset{\sim}{G}$. This means that:

$$\beta([A,B],C) + \beta(B,[A,C]) = 0 \qquad (3.6)$$

for $A,B,C \;\varepsilon\; \underset{\sim}{G}$.

Now, $\underset{\sim}{G}^1$ is a Lie subalgebra of $\underset{\sim}{G}$. Also,

$$[\underset{\sim}{G}^1,\underset{\sim}{G}^2] \subset \underset{\sim}{G}^2 \tag{3.7}$$

Relation 3.4 implies that:

$$\underset{\sim}{G}^1 \subset (\underset{\sim}{G}^2)^{\perp} \tag{3.8}$$

<u>Theorem 3.6</u>.

$$[\underset{\sim}{G}^1,(\underset{\sim}{G}^2)^{\perp}] \subset (\underset{\sim}{G}^2)^{\perp} \tag{3.9}$$

<u>Proof</u>. Using 3.6,

$$\beta([\underset{\sim}{G}^1,(\underset{\sim}{G}^2)^{\perp}],\underset{\sim}{G}^2) = \beta((\underset{\sim}{G}^2)^{\perp},[\underset{\sim}{G}^1,\underset{\sim}{G}^2])$$

$$\subset \beta((\underset{\sim}{G}^2)^{\perp},\underset{\sim}{G}^2) = 0$$

<u>Theorem 3.7</u>. Suppose $\underset{\sim}{G}$ is semi-simple. Then,

$$\dim \underset{\sim}{G}^2 \leq \dim \underset{\sim}{G}/\underset{\sim}{G}^1 \tag{3.10}$$

<u>Corollary to Theorem 3.7</u>. If $\underset{\sim}{G}$ is a semi-simple Lie algebra of vector fields on a manifold X, and if the filtration on $\underset{\sim}{G}$ is that defined by a point $x \in X$, then

$$\dim \underset{\sim}{G}^2 \leq \dim X \ . \tag{3.11}$$

<u>Remark</u>. For example, if X is two-dimensional, 3.11 is the bound for $\underset{\sim}{G}^2$ found by Lie by calculation.

Proof of Theorem 3.7. Since β is non-degenerate, we have (from linear algebra) that:

$$\dim \underset{\sim}{G}^2 = \dim \underset{\sim}{G} - \dim ((\underset{\sim}{G}^2)^{\perp}) . \qquad (3.12)$$

Now,

$$(\underset{\sim}{G}^2)^{\perp} \supset \underset{\sim}{G}^1 ,$$

using 3.4. Hence,

$$\dim (\underset{\sim}{G}^2)^{\perp} \geq \dim \underset{\sim}{G}^1 \qquad (3.13)$$

Combine 3.12 and 3.13

$$\dim \underset{\sim}{G}^2 \leq \dim \underset{\sim}{G} - \dim \underset{\sim}{G}^1 = \dim (\underset{\sim}{G}/\underset{\sim}{G}^1)$$

Proof of Corollary. $\underset{\sim}{G}^1$ consists of the vector fields in $\underset{\sim}{G}$ which vanish at x. Hence, $\underset{\sim}{G}$ is the kernel in the linear map

$$A \to A(x)$$

of $\underset{\sim}{G} \to X_x$, which shows that

$$\dim (\underset{\sim}{G}/\underset{\sim}{G}^1) \leq \dim X .$$

The argument given in the proof proves a bit more, namely:

Theorem 3.8. If $\underset{\sim}{G}$ is semi-simple, then $\dim (\underset{\sim}{G}^2) = \dim (G/G^1)$ if and only if

$$G^1 = (G^2)^{\perp} .$$

<u>Theorem 3.9</u>. Suppose $\underset{\sim}{G}$ is a semi-simple Lie algebra of vector fields on a manifold X. Let x be a point of X which is <u>not</u> a fixed point of $\underset{\sim}{G}$, but at which the linear stability algebra acts irreducibly on X_x. Then, either

$$\underset{\sim}{G}^2 = 0 ,$$

or

$$\dim \underset{\sim}{G}^2 = \dim X . \qquad (3.14)$$

<u>Proof</u>. If x is not a fixed point, $\underset{\sim}{G}(x) \subset X_x$ is a linear subspace invariant under the linear stability algebra, hence

$$\underset{\sim}{G}(x) = X_x ,$$

since the linear stability subalgebra is assumed to act irreducibly. Hence, X_x is identified with $\underset{\sim}{G}/\underset{\sim}{G}^1$, and the linear stability algebra is identified with Ad $(\underset{\sim}{G}^1)$ acting on $(\underset{\sim}{G}/\underset{\sim}{G}^1)$. Since

$$(\underset{\sim}{G}^2)^{\perp} \supset \underset{\sim}{G}^1, \quad \text{and} \quad [\underset{\sim}{G}^1, (\underset{\sim}{G}^2)^{\perp}] \subset (\underset{\sim}{G}^2)^{\perp},$$

the projection of $(\underset{\sim}{G}^2)^{\perp}$ on $(\underset{\sim}{G}/\underset{\sim}{G}^1)$ is a linear subspace invariant under the linear stability algebra. Hence, using irreducibility again, it must equal all of (G/G^1) (which would contradict semi-simplicity of $\underset{\sim}{G}$) or must be zero. The latter possibility leads to 3.14.

Remark. Again, this simple result may be used to replace difficult calculations in Lie's work. For example, if $\underset{\sim}{G}$ acts on $X = R^2$, "irreducibility of linear stability algebra" is equivalent to "primitivity", and 3.14 is indeed the result found by Lie in this case.

Theorem 3.10. If $\underset{\sim}{G}$ is a semi-simple Lie algebra of vector fields on an n-dimensional manifold X, and if the filtration is determined by a point of X_x at which $\underset{\sim}{G}$ acts transitively, then:

$$\dim \underset{\sim}{G} \leq 2n + n^2 \tag{3.15}$$

Proof. Since $\underset{\sim}{G}^3 = 0$, we have:

$$\begin{aligned} \dim \underset{\sim}{G} &= \dim \underset{\sim}{G}^2 + \dim (\underset{\sim}{G}^1/\underset{\sim}{G}^2) + \dim (\underset{\sim}{G}/\underset{\sim}{G}^1) \\ &\leq 2n + \dim (\underset{\sim}{G}^1/\underset{\sim}{G}^2) \quad , \end{aligned} \tag{3.16}$$

Now $\underset{\sim}{G}^2$ is an ideal in $\underset{\sim}{G}^1$. The linear stability algebra determines a Lie algebra homomorphism

$$\underset{\sim}{G}^1 \to (\text{linear maps on } X_x)$$

Since $\underset{\sim}{G}$ acts transitively, $\underset{\sim}{G}^2$ is the kernel of this homomorphism. Now, the dimension of the linear maps on an n-dimensional vector space is n^2, the dimension of the space of $n \times n$ matrices. Hence,

$$\dim (\underset{\sim}{G}^2/\underset{\sim}{G}^2) \leq n^2 .$$

This combined with 3.16, completes the proof.

Remark. If $n = 2$, then 3.15 implies that:

$$\dim \underset{\sim}{G} \leq 8 .$$

Now, SL(3,R) has dimension eight. It follows from Lie's results that it, or its subgroups, are the only semi-simple groups which may act on a two-dimensional manifold. We will now show how these inequalities may be combined with simple facts from the classification theory of Lie algebras to prove Lie's results in the case the algebra is semi-simple.

4. SEMI-SIMPLE LIE ALGEBRAS WHICH ACT PRIMITIVELY ON TWO-DIMENSIONAL MANIFOLDS

Let $\underset{\sim}{G}$ be a Lie algebra of vector fields on a two-dimensional manifold which acts primitively, i.e., leaves invariant (locally) no foliation.

Restrict attention to the non-singular points of the action of $\underset{\sim}{G}$, i.e., the points where the dimension of the orbits of $\underset{\sim}{G}$ are maximal.

Now, the orbits themselves form an invariant foliation, hence, $\underset{\sim}{G}$ must act transitively. Further, the linear isotropy algebra must be irreducible. (Otherwise, it would admit a one-dimensional invariant subspace, which would define a $\underset{\sim}{G}$-invariant, one-dimensional vector field system, which would <u>automatically</u> be integrable, hence would define an invariant foliation, which would contradict primitivity.)

Now, the only irreducible Lie algebras of linear maps on a two-dimensional real vector space are the following:

$SL(\underset{\sim}{2},\underset{\sim}{R})$, the 2×2 real matrices of trace zero. Its dimension is three.

$SO(\underset{\sim}{2},\underset{\sim}{R})$, the 2×2 skew-symmetric matrices. Its dimension is one.

$SO(\underset{\sim}{1},\underset{\sim}{1})$, again its dimension is one.

The algebras obtained by adding to these are the constant multiples of the identity

Hence, using Theorem 3.9 we have:

$$\dim \underset{\sim}{G} = 8,7,6, \text{ or } 5 \quad \text{if } \underset{\sim}{G}^2 \neq 0 \tag{4.1}$$

$$\dim \underset{\sim}{G} = 6,5, \text{ or } 3 \quad \text{if } \underset{\sim}{G}^2 = 0 \tag{4.2}$$

Let $\underset{\sim}{G}_c = \underset{\sim}{G} \otimes C$ be the complexification of $\underset{\sim}{G}$, i.e., the Lie algebra obtained by extending the ground field from the reals to the complexes.

The only possibilities for $\underset{\sim}{G}_c$ simple are:

$$\underset{\sim}{G}_c = SL(3, C) \quad \text{or} \quad SL(2, C) \quad .$$

(These are the only complex simple Lie algebras of dimension eight or less.) The only possibility for $\underset{\sim}{G}_c$ semi-simple is

$$\underset{\sim}{G}_c = SL(2, C) \oplus SL(2, C) = O(4, C) \quad .$$

Now, $\dim \underset{\sim}{G} = 3$ is incompatible with $\underset{\sim}{G}^2 \neq 0$. Hence,

$$\underset{\sim}{G}_c = SL(3, C) \quad \text{or} \quad O(4, C) \tag{4.3}$$

are the only possibilities of

$$G^2 \neq 0.$$

These groups correspond geometrically to projective and conformal geometry.

In case

$$\underset{\sim}{G}^2 = 0 \quad ,$$

one can show that $\underset{\sim}{G}^1$ is semi-simple and is the direct sum of a semi-simple ideal and a one-dimensional center. The results of Dynkin [1] on semi-simple subalgebras of the simple Lie algebras enable one now to finish the classification. I will not go into detail.

In case $\mathbf{G}^2 \neq 0$, proceed as follows. Write

$$\mathbf{G}^1 = \mathbf{R} \oplus \mathbf{S} \quad ,$$

where $\mathbf{R}$ is the radical of $\mathbf{G}^1$, and $\mathbf{S}$ is a maximal semi-simple subalgebra. It is clear that:

$$\mathbf{G}^2 \subset \mathbf{R} \quad .$$

Ad $\mathbf{S}$ acts irreducibly on

$$\mathbf{G}/\mathbf{G}^1 \quad \text{or} \quad \text{Ad } S = 0 \quad \text{on} \quad \mathbf{G}/\mathbf{G}^1 \quad .$$

The only possibility is then:

$$\dim \mathbf{S} = 3 \quad \text{or} \quad 0 \quad .$$

Hence,

$$\text{If} \quad \dim \mathbf{G} = 8, \quad \dim \mathbf{R} = 3$$

$$\text{If} \quad \dim \mathbf{G} = 6, \quad \mathbf{S} = 0$$

i.e., $\mathbf{G}^1$ is <u>solvable</u>.

In case $\dim \mathbf{G} = 6$, we can easily find $\mathbf{G}^1$ (hence also $\mathbf{G}$ as a Lie algebra of vector fields on X) by using Lie's theorem on solvable Lie algebras. Represent $\mathbf{G}$ by 4×4 complex orthogonal matrices, and use the fact that a asis can be chosen with respect to which $\mathbf{S}$ is in tri-ngular form. In fact, $\mathbf{S}_c$ is a <u>maximal</u> solvable subalgebra f $\mathbf{G}_0$, and Morosov's theorem (which is a generalization of ie's theorem) implies that $\mathbf{S}$ is uniquely determined.

Therefore, possibly after complexification of the variables, $\underset{\sim}{G}$ must be the Lie algebra of infinitesimal conformal transformations on R^2.

In case $\dim \underset{\sim}{G} = 8$, i.e., $\underset{\sim}{G}_c = SL(3,C)$, $\underset{\sim}{G}_c^1$ is also a maximal subalgebra, and Dynkin's classification might be applied. (Of course, that is using a cannon to kill a fly--it is easy to directly represent $\underset{\sim}{G}_c$ by 3×3 complex matrices of trace zero, and calculate what

$$G^1 = SL(2,C) \oplus \underset{\sim}{R}$$

might be.) The result is that G/G^1 is just $PS(C^3)$, the two-dimensional complex projective space.

In summary, we see that Lie's classification of Lie algebras $\underset{\sim}{G}$ of vector fields acting in R^2 may be readily reproduced using modern Lie algebra techniques in case $\underset{\sim}{G}$ is semi-simple, and acts primitively.

5. TRANSITIVE ACTIONS OF SEMI-SIMPLE LIE GROUPS

One can approach the problems discussed in the preceding sections in the following general way:

> Find the manifolds X which can be exhibited as coset spaces G/G^1, where G is a semi-simple Lie group and G^1 is a subgroup.

Now, the work of Dynkin [1] on the Lie subalgebras of semi-simple Lie algebras is, in a sense, a solution to this problem. However, it is not easy in practice to apply Dynkin's results in this way. In this section, I will present some general remarks about such transitive semi-simple group actions in terms of the theory of filtered Lie algebras.

For the sake of algebraic simplicity, I will work only with the case that X is a complex-analytic manifold, G is a complex semi-simple Lie group, and G^1 is a complex subgroup. $\underset{\sim}{G}$, its Lie algebra, is then a complex Lie algebra, and $\underset{\sim}{G}^1$ is a complex subalgebra. (As I have indicated in the preliminaries to Chapter D, a real transitive action can always be "complexified" in the natural way by letting the real analytic functions defining the group action take on complex values. Lie often, and implicitly, works in this way.)

$\underset{\sim}{G}^1$ defines a filtration of $\underset{\sim}{G}$:

$$\underset{\sim}{G} = \underset{\sim}{G}^0 \supset \underset{\sim}{G}^1 \supset \underset{\sim}{G}^2 \supset \cdots$$

Thus,

$$\underset{\sim}{G}^2 = \{A \in \underset{\sim}{G}^1 : [A, \underset{\sim}{G}] \subset \underset{\sim}{G}^1\}$$

$$\underset{\sim}{G}^3 = \{A \in \underset{\sim}{G}^1 : [A, [A, \underset{\sim}{G}] \subset \underset{\sim}{G}^1\}$$

and so forth.

This, of course, coincides with the "geometric" filtration obtained by regarding $\underset{\sim}{G}$ as a Lie algebra of vector fields on the coset space G/G^1.

Since $\underset{\sim}{G}$ is semi-simple we know from previous work that:

$$\underset{\sim}{G}^3 = 0$$

$$\dim \underset{\sim}{G} \leq n^2 + 2n \quad , \tag{5.1}$$

with $n = \dim X$.

<u>Remark</u>. Of course, here by "dimension" I mean the dimension of X as a complex analytic manifold, and the dimension of $\underset{\sim}{G}$ as a complex Lie algebra.

The elementary estimate 5.1 is surprisingly powerful in narrowing down the possibilities for $\underset{\sim}{G}$, at least in the case of low values of n. For example,

$$\text{If } n = 1,2,3, \text{ then} \quad \dim \underset{\sim}{G} \leq 3,8,15 \tag{5.2}$$

In turn, one can look at the classification of complex semi-simple Lie algebras and write down the possibilities for $\underset{\sim}{G}$:

$$\text{If } n = 1, \quad G = A_1 \equiv SL(2,C) = SO(3,C) = Sp(1,C)$$

$$\text{If } n = 2, \quad G = A_2 \equiv SL(3,C) \text{ or}$$

$$G = O_2 \equiv SO(4,C) = SO(3,C) \times SO(3,C)$$

If $n = 3$, $G = A_3 \equiv SL(4,C)$

$$G = B_2 \equiv SO(5,C)$$

$$G = G_2$$

plus the semi-simple possibilities.

(Here the equal sign between these groups means that they are <u>locally isomorphic</u>, or, alternatively, that their Lie algebras are isomorphic. Of course, the possibilities listed for $n = 1$ are also possibilities for $n = 2$, and so forth.)

<u>Case 1</u>. $\underset{\sim}{G}^1$ <u>is solvable</u>.

By the Lie-Morosov theorem, $\underset{\sim}{G}^1$ may then be transformed, within Ad G, to a subalgebra of a fixed maximal solvable subalgebra H of G. Further,

$$\dim(\underset{\sim}{G}/\underset{\sim}{H}) = \frac{1}{2}(\dim \underset{\sim}{G} - \text{rank } \underset{\sim}{G}) \qquad (5.3)$$

Hence, we have:

$$n \_ \frac{1}{2}(\dim \underset{\sim}{G} - \text{rank } \underset{\sim}{G}) \qquad (5.4)$$

If $n = 1$, we have equality in 5.4, i.e., $\underset{\sim}{G}^1$ is <u>maximal solvable</u>. If $n = 2$, $\dim \underset{\sim}{G} = 8$ or 6 or 3

rank = 2 or or 1 or 3

If $\dim \underset{\sim}{G} = 8$, identity 5.4 is impossible. If $\dim \underset{\sim}{G} = 6$, only equality is possible. Hence:

> The only two-dimensional manifold G/G^1, with $\underset{\sim}{G}$ semi-simple, $\underset{\sim}{G}^1$ solvable is: $G = SO(4,C)$, $\underset{\sim}{G}^1 =$ maximal solvable subalgebra.

As was already mentioned in the previous section, this possibility corresponds geometrically to conformal geometry.

> If $n = 3$ and G is simple,
> $\dim G = 15$ or 10, $\operatorname{rank} G = 3$ or 2

Neither possibility is compatible with inequality 5.4. Hence:

> A three-dimensional manifold is never the quotient G/G^1 of a simple group and a solvable subgroup.

Finally, the possibilities for G semisimple must be examined. Say, that:

$$G = G_1 \times \cdots \times G_m ,$$

with $G_1, \ldots, G_m$ simple. Let:

H_1 = maximal solvable subgroup of G_1

H_2 = maximal solvable subgroup of G_2

and so forth.

Then,

$$H_1 \times \cdots \times H_m$$

is a maximal solvable subgroup of G. Hence,

$$n \geq \dim (G_1/H_1) + \cdots + \dim (G_m/H_m)$$

Here are the possibilities for G semisimple:

If $n = 1$, G is simple, G^1 is maximal solvable.

If $n = 2$, $G = O(4,C)$, G^1 is maximal solvable.

If $n = 3$, $G = A_1 \times A$, or $A_1 \times A_1 \times A_1$
$G = O(4,C)$, $O(4,C) \times A_1$

<u>Case 2</u>. $\underset{\sim}{G}^1$ <u>is not solvable</u>, <u>but</u> $\underset{\sim}{G}^2 \neq 0$.

Then,

$$[\underset{\sim}{G}^2, \underset{\sim}{G}^1] \subset \underset{\sim}{G}^2$$

$$[\underset{\sim}{G}^2, \underset{\sim}{G}^2] = 0 \quad ,$$

i.e., $\underset{\sim}{G}^2$ is an abelian ideal of $\underset{\sim}{G}^1$. Let $\underset{\sim}{R}$ be the radical, i.e., <u>maximal solvable ideal</u>, of $\underset{\sim}{G}^1$, and $\underset{\sim}{S}$ a maximal semi-simple subalgebra. Then we have the <u>Levi-decomposition</u>, i.e.,

$$\underset{\sim}{G}^1 = \underset{\sim}{R} \oplus \underset{\sim}{S} \qquad \text{(as a vector space)}$$

Also,

$$\underset{\sim}{G}^2 \subset \underset{\sim}{R} .$$

Ad $\underset{\sim}{S}$, acting in $\underset{\sim}{G}/\underset{\sim}{G}^1$, is a faithful representation of $\underset{\sim}{S}$.

In previous sections we have worked out the possibilities for $n = 1,2$. (There are none for $n = 1$. For $n = 2$, only X = two-dimensional complex projective space, G = projective group.) Let us now do $n = 3$.

$\underset{\sim}{S}$ is semi-simple and admits a faithful three-dimensional representation. The only possibilities are:

$$S = SL(2,C) \quad \text{or} \quad SL(3,C)$$

Now, we can work out what the dimension of the radical $\underset{\sim}{R}$ must be, in the case where:

$$n = 3, \quad G \text{ is simple.}$$

Case a. $G = SL(4,C)$

$$\begin{aligned} 3 &= 15 - \dim S - \dim \underset{\sim}{R} \\ &= 15 - 3 - \dim \underset{\sim}{R} \quad \text{or} \quad 15 - 8 - \dim R \end{aligned}$$

Thus, $\dim \underset{\sim}{R} = 9$ or 4.

Now, Ad S acts in $\underset{\sim}{R}$, and preserves $\underset{\sim}{G}^2$, which is an abelian ideal of $\underset{\sim}{R}$ which is invariant under Ad $\underset{\sim}{S}$. I believe that all these facts can be put together readily to finally determine $\underset{\sim}{G}^1$, but I will not carry out the details here. (For example, the case $\dim \underset{\sim}{R} = 4$ corresponds to the case where $\underset{\sim}{G}/\underset{\sim}{G}^1$ = three-dimensional projective space.)

<u>Case b</u>. $G = SO(5,C)$

$$3 = 10 - \dim \underset{\sim}{S} - \dim \underset{\sim}{R}$$

$$= 10 - 3 - \dim \underset{\sim}{R} \quad \text{or} \quad 10 - 8 - \dim \underset{\sim}{R} .$$

Only:

$$\dim \underset{\sim}{R} = 4, \qquad S = SL(2,C)$$

is possible.

Again, I believe that working out these conditions will lead to the unique possibility that:

G/G^1 = space of conformal geometry in three variables.

The striking fact lying in the background here is the connection between semi-simple Lie group theory and the "classical" geometries.

6. CLASSIFICATION OF THE INFINITESIMAL ACTION OF THE ONE-VARIABLE AFFINE GROUP ON THE PLANE

It is important to notice that Lie put much more effort into the classification of intransitive group actions than into transitive ones. In the modern literature, this situation is reversed, probably because the intransitive situation is very messy and computational, while the classification of transitive actions lends itself to general statements. However, the intransitive case is much more important from the point of view of applications to the theory of differential equations.

In this paper, Lie classified intransitive actions by classifying foliations (i.e., differential equations) they left invariant. Before examining this viewpoint in detail, I want to consider the classification problem directly. In this section, we deal with the simplest non-abelian group, the group of affine transformations on R. (It is a two-dimensional solvable, but non-abelian group. All groups of this type are locally isomorphic.) In the next section we consider the simplest semi-simple group, namely SL(2,R).

Let $\underset{\sim}{G}$ be the two-dimensional real Lie algebra, spanned by two elements

$$A_1, A_2$$

such that:

$$[A_2, A_1] = A_1 \tag{6.1}$$

Let $\underset{\sim}{G}_1$ be the subalgebra spanned by A_1. It is the derived ideal of $\underset{\sim}{G}$. Since

$$[\underset{\sim}{G}_1, \underset{\sim}{G}_1] = 0 \ ,$$

$\underset{\sim}{G}$ is a solvable Lie algebra, and $\underset{\sim}{G}_1$ is the nilradical, i.e., the maximal nilpotent Lie ideal.

Suppose that $\underset{\sim}{G}$ is a Lie algebra of vector fields on a two-dimensional manifold X. If

$$A_1 \equiv 0 \quad \text{on} \quad X \ , \tag{6.2}$$

then $\underset{\sim}{G}/\underset{\sim}{G}_1$ acts on X. One-dimensional Lie algebras can be classified locally (at the non-singular points), so the classification of this situation can be considered as known.

Let us suppose then that 6.2 is not satisfied. Pick a point at which $A_1 \neq 0$, and a coordinate system (x,y) valid in a neighborhood of that point such that:

$$A_1 = \frac{\partial}{\partial x} \tag{6.3}$$

Since

$$(\text{Ad}\ A_1)^2(A_2) = 0 \tag{6.4}$$

the coefficients of A_2 must, as functions of x, be polynomials of degree at most one. Explicitly, A_2 must be of the following form:

$$A_2 = (a_0(y)+a_1(y)x)\frac{\partial}{\partial x} + (b_0(y)+b_1(y)x)\frac{\partial}{\partial y} \tag{6.5}$$

Hence, using 6.1,

$$[A_1,A_2] = a_1(y)\frac{\partial}{\partial x} + b_1(y)\frac{\partial}{\partial y}$$

$$= A_1 \equiv \frac{\partial}{\partial x} \quad ,$$

hence:

$$a_1 = 1, \qquad b_1 = 0 \quad ,$$

or:

$$A_2 \;=\; (a_0+x)\,\frac{\partial}{\partial x} + b_0(y)\,\frac{\partial}{\partial y} \tag{6.6}$$

We have made one "generaticity" assumption, which led to (6.3), namely that A_1 did not vanish at the point of X whose coordinates are $(0,0)$. Let us make another one, namely that

$$A_2(y)(0,0) \;\neq\; 0 \quad ,$$

i.e.,

$$b_0(y) \;\neq\; 0 \quad .$$

Then, we can suppose the coordinates (x,y) are chosen so that:

$$A_2 \;=\; (a_0(y)+x)\,\frac{\partial}{\partial x} + \frac{\partial}{\partial y} \tag{6.7}$$

This completely determines the action of $\underset{\sim}{G}$. It depends on "one arbitrary function $a_0(y)$ be of one variable".

Suppose now that we have two such actions, i.e., another pair (A_1',A_2') of vector fields which generate a Lie algebra $\underset{\sim}{G}'$ which is isomorphic to $\underset{\sim}{G}$, and which is transformable into $\underset{\sim}{G}$ by a local diffeomorphism. (This is what Lie means by "similar", or "ähnlich" in German. Precisely, we mean that there is an open subset $U \subset X$ of the point in question, and a diffeomorphism $\phi\colon U \to V$ such that

$$\phi_*(\underset{\sim}{G}') = \underset{\sim}{G} .$$

Thus, there are coordinates

$$(x',y')$$

such that A_1,A_2 take the following form:

$$A_1' = \frac{\partial}{\partial x'}$$

$$A_2' = (a_0'(y)+x')\frac{\partial}{\partial x'} + \frac{\partial}{\partial y'}$$

Further, we have real numbers $\lambda_1,\lambda_2,\lambda_3$ such that:

$$A_1 = \lambda_1 A_1'$$

$$A_2 = \lambda_2 A_2' + \lambda_3 A_1' .$$

Now,

$$\frac{\partial}{\partial x} = \frac{\partial x'}{\partial x}\frac{\partial}{\partial x'} + \frac{\partial y'}{\partial x}\frac{\partial}{\partial y'} ,$$

$$\frac{\partial}{\partial y} = \frac{\partial x'}{\partial y}\frac{\partial}{\partial x'} + \frac{\partial y'}{\partial y}\frac{\partial}{\partial y'}$$

hence,

$$A_1 = \frac{\partial}{\partial x} = \lambda_1\frac{\partial}{\partial x'} ,$$

which forces:

$$\frac{\partial y'}{\partial x} = 0$$

i.e., y' is a function $y'(y)$ of y alone

$$\frac{\partial x'}{\partial x} = \lambda, \quad \text{or} \quad x' = \lambda x + f(y)$$

$$(a_0+x)\lambda_1 \frac{\partial}{\partial x'} + \frac{df}{dy}\frac{\partial}{\partial x'} + \frac{dy'}{dy}\frac{\partial}{\partial y'} = \lambda_2\left((a_0'+x')\frac{\partial}{\partial x'} + \frac{\partial}{\partial y'}\right)$$

$$+ \lambda_3 \frac{\partial}{\partial x'}$$

Thus,

$$a_0 - f(y) + \frac{df}{dy} = \lambda_2 a_0' + \lambda_3 \qquad (6.8)$$

$$\lambda_2 = 1$$

$$\frac{dy'}{dy} = \lambda_2 \quad ,$$

or

$$y' = \lambda_2 y \quad . \qquad (6.9)$$

When 6.9 is substituted into 6.8, we see that 6.8 is a differential equation for $f(y)$, which can be solved (using the classical terminology) "by quadratures". Hence, we have proved:

> There is just one action (up to local similarity) of $\underset{\sim}{G}$ on R^2 which is "generic" in the sense described above.

This example is interesting methodologically. From the point of view of general Lie theory, it is somewhat trivial. For our generaticity assumption implies that

$$\underset{\sim}{G}(x) = X_x$$

at the chosen point x of X.

General Lie theory implies that there is but one action--left translation on the group manifold.

7. LIE'S CLASSIFICATION PROBLEMS AND LIE ALGEBRA COHOMOLOGY THEORY

There is an abstract, algebraic feature to Lie's work in this paper which is worthwhile investigating for two reasons: First, for the general insights it might give into Lie's specific, computational work, and, second, as a link to the modern theory of deformations of algebraic and geometric structures, and related Lie algebra cohomological ideas. (Some modern names here are Kodaira, Spencer, Gerstenhaber, Nijhuis, Richardson, Guillemin and Steinberg, Gelfand and Fuks. See the treatise by Kumpera and Spencer for full references. My papers "Analytic continuation of group representation" and "Linearization of Lie algebras of vector fields near invariant submanifolds" are also relevant.)

The general deformation theory deals with differentiable families of algebraic and geometric structures and their realizations. Usually, it attempts to classify them up to some sort of "equivalence". Typically, "cohomology" of the algebraic structure appears as one of the first invariants describing the "equivalence". It should be clear to the reader who has carried on this far that Lie's classification of various sorts of Lie algebras acting on one and two dimensional manifolds involves examples of this sort.

I will now attempt to explain some of these connections with deformation theory in a relatively simple-minded way.

Let us first restate Lie's fundamental problem.

Definition. Let $\underset{\sim}{G}$ be a Lie algebra of vector fields on a manifold X, and let $\underset{\sim}{G}'$ be a Lie algebra of vector fields on a manifold X'. Then, $(\underset{\sim}{G},X)$ and $(\underset{\sim}{G}',X')$ are locally equivalent (or equivalent in the sense of Lie) if the following condition is satisfied:

There are points $x \in X$, $x' \in X'$, and open neighborhoods

$$x \in U \subset X$$

$$x' \in U' \subset X' \quad ,$$

and a diffeomorphism

$$\phi : U \to U'$$

such that:

$$\phi_*(\underset{\sim}{G}) = \underset{\sim}{G}' .$$

This is, of course, nothing but a formalization of the idea that the vector fields of G may be converted into the vector fields in G' by a change of coordinates, i.e., that the two Lie algebras are "similar" ("ähnlich", in German).

There are now two distinct deformation problems which are related to Lie's problem. The first is:

Deform a Lie algebra homomorphism

$$\phi : \underset{\sim}{G} \to V(X)$$

of a given Lie algebra $\underset{\sim}{G}$ *into the Lie algebra of vector fields on a manifold* X.

The second is:

Deform Lie subalgebras of $V(X)$.

The first problem is simpler conceptually and computationally, hence will be the one considered here.

Let $\underset{\sim}{G}$ be a real Lie algebra, X a manifold,

$$\alpha: \underset{\sim}{G} \to V(X)$$

a Lie algebra homomorphism. Let λ be a real parameter, varying over

$$0 \leq \lambda \leq 1 \quad .$$

A <u>deformation</u> of ϕ is a one-parameter family

$$\lambda \to \alpha_\lambda: \underset{\sim}{G} \to V(X)$$

of Lie algebra homomorphisms, reducing to ϕ at $\lambda = 0$.

For $A \in \underset{\sim}{G}$, set:

$$\theta(A) = \frac{\partial}{\partial t} \alpha_\lambda(A)|_{\lambda=0} \tag{7.1}$$

The map

$$\theta: A \to \theta(A) \quad \text{of} \quad \underset{\sim}{G} \to V(X)$$

is R-linear. Let

$$\rho: \underset{\sim}{G} \to L(V(X))$$

be the representation of $\underset{\sim}{G}$ by linear maps in $V(X)$ defined as follows:

$$\rho(A)(B) = [\alpha(A), B] \tag{7.2}$$

$$\text{for} \quad A \in \underset{\sim}{G}, \quad B \in V(X)$$

Then, θ, defined by 7.1, is a one-cochain of $\underset{\sim}{G}$, with coefficients in the vector space (i.e., $V(X)$) in which the representation ρ takes place.

Now, we are <u>given</u> that, for each λ, ϕ_λ is a Lie algebra homomorphism. Thus,

$$\alpha_\lambda([A,B]) = [\alpha_\lambda(A), \alpha_\lambda(B)] \qquad (7.3)$$

for $A,B \in \underset{\sim}{G}$.

Differentiate both sides of 7.3 with respect to λ, and set $\lambda = 0$:

$$\theta([A,B]) = [\theta(A), \alpha(B)] + [\alpha(A), \theta(B)] \quad ,$$

or

$$\rho(A)(\theta(B)) - \rho(B)(\theta(A)) - \theta([A,B]) = 0 \quad (7.4)$$

This is seen to be the condition that θ defines a <u>one-cocycle of</u> $\underset{\sim}{G}$, <u>with coefficients in the representation</u> ρ.

What is the meaning in terms of Lie theory, of the condition that this one-cocycle be a coboundary? To see this, suppose that each of the Lie algebras

$$\alpha_\lambda(\underset{\sim}{G})$$

of vector fields are mutually Lie equivalent. Let us not worry about the local aspects, but suppose this equivalence is realized by a one-parameter family

$$\lambda \to \phi_\lambda : X \to X$$

of diffeomorphisms of X, i.e., by a flow. Let $A \in V(X)$ be the infinitesimal transformation corresponding to this flow.

We then have, as the definition of "Lie equivalence":

$$\phi_{\lambda *}(\alpha_0(\underset{\sim}{G})) = \alpha_\lambda(\underset{\sim}{G}) .$$

Then, for each $A \in \underset{\sim}{G}$, each λ, there is an element

$$\beta_\lambda(A) \in \underset{\sim}{G}$$

such that:

$$\phi_{\lambda *}(\alpha_0(B)) = \alpha_\lambda(\beta_\lambda(B)) \tag{7.5}$$

for all $B \in \underset{\sim}{G}$.

Differentiating 7.5 with respect to λ, and setting $\lambda = 0$, we have:

$$[\alpha_0(B),A] = \theta(B) + \alpha_0(\gamma(B)) \tag{7.6}$$

where

$$\gamma(B) = \frac{d}{d\lambda} \beta_\lambda(B)\Big|_{\lambda=0} \tag{7.7}$$

Now, it is readily seen from 7.5 that β_λ is an automorphism of the Lie algebra $\underset{\sim}{G}$. This implies that γ is a derivation of the Lie algebra $\underset{\sim}{G}$. "A" determines a zero-cocycle of the representation ρ, i.e., an element of the vector space in which ρ operates. Relation 7.6 then expresses the one-cocycle θ as the cohomology of "A" plus the one-cocycle

$$B \to \alpha_0(\gamma(B)) .$$

The **existence** of such an "A" and "γ" is then a **necessary** condition that the continuous family

$$\lambda \to \alpha_\lambda(\underset{\sim}{G})$$

of Lie algebras of vector fields be mutually Lie equivalent. In particular, the condition:

$$H^1(\underset{\sim}{G},\rho) = 0$$

is often indicative of the fact that $\underset{\sim}{G}$ can only act in essentially one Lie equivalent way on the manifold X.

Rather than go into an inevitably lengthly and complicated further discussion of the general deformation-cohomology theory, I prefer to calculate some of these cohomology groups in the typical examples considered by Lie in this paper.

8. COHOMOLOGY OF FINITE DIMENSIONAL LIE ALGEBRAS ACTING ON R.

Let $\underset{\sim}{G}$ be a finite dimensional Lie algebra of vector fields on a one-dimensional manifold X. Suppose the coordinate variable on X is x.

Let

$$\theta \colon \underset{\sim}{G} \to V(X)$$

be a one-cocycle of $\underset{\sim}{G}$. Then,

$$\theta(A) = f_A(x)\,\frac{d}{dx} \tag{8.1}$$

The cocycle condition, 7.4, is then:

$$\left[A, f_B \frac{d}{dx}\right] - \left[B, f_A \frac{d}{dx}\right] = f_{[A,B]} \frac{d}{dx}$$

for $A,B \in G$.

Suppose:

$$A = a_A \frac{d}{dx} .$$

Then, 8.1 takes the following form:

$$a_A \frac{df_B}{dx} - f_B \frac{da_A}{dx} - a_B \frac{df_A}{dx} + f_A \frac{df_B}{dx} = f_{[A,B]} \qquad (8.2)$$

We know that there are only three possibilities of $\underset{\sim}{G}$, one-dimensional abelian, two-dimensional solvable, and three-dimensional simple. Let us consider these possibilities:

Case 1. $\underset{\sim}{G}$ is two-dimensional.

$\underset{\sim}{G}$ is then spanned by two vector fields

$$A_1 = \frac{d}{dx}$$

$$A_2 = x \frac{d}{dx} ,$$

satisfying the following commutation relation:

$$[A_1, A_2] = A_1 . \qquad (8.3)$$

Set:

$$f_1 = f_A, \qquad f_2 = f_{A_2} \quad .$$

Then 8.2 takes the following form:

$$\frac{df_2}{dx} - x\frac{df_1}{dx} + f_1 = f_1 \quad ,$$

or

$$\frac{df_2}{dx} = x\frac{df_1}{dx} \qquad (8.4)$$

In summary, the vector space of one-cocycles of $\underset{\sim}{G}$ with coefficients in $V(X)$ is isomorphic to the pairs

$$(f_1, f_2)$$

of functions f on real variable x satisfying condition 8.4.

We shall now determine the subspace of one-cocycles which cobound. A zero-cochain is determined by a single vector field

$$C = h\frac{d}{dx} \quad .$$

Its coboundary θ is the one-cochain $\theta\colon \underset{\sim}{G} \to V(X)$ defined as follows

$$\theta(A) = [A,C] \quad .$$

Thus,

$$\theta(A) = \left(a_A \frac{dh}{dx} - h\frac{da_A}{dx}\right)$$

Suppose θ is defined by 8.1. Then, we have:

$$f_1 = \frac{dh}{dx}$$
$$f_2 = x\frac{dh}{dx} - h \qquad (8.5)$$

Thus, the existence of h satisfying 8.5 is the condition that the one-cocycle θ also cobound.

We shall now prove that such an h exists, if (f_1, f_2) satisfy condition 8.4. Suppose that h is chosen so that:

$$\frac{dh}{dx} = f_1 ,$$

i.e., h is an indefinite integral of f_1. Then,

$$\frac{d}{dx}\left(f_2 - x\frac{dh}{dx} + h\right) = \frac{df_2}{dx} - \frac{dh}{dx} - x\frac{df_1}{dx} + f_1 = 0$$

as a consequence of the cocycle condition 8.4. Then, we can adjust the choice of arbitrary constant in h to satisfy both conditions 8.5, i.e., the one-cocycle θ cobounds. Let us sum up this result as follows:

Theorem 8.1. Let $\underset{\sim}{G}$ be a two-dimensional solvable subalgebra of $V(R)$. Let

$$\rho: \underset{\sim}{G} \rightarrow V(R)$$

be defined by the following formula.

$$\rho(A)(B) = [A,B]$$

$$\text{for } A \in \underset{\sim}{G}, \quad B \in V(R) \quad .$$

Then, the first cohomology group of $\underset{\sim}{G}$, with coefficients determined by ρ, is zero. In symbols,

$$H^1(\underset{\sim}{G},\rho) = 0 \tag{8.6}$$

Case 2. $\underset{\sim}{G}$ is three-dimensional and simple

In addition to A_1, A_2, as in Case 1, $\underset{\sim}{G}$ contains a third vector field

$$A_3 = \frac{1}{2} x^2 \frac{d}{dx} \quad ,$$

satisfying the following commutation relations:

$$[A_1,A_3] = A_2$$

$$[A_2,A_3] = A_3$$

Let

$$\theta: \underset{\sim}{G} \to V(R)$$

be a one-cocycle. Then, θ restricted to the two-dimensional subalgebra spanned by A_1, A_2 is also a cocycle. Hence, there is a $C \in V(R)$ such that:

$$\theta(A_1) = [A_1,C]$$

$$\theta(A_2) = [A_2,C] \quad .$$

We will now examine

$$\theta(A_3) - [A_3,C] = D$$

Now,

$$[A_1,D] = [A_1,\theta(A_3)] - [A_1,[A_3,C]]$$

$=$ using the condition that θ is a one-cocycle on $\underset{\sim}{G}$,

$$[A_3,\theta(A_1)] - \theta[A_1,A_3]) - [[A_1,A_3],C]$$

$$- [A_3,[A_1,C]]$$

$$= [A_3,[A_1,C]] - \theta(A_2) - [A_2,C]$$

$$- [A_3,[A_1,C]]$$

$$= 0 \quad .$$

Hence, there is a constant c such that:

$$D = cA_1 \quad .$$

By a similar calculation,

$$[A_2,D] = 0 \quad .$$

Since $[A_2,A_1] \neq 0$, this forces

$$c = 0 \quad ,$$

hence,

$$D = 0 \quad ,$$

hence,

$$\theta(A_3) = [A_3, C]$$

We have proved that θ cobounds <u>as a one-cocycle on</u> $\underset{\sim}{G}$. This proves:

<u>Theorem 8.2</u>. If $\underset{\sim}{G}$ is the three-dimensional simple subalgebra of $V(R)$, and if ρ is the representation $: \underset{\sim}{G} \to L(V(R))$ defined above then,

$$H^1(\underset{\sim}{G}, \rho) = 0 .$$

<u>Remark</u>. Theorems 8.1 and 8.2 reflect the fact that there is but one Lie equivalence class of $\underset{\sim}{G}$ acting on R. In a sense, they are the "infinitesimal" versions of the fact that there is but one Lie equivalence class.

This example suggests a natural:

<u>Conjecture</u>. If $\underset{\sim}{G}$ is a finite (or certain types of infinite (?)) dimensional Lie algebra of vector fields on a manifold X, and if

$$\rho: \underset{\sim}{G} \to L(V(X))$$

is the linear representation defined as follows:

$$\rho(A)(B) = [A, B] ,$$

<u>and</u> if

$$H^1(\underset{\sim}{G}, \rho) = 0$$

then the Lie equivalence classes of $\underset{\sim}{G}$ acting on X form a discrete set.

In fact, in general deformation theory one is used to the vanishing of the "first obstruction cohomology group" implying the "rigidity" or "triviality" of the deformation. What makes this problem not of the standard type is that the vector space $V(X)$ is <u>infinite dimensional</u>. One can use standard techniques to prove that $H^1(\underset{\sim}{G},p) = 0$ implies the discreteness of the <u>formal power series Lie equivalence classes</u>. (See Chapter D.)

9. DEFORMATIONS OF LIE ALGEBRA HOMOMORPHISMS AND CLASSIFICATION OF LIE ALGEBRAS OF VECTOR FIELDS TANGENT TO FOLIATIONS

In my Comments to Section 7 I have briefly described relations between Lie algebra deformation theory and some of Lie's classification problems. In this section I plan to go into this in more detail.

Let X and Y be connected manifolds. Denote a typical point of X by x, a typical point of Y by y. Let:

$$Z = X \times Y \quad .$$

For $x \in X$, let $Y(x)$ be the following submanifold of Z:

$$Y(x) = \{(x,y)\} \quad .$$

$Y(x)$ is diffeomorphic to Y. Let

$$\pi : Z \to X$$

be the Cartesian projection map, i.e.,

$$\pi((x,y)) = x \quad .$$

Then, we have:

$$Y(x) = \pi^{-1}(x) \quad ,$$

i.e., $Y(x)$ is the *fiber* of π above the point x.

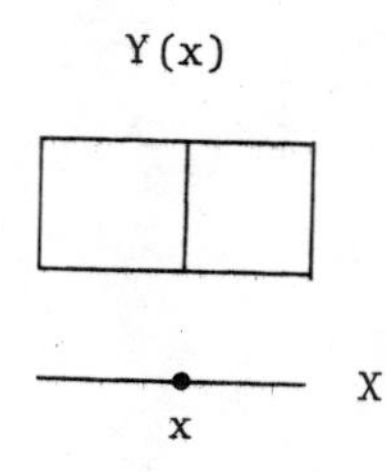

π is a submersion mapping, and its fibers determine a *foliation* V of Z. The vector fields $A \in V(Z)$ which lie in V are those which are first order differential operators on the variables of Y, with the *variables of X appearing as parameters in the coefficients*. For example, if

$$X = R = Y \quad ,$$

hen

$$V = \left\{ a(x,y) \frac{\partial}{\partial y} \right\} \quad .$$

Let $\underset{\sim}{G}$ be a <u>Lie algebra of vector fields on</u> Z <u>which is tangent to the fibers of this foliation</u>. This means that

$$\underset{\sim}{G} \subset V \quad .$$

For each $x \in X$, there is then a Lie algebra homomorphism

$$\phi_x : \underset{\sim}{G} \to V(Y)$$

such that:

$$A(x,y) = \phi_x(A)(y) \tag{9.1}$$

$$\text{for} \quad A \in \underset{\sim}{G}, \quad x \in X, \quad y \in Y \quad .$$

<u>Remark</u>. To interpret the right hand side of 9.1 as a tangent vector to $(x,y) \in Z$, identify

$$Z_{(x,y)} \quad \text{with} \quad X_x \oplus Y_y$$

in the obvious way.

Now, the assignment

$$x \to \phi_x$$

defines a typical "deformation" of an algebraic structure--in this case a Lie algebra homomorphism. We know that their "classification" involves certain algebraic techniques, such as Lie algebra cohomology theory. Let us examine the relation to the classification <u>in the sense of deformation theory</u> and in the sense of Lie.

Let $\underset{\sim}{G}'$ be another Lie algebra of vector fields on Z which is tangent to the fibers of V. Here is the definition of "equivalence" that Lie always uses:

<u>Definition</u>. $\underset{\sim}{G}$ is (globally) equivalent to $\underset{\sim}{G}'$ (in the sense of Lie) if there is a diffeomorphism

$$\alpha: Z \to Z$$

and a Lie algebra isomorphism

$$\beta: \underset{\sim}{G} \to \underset{\sim}{G}'$$

such that:

$$\alpha_*(A) = \beta(A) \tag{9.2}$$

for all $A \in G$.

<u>Definition</u>. A diffeomorphism

$$\alpha: Z \to Z$$

is an <u>automorphism</u> of the foliation V if α preserves the leaves of V.

We can determine such α's by analytical conditions. Since the points of X parameterize the leaves of V, there is a diffeomorphism:

$$\alpha^X: X \to X$$

such that:

$$\alpha(Y_x) = Y_{\alpha^X(x)} \tag{9.3}$$

For each $x \in X$, there is then a diffeomorphism

$$\alpha_x \colon Y \to Y$$

such that:

$$\alpha(x,y) = (\alpha^X(x), \alpha_x(y)) \tag{9.4}$$

We see that α is determined uniquely by the pair

$$(\alpha^X, x \to \alpha_x)$$

of the diffeomorphism $\alpha^X \colon X \to X$ and the mapping $x \to \alpha_x$ of X

$$X \to (\mathrm{DIFF}\ (Y)) \quad .$$

Remark. Physicists often give the name gauge transformations to objects like $x \to \alpha_x$.

Suppose now that $\underset{\sim}{G}, \underset{\sim}{G}'$ are two Lie algebras of vector fields on Z which are tangent to the foliation V. Suppose that $\underset{\sim}{G}$ is determined by a map

$$x \to \phi_x$$

of $X \to \mathrm{Hom}\ (\underset{\sim}{G}, V(Y))$ and $\underset{\sim}{G}'$ is determined by a map

$$x \to \phi'_x$$

of $X \to \mathrm{Hom}\ (\underset{\sim}{G}', V(Y))$. Suppose also that $\underset{\sim}{G}, \underset{\sim}{G}'$ are equivalent, in Lie's sense, via a pair (α, β), with:

$$\beta \colon \underset{\sim}{G} \to \underset{\sim}{G}'$$

a Lie algebra isomorphism

$$\alpha: Z \to Z$$

a diffeomorphism which is an isomorphism of the foliation, described analytically by

$$(\alpha^X, x \to \alpha_x)$$

as suggested in the preceding discussion. We shall work out the conditions this implies on $x \to \phi_x$, $x \to \phi'_x$.

For $A \in \underset{\sim}{G}$, $(x,y) \in Z$,

$$A(x,y) = \phi_x(A)(y) \quad .$$

Hence,

$$\begin{aligned} \alpha_*(A(x,y)) &= (\alpha_x)_*(\phi_*(A)(y)) \\ &= \beta(A)(\alpha^X(x),\alpha_x(y)) \\ &= \phi'_{\alpha^X(x)}(\beta(A))(\alpha_x(y)) \\ &= (\alpha_x)_*(\alpha_x^{-1})_*(\phi'_{\alpha^X(x)}(\beta(A))(y)) \end{aligned}$$

This relation implies the following one:

$$\phi_x(A) = (\alpha_x^{-1})_*(\phi'_{\alpha^X(x)}(\beta(A))) \quad ,$$

or

$$\phi_x = (\alpha_x^{-1})_*\phi'_{\alpha^X(x)}\beta \tag{9.5}$$

for all $x \in X$.

This is the natural notion of equivalence, in terms of deformation theory, for maps

$$X \to \text{Hom}\,(\underset{\sim}{G}, V(Y))$$

All these steps are reversible, showing that:

For the class of vector fields which are tangent to the fibers of the foliation V, *Lie's notion of equivalence coincides with the natural definition of deformation theory when one restricts the diffeomorphism* α *in Lie's definition to be those which preserve the foliation* V.

In the "transitive case", one can supplement this by showing that Lie's equivalence *coincides* with the deformation-theory one, without restriction on the diffeomorphism α.

Definition. A Lie algebra $\underset{\sim}{G}$ of vector fields on Z which is tangent to the fibers of the foliation V is said to be *transitive on fibers of* V if, for each $(x,y) \in Z \equiv X \times Y$, the values

$$\underset{\sim}{G}(x,y)$$

of the vector fields of $\underset{\sim}{G}$ at (x,y) fill up the tangent space to the leaf $Y(x)$, i.e., *if the smallest* $F(X)$-*submodule of* $V(Z)$ *containing* $\underset{\sim}{G}$ *is* V *itself*.

Theorem 9.1. Let $\underset{\sim}{G}, \underset{\sim}{G}'$ be Lie algebras of vector fields on Z which are transitive on the fibers of V. Suppose $\underset{\sim}{G}$

and $\underset{\sim}{G}'$ are equivalent, in the sense of Lie, via a diffeomorphism

$$\alpha: Z \to Z \quad .$$

Then, α preserves the foliation V.

<u>Proof</u>. It should be geometrically obvious. α must satisfy

$$\alpha_*(\underset{\sim}{G}(z)) = \underset{\sim}{G}'(\alpha(z))$$

$$\text{for all} \quad z \in Z \quad .$$

By definition of "transitive" $\underset{\sim}{G}(z)$ and $\underset{\sim}{G}'(\alpha(z))$ fill up the tangent spaces to the leaves of V at z and $\alpha(z)$. Hence, α maps a leaf of V onto another leaf, which is what one means by "α preserving the foliation V."

<u>Remark</u>. Already, we have clues to why Lie obtained the results listed in Theorem 14.4.1. In this case,

$$X = Y = R \quad .$$

If $\underset{\sim}{H}$ is a two- or three-dimensional Lie algebra of vector fields on $Y \equiv R$, we know that:

$$H^1(\underset{\sim}{H}, \rho) = 0 \quad ,$$

where

$$\rho(A)(B) = [A,B]$$

$$\text{for} \quad A \in \underset{\sim}{H}, \quad B \in V(Y)$$

Let

$$x \to \phi_x$$

be a mapping $X \to \text{Hom}(\underset{\sim}{G}, V(Y))$. Assume that the situation is "generic" in the sense that:

For all $x \in X$, $\phi_x(\underset{\sim}{G})$ are isomorphic to a Lie algebra $\underset{\sim}{H}$.

(Lie implicitly makes this assumption. If it is not satisfied, one must proceed to use a more general deformation-of-Lie-subalgebra theory.)

Now, if $\underset{\sim}{H}$ is semi-simple, the homomorphism $\phi_x\colon \underset{\sim}{G} \to \phi_x(\underset{\sim}{G})$ "splits", by the Levi-Malcev theorem. (Basically, this involves the J. H. C. Whitehead theorems on vanishing of the first and second cohomology groups of a semi-simple Lie algebra.) In other words, there are Lie subalgebras

$$\underset{\sim}{S}_x, \underset{\sim}{K}_x \subset \underset{\sim}{G}$$

such that:

$$\underset{\sim}{G} = \underset{\sim}{S}_x \oplus \underset{\sim}{K}_x$$

$$\underset{\sim}{K}_x = \text{kernel } \phi_x$$

$$\underset{\sim}{S}_x \text{ is isomorphic to } \underset{\sim}{H}.$$

Now, the subalgebra system $\{\underset{\sim}{S}_x\}$ is "rigid", because $\underset{\sim}{H}$ is semi-simple. This means that we can arrange (by suitable choice of the isomorphism β) that

$\underset{\sim}{S}_x$ is independent of x.

In fact, we can suppose that:

$\underset{\sim}{S}_x$ is a <u>maximal</u> semi-simple subalgebra of $\underset{\sim}{G}$.

By the Levi-Malcev theorem,

$$\underset{\sim}{K}_x = \text{Radical of } \underset{\sim}{G} \equiv \text{maximal solvable ideal.}$$

In particular,

$\{\underset{\sim}{K}_x\}$ <u>is independent of</u> x.

Now, we are <u>given</u> that the map

$$A \to (0, \phi_x(A))$$

of $\underset{\sim}{G} \to V(Z)$ is <u>one-one</u>. (This is built into Lie's problem--classify "r-term groups".) Hence,

$$\phi_x(A) = 0$$

for <u>all</u> $x \in X$, $A \in \underset{\sim}{K}_x$.

This forces:

$$\underset{\sim}{K}_x = 0 \ .$$

This is what is indicated in the third box of the results of Theorem 14.4.1. When $\underset{\sim}{G}$ is semi-simple, it can act (up to Lie's equivalence) in only <u>one</u> way on R^2, in such a way that it is tangent to (and transitive on the leaves of) a one-dimensional foliation of R^2.

In case $\underset{\sim}{G}$ is not semi-simple, (but is finite dimensional) a decomposition can be made on the basis of the Levi-Malcev theorem. In the case

$$\dim X = \dim Y = 1 ,$$

one sees readily that $\underset{\sim}{G}$ must be <u>solvable</u>. $\underset{\sim}{K}_x$, the kernel of ϕ_x, is then an ideal such that:

$$\dim (\underset{\sim}{G}/\underset{\sim}{K}_x) = 1 \text{ or } 2 .$$

Again, this situation may be analyzed and classified using Lie algebra cohomology. Presumably, the results obtained by Lie and listed in the first two boxes of Theorem 14.4.1 may be obtained in this way.

It would be very interesting to consider generalizations to higher dimensional situations. I have written enough to show the reader that the techniques of deformation theory are available to develop these higher dimensional generalizations of Lie's work. A key problem then is the following:

> Given a manifold Y, and a Lie algebra $\underset{\sim}{H}$ of vector fields on Y, let $\rho: H \to L(Y)$ be the linear map such that $\rho(A)(B) = [A,B]$. Compute $H^1(\underset{\sim}{H},\rho)$.

We have seen that $H^1(\underset{\sim}{H},\rho) = 0$ if $\dim \underset{\sim}{H} = 1$ or 2, and $\dim Y = 1$. This plays a key qualitative role in Theorem 14.4.1.

Remark. I believe that the "unitary trick" developed by H. Weyl enables one to prove that

$$H^1(\underset{\sim}{H},\rho) = 0$$

if $\underset{\sim}{H}$ is a semi-simple, real analytic Lie algebra of vector fields on a general manifold Y. Alternately, this might be provable purely algebraically by Invariant Theory, at least in certain common situations.

It might be interesting to develop an algebraic-geometric setting for these ideas. X and Y could be algebraic varieties (say, with the complex numbers as scalar field). V(Y) could be the vector fields with rational functions as coefficients. The "inner automorphisms" of V(Y) could be those defined by birational transformations of Y. This set-up leads to many interesting problems on the frontier between algebra, geometry and analysis!

10. SOME GENERAL FACTS ABOUT NON-SEMI-SIMPLE LIE ALGEBRAS OF VECTOR FIELDS

Here are some introductory remarks. We have seen that there are various general and modern techniques for studying semi-simple Lie algebras of vector fields. These results reproduce--with much less computational effort and greater geometrical and algebraic insight--a part of Lie's results. However, the greater part of his list of possibilities of Lie algebras in the plane is concerned with the non-semi-simple ones. Accordingly, I will now attempt to develop general tools to deal with this possibility. First, some results which hold for arbitrary finite dimensional Lie algebras of vector fields.

Theorem 10.1. Let $\underset{\sim}{G} \subset V(X)$ be a finite dimensional Lie algebra of vector fields on a connected manifold X which acts transitively at each point of X. Let x be a point of X, and let $\underset{\sim}{G}^1$ be the stability subalgebra of $\underset{\sim}{G}$ at x. Then, $\underset{\sim}{G}^1$ contains no non-zero ideal of $\underset{\sim}{G}$.

Proof. Suppose otherwise, i.e.

$$\underset{\sim}{H} \subset \underset{\sim}{G}^1 \subset \underset{\sim}{G} \quad ,$$

with

$$[\underset{\sim}{G},\underset{\sim}{H}] \subset \underset{\sim}{H} \quad .$$

By the definition of $\underset{\sim}{G}^1$ as stability subalgebra,

$$\underset{\sim}{H}(x) = 0 \quad .$$

Let A be an arbitrary element of $\underset{\sim}{G}$, and let

$$t \to x(t)$$

be the orbit curve of A which begins at x. Let $t \to g(t)$ be the one-parameter group of diffeomorphisms generated by A. Then,

$$x(t) = g(t)(x)$$

Hence,

$$\underset{\sim}{H}(x(t)) = g(t)_*(\exp\,(\mathrm{Ad}(-At))(\underset{\sim}{H})(x))$$

$$= g(t)_*(\underset{\sim}{H}(x) - t[A,\underset{\sim}{H}](x) + \cdots)$$

$$= 0 \quad ,$$

since $\underset{\sim}{H}$ is an ideal.

Since $\underset{\sim}{G}$ acts transitively at x (i.e., $\underset{\sim}{G}(x) = X_x$), we see that the orbits of elements of $\underset{\sim}{G}$ fill up an open neighborhood of x, hence $\underset{\sim}{H}$ vanishes in a neighborhood of x.

Using the hypothesis that $\underset{\sim}{G}$ acts transitively at each point of X, we see that the set of all points of X at which $\underset{\sim}{H}$ vanishes is open. It is also closed, by continuity. Since X is connected, it equals X, i.e., $\underset{\sim}{H} \equiv 0$.

This standard result relating transitivity and an algebraic property of the stability subalgebra suggests the following general:

Definition. Let $\underset{\sim}{G} = \underset{\sim}{G}^0 \supset \underset{\sim}{G}^1 \supset \cdots$ be a filtered Lie algebra. It is said to be an effective filtered Lie algebra if $\underset{\sim}{G}^1$ contains no non-zero ideal of $\underset{\sim}{G}$.

Here is a geometric property of Lie algebras of vector fields that is often useful. (For example, it played an important role in the classification of vector fields on R.)

<u>Theorem 10.2</u>. Let $\underset{\sim}{G}$ be a Lie algebra of vector fields on a manifold X which has the property that no non-zero element of $\underset{\sim}{G}$ vanishes in an open subset of X. Let x be a point of X and let $\underset{\sim}{A}$ be an abelian Lie subalgebra of $\underset{\sim}{G}$ such that

$$\underset{\sim}{A}(x) = X_x .$$

Then, $\underset{\sim}{A}$ is a maximal abelian subalgebra of $\underset{\sim}{G}$.

<u>Proof</u>. This result is basically a continuation of the development begun in Section 2.

Suppose $\underset{\sim}{A}$ were <u>not</u> maximal abelian in $\underset{\sim}{G}$. Let B be a non-zero element of $\underset{\sim}{G}$ such that:

$$[B,\underset{\sim}{A}] = 0$$

By the argument given in Section 2, B is linearly dependent on the elements of A in some neighborhood of x. Our hypothesis that no non-zero element of $\underset{\sim}{G}$ vanishes in an open subset implies that $B \in \underset{\sim}{A}$.

<u>Theorem 10.3</u>. Let $\underset{\sim}{G}$ be a finite dimensional Lie algebra of vector fields on a connected manifold X which acts

transitively at each point of X. Then, X admits a real analytic manifold structure such that the elements of $\underset{\sim}{G}$ are real-analytic vector fields. In particular, if an element of $\underset{\sim}{G}$ vanishes in an open subset of X it vanishes identically.

<u>Proof</u>. To give a real analytic structure means that one picks out a set of C^∞ functions in a neighborhood of each point of X which provide local coordinate systems and which transform in a real-analytic way in the overlap of two such neighborhoods.

In this situation, let us say that a function

$$f: U \to R$$

defined in an open set $U \subset X$ is <u>real analytic</u>, if, for each orbit curve

$$t \to x(t)$$

of G, the function

$$t \to f(x(t))$$

is real analytic.

The details are left to the reader, but it is routine to show that this class provides a real-analytic structure for X, such that the vector fields in $\underset{\sim}{G}$ are themselves real analytic.

With these simple, general results at hand, let us turn to the study of non-semi-simple Lie algebras of vector fields. Let X be a connected manifold, and let $\underset{\sim}{G}$ be a finite dimensional Lie algebra of vector fields on X such that no non-zero element of $\underset{\sim}{G}$ vanishes in an open subset of X.

To say that $\underset{\sim}{G}$ is non-semi-simple is to say that $\underset{\sim}{G}$ has a non-zero abelian Lie ideal $\underset{\sim}{A}$.

<u>Theorem 10.4</u>. Suppose that x is a point of X such that:

$$\underset{\sim}{A}(x) = X_x \quad . \tag{10.1}$$

Let $\underset{\sim}{G}^1$ be the stability subalgebra of $\underset{\sim}{G}$, and let $\underset{\sim}{G}^2$ be the set of vector fields which vanish to the second order at x. Then,

$$\underset{\sim}{G} = \underset{\sim}{G}^1 \oplus \underset{\sim}{A} \tag{10.2}$$

and

$$\underset{\sim}{G}^2 = 0 \tag{10.3}$$

<u>Proof</u>. We know (see Section 2) that 10.1 implies that the assignment

$$A \to A(x)$$

of $\underset{\sim}{A} \to X_x$ is a vector space isomorphism. 10.2 follows from linear algebra since $\underset{\sim}{G}^1$ is the kernel of the evaluation map

$\underset{\sim}{G} \to X_x$. (The $\oplus$ sign in 10.2 means vector space direct sum, not Lie algebra direct sum. Of course, $\underset{\sim}{G}$ is a semidirect sum of the Lie ideal $\underset{\sim}{G}^1$ and the Lie algebra $\underset{\sim}{A}$)

To prove 10.3, let

$$\rho: \underset{\sim}{G} \to L(X_x)$$

be the linear stability representation. We know that $\underset{\sim}{G}^2$ is the kernel of ρ.

Suppose $\underset{\sim}{G}^2$ were non-zero. Let $B \in \underset{\sim}{G}^2$ be a non-zero element. Then, $\rho(B) = 0$ means that:

$$[B,\underset{\sim}{A}] \subset \underset{\sim}{G}^1 \quad .$$

But, $\underset{\sim}{A}$ is an ideal, hence:

$$[B,\underset{\sim}{A}] \subset \underset{\sim}{A} \quad .$$

This forces

$$[B,\underset{\sim}{A}] = 0 \quad .$$

Theorem 10.2 implies that

$$B \in \underset{\sim}{A} \quad ,$$

hence,

$$B = 0 \quad ,$$

which finishes the proof of relation 10.3.

We can now polish off the classification of the primitive non-semi-simple vector field Lie algebras.

Theorem 10.5. Suppose that $\underset{\sim}{G}$ acts *primitively* on X, in the sense that $\underset{\sim}{G}$ acting in each open subset of X leaves invariant no non-singular foliation. Suppose that $\underset{\sim}{G}$ is not semisimple. Then, $\underset{\sim}{G}$ is a semidirect sum:

$$\underset{\sim}{G} = \underset{\sim}{H} \oplus \underset{\sim}{A} \ ,$$

$$[\underset{\sim}{H}, \underset{\sim}{A}] \subset \underset{\sim}{A}$$

$$[\underset{\sim}{H}, \underset{\sim}{H}] \subset \underset{\sim}{H}$$

$$[\underset{\sim}{A}, \underset{\sim}{A}] = 0$$

of a subalgebra $\underset{\sim}{H}$ and a abelian ideal $\underset{\sim}{A}$. $\underset{\sim}{H}$ is the isotropy subalgebra of $\underset{\sim}{G}$ at a point of X. The linear isotropy representation of it is equivalent to Ad $\underset{\sim}{H}$ acting in $\underset{\sim}{A}$, which is *irreducible* and *faithful*. In particular, $\underset{\sim}{H}$ is a *reductive* Lie algebra, and is either semisimple or is the direct sum of a semisimple ideal and a one-dimensional ideal.

Proof. Let X' be the following subset of X:

$$X' = \{x' \varepsilon X: \dim (\underset{\sim}{G}(x')) \geq \dim G(x) \quad \text{for all} \quad x \varepsilon X\} \ .$$

In words, X' is the set of *maximal points* of $\underset{\sim}{G}$, where the maximal number of the vector fields in $\underset{\sim}{G}$ are linearly independent) X' is an *open* subset of X.

The vector fields of $\underset{\sim}{G}$ then define a non-singular foliation of X'. Clearly, this foliation is invariant under $\underset{\sim}{G}$ itself. Since $\underset{\sim}{G}$ is, by hypothesis, primitive, the leaves of this foliation must be open subsets, i.e., $\underset{\sim}{G}$ acts transitively on X'.

Let $\underset{\sim}{A}$ be any abelian ideal of $\underset{\sim}{G}$. Let X" be the set of maximal points of $\underset{\sim}{A}$ acting in X'. Again, X" is an open subset of X, and the vector fields of $\underset{\sim}{A}$ define a foliation of X" which is invariant under $\underset{\sim}{G}$. By primitivity again, $\underset{\sim}{A}$ must be transitive on X". Choose x to be a point of X". Theorem 10.4 applies and the conclusions of Theorem 10.5 follow readily.

In case

$$\dim X = 2 \quad ,$$

we have seen that the cases $\underset{\sim}{G}$ semi-simple, acting primitively, are readily classified. These results, together with Theorem 10.5, then complete the classification of <u>primitive</u> Lie algebras acting in the plane.

Now, let us go to work on the imprimitive ones. Let $\underset{\sim}{G}$ be a non-semi-simple Lie algebra of vector fields acting on a manifold X. Let $\underset{\sim}{A}$ be a <u>maximal abelain ideal</u> of $\underset{\sim}{G}$. Set:

$$X' = \text{set of maximal points of } \underset{\sim}{A}.$$

X' is an open subset of X.

<u>Theorem 10.6</u>. Suppose that there is a point $x \in X'$ such that:

$$\underset{\sim}{A}(x) = X_x .$$

Let $\underset{\sim}{G}^1$ be the isotropy subalgebra of $\underset{\sim}{G}$ at x. Then,

$$\underset{\sim}{G}^2 = 0 ,$$

and $\underset{\sim}{G}$ is the semidirect sum of $\underset{\sim}{G}^1$ and $\underset{\sim}{A}$. In the neighborhood of x, X and $\underset{\sim}{G}$ acting on X are determined (up to Lie equivalence) by the Lie algebra $\underset{\sim}{G}^1$, and the linear isotropy representation on the vector space $\underset{\sim}{A}$.

<u>Proof</u>. The methods are the same as those used to prove Theorem 10.5. It is known from Section 2 that $\underset{\sim}{A}$ is a maximal abelian subalgebra of $\underset{\sim}{G}$, and that

$$\underset{\sim}{A} \cap \underset{\sim}{G}^1 = (0) ,$$

which implies $\underset{\sim}{G}^2 = 0$, and

$$\underset{\sim}{G} = \underset{\sim}{G}^1 \oplus \underset{\sim}{A} .$$

The only real complication in studying non semi-simple Lie algebras of vector fields on the manifold X comes when the orbits of $\underset{\sim}{A}$ generate a foliation of X.

In the next section we study this geometric situation in more detail.

11. FOLIATIONS LEFT INVARIANT BY LIE ALGEBRAS OF VECTOR FIELDS WHICH HAVE SUBALGEBRAS WHICH ACT TRANSITIVELY ON THE FIBERS

Here is the geometric situation to be studied in this section. Z and X are manifolds

$$\pi: Z \to X$$

is a submersion mapping, whose fibers are the leaves of a foliation of Z.

$\underset{\sim}{G}$ is a finite dimensional Lie algebra of vector fields on Z which leaves invariant this foliation. This is equivalent to requiring that each vector field $A \in \underset{\sim}{G}$ is projectable under π, i.e.,

$$\pi_*(\underset{\sim}{G})$$

is a Lie algebra of vector fields on X.

$$\pi_*: \underset{\sim}{G} \to \pi_*(\underset{\sim}{G})$$

is a Lie algebra homomorphism. Let $\underset{\sim}{H}$ be its kernel, a Lie ideal of $\underset{\sim}{G}$. $\underset{\sim}{H}$ consists of the vector fields in $\underset{\sim}{G}$ which are <u>tangent</u> to the fibers of π.

Let us now make the following assumptions:

$$\underset{\sim}{G} \text{ acts transitively on } Z \tag{11.1}$$

$$\underset{\sim}{H} \text{ acts transitively on the fibers of } \pi \tag{11.2}$$

$$\pi_*(\underset{\sim}{G}) \text{ acts transitively on } X \tag{11.3}$$

Here is a useful, purely algebraic fact.

<u>Theorem 11.1</u>. Let $\underset{\sim}{G}$ be a finite dimensional Lie algebra, and let $\underset{\sim}{S}$ be a <u>semi-simple</u> Lie ideal. Then, there is another Lie ideal $\underset{\sim}{G}'$ of $\underset{\sim}{G}$ such that:

$$\underset{\sim}{G} = \underset{\sim}{G}' \oplus \underset{\sim}{S}, \qquad [\underset{\sim}{G}', \underset{\sim}{S}] = 0 ,$$

i.e., $\underset{\sim}{G}'$ is the direct sum of $\underset{\sim}{G}'$ and $\underset{\sim}{S}$.

<u>Proof</u>. Let $\underset{\sim}{R}$ be the radical of $\underset{\sim}{G}$, i.e., the maximal solvable ideal. By the Levi-Malcev theorem there is a semi-simple Lie subalgebra $\underset{\sim}{S}'$ such that:

$$\underset{\sim}{G} = \underset{\sim}{R} \oplus \underset{\sim}{S}' ,$$

$$[\underset{\sim}{R}, \underset{\sim}{S}'] \subset \underset{\sim}{S}'$$

$$\underset{\sim}{S} \subset \underset{\sim}{S}' .$$

In particular, $\underset{\sim}{S}$ is an ideal of $\underset{\sim}{S}'$. Using the Killing form, one can write S' as a direct sum of $\underset{\sim}{S}$ and an ideal $\underset{\sim}{S}''$. Since $\underset{\sim}{S}$ and $\underset{\sim}{R}$ are both ideals of $\underset{\sim}{G}$,

$$[\underset{\sim}{R}, \underset{\sim}{S}] \subset \underset{\sim}{S}$$

and

$$[\underset{\sim}{R}, \underset{\sim}{S}] \subset \underset{\sim}{R}$$

These identities force:

$$[\underset{\sim}{R},\underset{\sim}{S}] = 0 .$$

Since $\underset{\sim}{S}'$ is a direct sum of $\underset{\sim}{S}$ and $\underset{\sim}{S}''$,

$$[\underset{\sim}{S},\underset{\sim}{S}''] = 0 .$$

Hence,

$\underset{\sim}{G}$ is the direct sum of $\underset{\sim}{R} + \underset{\sim}{S}''$ and $\underset{\sim}{S}$.

$\underset{\sim}{R} + \underset{\sim}{S}''$ is an ideal of $\underset{\sim}{G}$. If we define $\underset{\sim}{G}'$ to be $\underset{\sim}{R} + \underset{\sim}{S}''$, Theorem 11.1 is proved.

Now, return to the case where $(\underset{\sim}{G},\underset{\sim}{H},\pi_*(\underset{\sim}{G}))$ satisfy 11.1-11.3.

<u>Theorem 11.2</u>. Suppose that $\underset{\sim}{H}$ is semi-simple. Let $\underset{\sim}{G}'$ be the Lie ideal of $\underset{\sim}{G}$ provided by Theorem 11.1, such that:

$$\underset{\sim}{G} = \underset{\sim}{G}' \oplus \underset{\sim}{H} ,$$

Let $\underset{\sim}{H}^1$ be the Lie subalgebra of $\underset{\sim}{H}$ consisting of the elements of $\underset{\sim}{H}$ which vanish at one point of Z. Suppose the following condition is satisfied:

> There is no non-zero element $A \in \underset{\sim}{H}$ such that
>
> $$[\underset{\sim}{A},\underset{\sim}{H}^1] \subset \underset{\sim}{H}^1 , \qquad (11.4)$$
>
> i.e., H^1 is its own normalizer in $\underset{\sim}{H}$

Then, $\underset{\sim}{G}$ acting on Z is Lie equivalent to the product of $\pi_*(\underset{\sim}{G}')$ acting (transitively) on X and $\underset{\sim}{G}'$ acting transitively on a manifold Y.

<u>Proof</u>. Let z be a point of Z. We shall prove that hypothesis 11.4 guarantees that:

$$\underset{\sim}{G}'(z) \cap \underset{\sim}{H}(z) = 0 \tag{11.5}$$

If, in fact, 11.5 is <u>not</u> satisfied, there is an element $A \in \underset{\sim}{G}'$ and an element $B \in H$ such that:

$$A - B \in \underset{\sim}{G}^1 , \quad \text{but} \quad B(z) \neq 0$$

Now,

$$[\underset{\sim}{H}^1, \underset{\sim}{G}^1] \subset \underset{\sim}{H}^1 ,$$

since $\underset{\sim}{H}$ is an ideal of $\underset{\sim}{G}$. Thus,

$$[A-B, \underset{\sim}{H}^1] \subset \underset{\sim}{H}^1 .$$

But, since $[B, \underset{\sim}{H}] = 0$,

$$[A-B, \underset{\sim}{H}^1] = [B, \underset{\sim}{H}^1] ,$$

hence,

$$[B, \underset{\sim}{H}^1] \subset \underset{\sim}{H}^1$$

Condition 11.4 forces:

$$B \in H^1, \quad \text{i.e.,} \quad B(z) = 0 ,$$

contradiction.

To finish the proof of Theorem 11.2, pick a point $z \in Z$. Let X' denote the orbit of $\underset{\sim}{G}'$ passing through

z, Y the orbit of $\underset{\sim}{H}$ through z. (See diagram.)

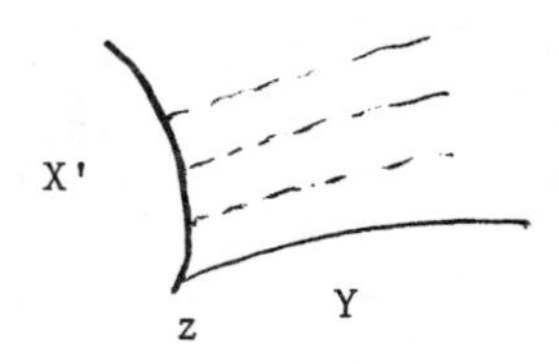

Moving this orbit Y, parameterized by points of X' gives the coordinate system required for Z, in which $\underset{\sim}{G}'$ and $\underset{\sim}{H}$ act independently to make up the action of $\underset{\sim}{G}$.

Remark. From what we already know about actions of semi-simple Lie algebras on one-dimensional manifolds, condition 11.4 is automatically satisfied if the fibers of π are one-dimensional. (In fact, I believe it is satisfied also if they are two-dimensional.) In particular, if:

$$\dim Z = 2 ,$$

and if $\underset{\sim}{G}$ is a transitive Lie algebra of vector fields with a semi-simple ideal $\underset{\sim}{H}$, then this argument proves that either H itself acts transitively, or the action of $\underset{\sim}{G}$ on Z is equivalent to the product of actions on one-dimensional manifolds, which we have already classified.

Having studied the case where the ideal $\underset{\sim}{H}$ (which is tangent to the foliation invariant under $\underset{\sim}{G}$) is semi-simple, let us study the opposite extreme case where it is abelian, i.e.,

$$[\underset{\sim}{H},\underset{\sim}{H}] = 0 \tag{11.6}$$

Suppose that the fibers of the foliation are m-dimensional. Since $\underset{\sim}{H}$ acts transitively on the fibers, we can find r elements

$$A_1,\ldots,A_m \in \underset{\sim}{H}$$

which are linearly independent in an open set of Z. The coordinates

$$(x^i,y^a) \; ; \qquad 1 \le i,j \le n, \quad 1 \le a,b \le m$$

can be chosen so that:

$$A_a = \frac{\partial}{\partial y^a} \tag{11.7}$$

Let

$$A_{m+1},\ldots,A_r$$

be the remaining linearly independent elements of $\underset{\sim}{H}$. Then, there are relations of the following form:

$$\begin{aligned} A_{m+1} &= f^a_{m+1}(x)\frac{\partial}{\partial y^a} \\ &\vdots \\ A_r &= f^a_r(x)\frac{\partial}{\partial y^a} \end{aligned} \tag{11.8}$$

Any other element $B \in \underset{\sim}{G}$ can be written in these coordinates in the following form:

$$B = h^a \frac{\partial}{\partial y^a} + h^i \frac{\partial}{\partial x^i} \tag{11.9}$$

Since $\underset{\sim}{H}$ is an ideal,

$$[B,\underset{\sim}{H}] \subset \underset{\sim}{H}$$

Since $\underset{\sim}{H}$ is <u>abelian</u>,

$$[\underset{\sim}{H}[\underset{\sim}{H},B]] = 0 \tag{11.10}$$

(This relation says that Ad $\underset{\sim}{H}$, acting in $\underset{\sim}{G}$, is <u>nilpotent</u>.) In particular, since $\partial/\partial y^a \in \underset{\sim}{H}$, we have:

$$\left[\frac{\partial}{\partial y^a}, \left[\frac{\partial}{\partial y^b}, B\right]\right] = 0 ,$$

or

$$\frac{\partial^2 h^c}{\partial y^a \partial y^b} = 0 = \frac{\partial^2 h^i}{\partial y^a \partial y^b}$$

In particular, h^c and h^i are polynomials of degree at most <u>one</u> in the variables y, with coefficients which are functions of x. In particular, B is of the following form:

$$B = (\beta^a(x)+\beta^a_b(x)y^b)\frac{\partial}{\partial y^a} + (\beta^i(x)+\beta^i_b(x)y^b)\frac{\partial}{\partial x^i} \tag{11.10}$$

We can then sum up these results as follows:

<u>Theorem 11.3</u>. Let $\underset{\sim}{G}$ be a finite dimensional Lie algebra of vector fields on a manifold Z, with an abelian ideal $\underset{\sim}{H}$. Suppose $\dim \underset{\sim}{H}(z)$ is constant as z ranges over Z.

(For example, this will be so if $\underset{\sim}{G}$ acts transitively on Z.) Then, each point of Z has a coordinate system

$$(x^i, y^a)$$

such that:

$$\frac{\partial}{\partial y^a} \in \underset{\sim}{H} \ .$$

Each element B of $\underset{\sim}{G}$ has the form 11.10.

I have now temporarily come to the end of my energy to work out the general setting for Lie's classification of finite dimensional Lie algebras of vector fields. I hope I have given enough detail to convince the reader that this is a very fruitful area for further research, utilizing to the full ideas of Lie algebra theory, differential geometry, differential topology, and so forth. In the next chapter we deal more specifically and directly with the main problems dealt with by Lie in this paper.

Chapter F

CLASSIFICATION OF IMPRIMITIVE FINITE DIMENSIONAL LIE ALGEBRAS ACTING ON 2-DIMENSIONAL MANIFOLDS

1. INTRODUCTION

Let Z be a two dimensional manifold, and let $\underset{\sim}{G} \subset V(Z)$ be a finite dimensional Lie algebra of vector fields on Z. The cases where $\underset{\sim}{G}$ acts primitively on Z may be readily classified by the general techniques described in Chapter E. Hence, one can regard the classification of the imprimitive actions as the Main Result of this paper. In this chapter, I will develop this classification, adhering more closely to Lie's methods.

Remark. The treatment of this topic in Volume III of "Transformationsgruppen" by Lie and Engel is considerably clearer and more concise than in the "Mathematische Annallen" paper being translated here.

Here is the general setting. Let Z and X be manifolds, with:

$$\dim Z = 2$$

$$\dim X = 1 .$$

Let

$$\pi: Z \to X$$

be a submersion mapping.

Let $\underset{\sim}{G}$ be a finite dimensional Lie algebra of vector fields on Z. Let $\underset{\sim}{L}$ be a finite dimensional Lie algebra of vector fields on X. Suppose that:

$$\pi_*(\underset{\sim}{G}) = \underset{\sim}{L}$$

Set

$$\underset{\sim}{H} = \{A \varepsilon G : \pi_*(A) = 0\} .$$

Then, <u>algebraically</u>, $\underset{\sim}{H}$ is a Lie ideal of $\underset{\sim}{G}$, and

$$\underset{\sim}{L} = \underset{\sim}{G}/\underset{\sim}{H} .$$

<u>Geometrically</u>, $\underset{\sim}{H}$ consists of the vector fields in $\underset{\sim}{G}$ which are tangent to the fibers of π. One says that:

$\underset{\sim}{G}$ <u>is an extension of</u> $\underset{\sim}{L}$ <u>by</u> $\underset{\sim}{H}$.

One knows (from the classification of Lie algebras acting on one dimensional manifolds) that:

$$\dim \underset{\sim}{L} = 0,1,2, \text{ or } 3 \qquad (1.1)$$

In principle, one might think that Lie's classification problem should be divided into an "algebraic" and a "geometric" part. The "algebraic" one is to classify all possibilities for $(\underset{\sim}{G},\underset{\sim}{H})$, with

$$\dim (\underset{\sim}{G}/\underset{\sim}{H}) = 0,1,2, \text{ or } 3 ,$$

and then to solve the problem of classifying the possible ways these algebraic possibilities may be realized "geometrically". In fact, Lie's method mixes up the algebraic and geometric features of the problem in a very interesting way!

Let x denote a coordinate for X. We do not make any notational distinction between x and $\pi^*(x)$. Thus, if y is a function on Z such that

$$(x,y) \equiv (\pi^*(x),y)$$

is a coordinate system for Z, y may be regarded as a choice of coordinate system for the fibers of π. x is called the coordinate of the base.

One now has a natural preliminary classification by two integers, the dimension of $\underset{\sim}{H}$ and the dimension of $\underset{\sim}{G}/\underset{\sim}{H}$. Another classification possibility is provided by the algebraic structure of the Lie algebra $\underset{\sim}{H}$. Our method will be to examine each of these possibilities in turn.

2. $\underset{\sim}{H}$ ABELIAN AND $\dim(\underset{\sim}{G}/\underset{\sim}{H}) = 0$

This is of course the simplest case. Another way of stating the condition $\dim(\underset{\sim}{G}/\underset{\sim}{H}) = 0$ is that:

$$\underset{\sim}{G} = \underset{\sim}{H} \quad ,$$

i.e., each vector field in $\underset{\sim}{G}$ is tangent to the fibers of π.

<u>Definition</u>. A coordinate system (x,y) for Z is said to be <u>adapted</u> to $\underset{\sim}{H}$ if each vector field $A \in \underset{\sim}{H}$ is of the form

$$A = a(x) \frac{\partial}{\partial y} \tag{2.1}$$

<u>Theorem 2.1</u>. Such adapted coordinate systems exist locally. If $\underset{\sim}{H} \neq 0$, x is base-like, and y is a coordinate system for the fibers of π.

<u>Proof</u>. If $\underset{\sim}{H} = 0$, everything is trivial. Suppose then that $\underset{\sim}{H} \neq 0$, and picks a vector field

$$B \in \underset{\sim}{H} \quad .$$

which is non-zero. Pick a coordinate system (x,y) such that:

$$B = \frac{\partial}{\partial y} \quad .$$

B is then tangent to the fibers of π. Since these fibers are 1-dimensional, each $A \in \underset{\sim}{H}$ is of the form

$$A = a(x,y)B$$

But, the condition

$$[A,B] = 0$$

implies

$$\frac{\partial a}{\partial y} = 0 \quad ,$$

which proves that A is of form 2.1.

Thus, in terms of an adapted coordinate system, $\underset{\sim}{H}$ has the following canonical form:

There is an R-linear map

$$A \to a(x)$$

of $\underset{\sim}{H} \to F(X)$ such that:

$$A = a(x) \frac{\partial}{\partial y} \tag{2.2}$$

for all $A \in \underset{\sim}{H}$.

Let us examine the conditions that two Lie algebras $\underset{\sim}{H}, \underset{\sim}{H}'$ of the form 2.2 be Lie equivalent. Suppose $\underset{\sim}{H}'$ is given in terms of adapted coordinates

$$(x',y') \quad .$$

Then,

x is a function of x' alone

y is a function of (x,y').

Hence,

$$\frac{\partial}{\partial y} = \frac{\partial x'}{\partial y} \frac{\partial}{\partial x'} + \frac{\partial y'}{\partial y} \frac{\partial}{\partial y'}$$

$$= 0 + \frac{\partial y'}{\partial y} \frac{\partial}{\partial y'} \quad .$$

In order that this change of variable send $\underset{\sim}{H}$ into $\underset{\sim}{H}'$, the formulas must take the following form:

$$y' = f(x)y + g(x) \quad (2.3)$$

This immediately gives us the Lie equivalence classes.

Definition. Let us say that two R-linear maps

$$\alpha: \underset{\sim}{H} \to F(X)$$

$$\alpha': \underset{\sim}{H}' \to F(X)$$

are Lie equivalent if there is a non-zero function $f \in F(X)$ and an R-linear isomorphism

$$\phi: \underset{\sim}{H} \to \underset{\sim}{H}'$$

such that:

$$\alpha(A) = f\alpha'(\phi(A)) \quad (2.4)$$

for all $A \in \underset{\sim}{H}$

Here is the main result of this section.

Theorem 2.2. The Lie equivalence classes of n-dimensional abelian Lie algebras of vector fields on Z which are tangent to the fibers of π are in one-one correspondence with Lie equivalence classes (as defined above) of one-one R-linear maps

$$\alpha: R^n \to F(X) \quad .$$

Remark. Again it is worth pointing out that this material--and its various generalizations--is in close relation to what the physicists call gauge transformations.

Now, let us turn to more complicated algebraic situations.

3. GENERAL ALGEBRAIC PROPERTIES OF LIE ALGEBRAS OF VECTOR FIELDS WHICH ACT IMPRIMITIVELY

Since it is no extra work, we may temporarily consider the general case. In this section Z and X are manifolds of arbitrary dimension,

$$\pi: Z \to X$$

is a submersion mapping. $\mathbf{G}$ is a finite dimensional Lie algebra of vector fields on Z. $\mathbf{L}$ is a finite dimensional Lie algebra of vector fields on X, such that:

$$\pi_*(\mathbf{G}) = \mathbf{L} .$$

$\mathbf{H}$ is the kernel of π_*, and consists of the vector fields of $\mathbf{G}$ which are tangent to the fibers of π.

<u>Definition</u>. Let $\mathbf{K}$ be a Lie algebra. Set

$$\mathbf{K}_1 = [\mathbf{K},\mathbf{K}], \ [\mathbf{K}_1,\mathbf{K}_1] = \mathbf{K}_2,$$

and so forth. $\mathbf{K}$ is said to be <u>n-step solvable</u> if

$$\mathbf{K}_n = 0 .$$

(For example, "abelian" is 1-step solvable, a non-abelian

solvable Lie algebra of vector fields on a 1-dimensional manifold is 2-step solvable, etc.)

<u>Theorem 3.1</u>. Suppose $\underset{\sim}{H}$ restricted to each fiber of π is n-step solvable. Then $\underset{\sim}{H}$ is n-step solvable.

<u>Proof</u>. Let x be a point of X, and let

$$Y(x) \equiv \pi^{-1}(x)$$

be the fiber of π. It is a submanifold of Z. Our definition of $\underset{\sim}{H}$ means that each vector field in $\underset{\sim}{H}$ is tangent to $Y(x)$.

Let $\underset{\sim}{H}(x)$ denote the set of vector fields on $Y(x)$, obtained by restricting the vector fields in $\underset{\sim}{H}$ to $Y(x)$. We then have:

$$[\underset{\sim}{H},\underset{\sim}{H}](x) = [\underset{\sim}{H}(x),\underset{\sim}{H}(x)] \quad ,$$

i.e.,

$$\underset{\sim}{H}_1(x) = (\underset{\sim}{H}(x))_1 \quad .$$

If $\underset{\sim}{H}(x)$ is n-step solvable for <u>each</u> $x \in X$, we see that vector fields in

$$\underset{\sim}{H}_n$$

are zero when <u>restricted to each fiber of</u> π. This implies (since $\pi_*(\underset{\sim}{H}) = 0$) that

$$\underset{\sim}{H}_n = 0 \quad ,$$

i.e., $\underset{\sim}{H}$ is n-step solvable.

Corollary to Theorem 3.1. If the fibers of π are 1-dimensional, and if $\underset{\sim}{H}$ restricted to the fibers of π is not simple, then $\underset{\sim}{H}$ is 2-step solvable.

Theorem 3.2. If $\underset{\sim}{H}$ is n-step solvable, and if $\underset{\sim}{L}$ is m-step solvable, then $\underset{\sim}{G}$ is

(n+m)-step solvable.

Proof. This is purely algebraic. Since $\underset{\sim}{L}_m = 0$, we have:

$$\pi_*(\underset{\sim}{G}_m) = 0 ,$$

i.e., $\underset{\sim}{G}_m \subset \underset{\sim}{H}$. Hence,

$$\underset{\sim}{G}_{m+n} = (\underset{\sim}{G}_m)_n \subset \underset{\sim}{H}_n = 0 .$$

Theorem 3.3. Suppose that $\underset{\sim}{H}$ restricted to each fiber of π is semisimple. Then, $\underset{\sim}{H}$ is semisimple.

Proof. Suppose $\underset{\sim}{H}$ is not semisimple. It has then a non-zero abelian ideal $\underset{\sim}{A}$. But, for each $x \in X$, $\underset{\sim}{A}(x)$ is an ideal of $\underset{\sim}{H}(x)$, hence is zero because $\underset{\sim}{H}(x)$ (the vector fields in $\underset{\sim}{H}$ restricted to the fiber above x) is semisimple. This implies that

$$\underset{\sim}{A} = 0 ,$$

contradiction.

4. CLASSIFICATION OF $\dim(\underset{\sim}{G}/\underset{\sim}{H}) = 1$, $\underset{\sim}{H}$ ABELIAN

Let (x,y) be a coordinate system which is adapted to $\underset{\sim}{H}$, and let B be an element of $\underset{\sim}{G}$ which is not in $\underset{\sim}{H}$. We can suppose the base coordinate (x) is chosen so that:

$$\pi_*(B) = \frac{\partial}{\partial x} .$$

Hence, B is of the following form:

$$B = \frac{\partial}{\partial x} + b(x,y)\frac{\partial}{\partial y} .$$

Each $A \in \underset{\sim}{H}$ is of the form:

$$A = a(x)\frac{\partial}{\partial y} .$$

Hence,

$$[A,B] = \frac{da}{dx}\frac{\partial}{\partial y} + \frac{\partial b}{\partial y}\frac{\partial}{\partial y} .$$

This must belong to $\underset{\sim}{H}$, which implies that:

$$\frac{\partial b}{\partial y} \text{ is a function of } x \text{ alone.}$$

Hence, B is of the following form:

$$B = \frac{\partial}{\partial x} + (b(x)y+b_1(x))\frac{\partial}{\partial y} . \tag{4.1}$$

Let us now make what the physicists would call a <u>gauge transformation</u>:

$$x_1 = x$$

$$y_1 = f(x) + fg(x)$$

$$\frac{\partial}{\partial y} = \frac{\partial y_1}{\partial y}\frac{\partial}{\partial y_1} = f\frac{\partial}{\partial y_1}$$

$$\frac{\partial}{\partial x} = \frac{\partial y_1}{\partial x}\frac{\partial}{\partial y_1} + \frac{\partial x_1}{\partial x}\frac{\partial}{\partial x_1}$$

$$= \left(\frac{df}{dx}y + \frac{dg}{dx}\right)\frac{\partial}{\partial y_1} + \frac{\partial}{\partial x_1}$$

Substitute these values into 4.1:

$$B = \frac{\partial}{\partial x_1} + \left(\frac{df}{dx}y + \frac{dg}{dx} + f(by+b_1)\right)\frac{\partial}{\partial y_1} \tag{4.2}$$

We can now choose f and g so that the term in parentheses on the right hand side of 4.2 vanishes. (f and g are then determined up to arbitrary constants.) Let us sum up as follows:

Theorem 4.1. Let $\underset{\sim}{G}$ such that $\dim(\underset{\sim}{G}/\underset{\sim}{H}) = 1$, and $\underset{\sim}{H}$ is abelian. There then exists a coordinate system (x,y) for Z such that:

a) x is base-like

b) Each $A \in \underset{\sim}{H}$ is of the form

$$a(x)\frac{\partial}{\partial y},$$

for some base-like function $a(x)$

c) As a vector space, $\underset{\sim}{G}$ is a direct sum of $\underset{\sim}{H}$ and $\partial/\partial x$.

$\underset{\sim}{H}$ is then determined by an R-linear, one-one map

$$\alpha\colon \underset{\sim}{H} \to F(X) \quad ,$$

namely:

$$\alpha(A) = a$$

$\underset{\sim}{H}$ is an ideal of $\underset{\sim}{G}$. The condition that

$$\left[\frac{\partial}{\partial x}, \underset{\sim}{H}\right] \subset \underset{\sim}{H}$$

determines an ordinary differential equation for α.

Specifically, let

$$\begin{aligned} A_1 &= a_1(x)\frac{\partial}{\partial y} \\ &\vdots \\ A_n &= a_n(x)\frac{\partial}{\partial y} \end{aligned} \qquad (4.3)$$

be a basis for $\underset{\sim}{H}$. Choose indices and summation convention as follows:

$$1 \leq i,j \leq n \quad .$$

Suppose that c_i^j are constants such that

$$\left[\frac{\partial}{\partial x}, A_j\right] = c_i^j A_j \qquad (4.4)$$

Conditions 4.4 require that:

$$\frac{da_i}{dx} = c_i^j a_j$$

We can write this in terms of vectors and matrices as follows:

$$\underline{a} = \begin{pmatrix} a_1 \\ \vdots \\ a_n \end{pmatrix}$$

$$\underline{c} = \begin{pmatrix} c_1^1 & \cdots & c_1^n \\ \vdots & & \\ c_n^1 & \cdots & c_n^n \end{pmatrix}$$

$$\frac{d\underline{a}}{dx} = \underline{ca} \tag{4.6}$$

Equation 4.6 is an ordinary, constant coefficient differential equation for $\underline{a}$, considered as a map

$$R \to R^n \quad .$$

It can be solved as follows:

$$\underline{a}(x) = e^{\underline{c}x} \underline{b} \quad , \tag{4.7}$$

where:

$$\underline{b} \in R^n \quad .$$

This determines the <u>canonical form</u> for the Lie algebra $\underset{\sim}{G}$. Each canonical form is parameterized by the $n \times n$ matrix $\underline{c}$ and the column n-vector $\underline{b}$.

When are two such canonical forms Lie equivalent? To answer this, let

$$(x',y')$$

be another coordinate system, $(\underset{\sim}{G}',\underset{\sim}{H}')$ another Lie algebra of the same type which takes its canonical form in the (x',y') coordinates.

$$\underset{\sim}{H}' = \left(\underline{a}'(x) \frac{\partial}{\partial y'}\right)$$

$$\underset{\sim}{G}' = \underset{\sim}{H}' + \frac{\partial}{\partial x'}$$

$$(x)\underline{a}' = e^{\underline{c}'x}\, \underline{b}' \quad .$$

Lie equivalence requires that there be a Lie algebra isomorphism

$$\phi\colon G' \to G$$

which is realized by the change of variable. The purely algebraic condition that ϕ be an isomorphism requires that:

$$\phi(\underset{\sim}{H}') = \underset{\sim}{H} \tag{4.8}$$

$$\phi\left(\frac{\partial}{\partial x'}\right) = \lambda \frac{\partial}{\partial x} + \lambda^i a_i(x) \frac{\partial}{\partial y} \quad , \tag{4.9}$$

where (λ,λ^i) are real constants.

These conditions are readily worked out. It is readily seen that

$$x = \lambda x' + \gamma \quad , \tag{4.10}$$

where γ is another real constant.

$$y = fy' + g \quad , \tag{4.11}$$

where $f(x), g(x)$ are arbitrary functions of the variable x.

Formulas 4.10, 4.11 determine what the physicists would call the gauge transformations of the canonical form. With them, the Lie equivalence classes can be readily described explicitly.

5. $\underset{\sim}{H}$ SOLVABLE, NON-ABELIAN, $\underset{\sim}{G} = \underset{\sim}{H}$

By the general remarks of Section 3, $\underset{\sim}{H}$ is 2-step solvable. It can be readily shown that there is a coordinate system (x,y) for Z, adapted to the foliation defined by the fibers π, such that, for each $A \in \underset{\sim}{H}$, there are base-like functions

$$a(x),\ b(x)$$

such that:

$$A = (a(x)+b(x)y)\,\frac{\partial}{\partial y} \quad . \tag{5.1}$$

Thus, $\underset{\sim}{H}$ as a Lie algebra of vector fields on Z which is tangent to the fibers of the foliation π, is determined by an n-dimensional R-linear subspace of

$$F(X) \oplus F(X) \quad ,$$

where

$$n = \dim \underset{\sim}{H} \quad .$$

(To see this, assign $(a,b) \in F(X) \oplus F(X)$ to each $A \in \underset{\sim}{H}$.)

However, not every such linear subspace can appear in this way. To determine what conditions it must satisfy, we follow Lie and Engel, "Transformationsgruppen", Vol. III, page 40.

Suppose

$$A' = (a' + b'y) \frac{\partial}{\partial y} \tag{5.2}$$

is another element of $\underset{\sim}{H}$, with $(a',b') \in F(X) \oplus F(X)$. Then,

$$[A,A'] = (a'b - ba') \frac{\partial}{\partial y}$$

In particular, suppose

$$b' = 0 \quad .$$

Then,

$$[A,A'] = ba' \frac{\partial}{\partial y} \quad .$$

Hence,

$$(\text{Ad } A)^n(A') = (b^n a') \frac{\partial}{\partial y} \tag{5.3}$$

At this point we can use the hypothesis that $\underset{\sim}{G}$ is <u>finite dimensional</u>. The Cayley-Hamilton theorem then asserts that Ad A acting in $\underset{\sim}{H}$ satisfies a polynomial equation. Then, there is a relation of the following form:

$$b^n a' + \lambda_{n-1} b^{n-1} a' + \cdots + \lambda_0 a' = 0 \qquad (5.4)$$

for some choice of real constants $\lambda_0, \ldots, \lambda_{n-1}$.

A' can be chosen so that a' is not identically zero. It can then be factored out of 5.4, resulting in a <u>polynomial equation for</u> b, <u>with constant coefficients</u>. This is only possible if:

$$b \equiv \text{constant} .$$

Let us sum up as follows:

<u>Theorem 5.1</u>. Let $\underset{\sim}{H}$ be a solvable, non-abelian, finite dimensional Lie algebra of vector fields on Z, which is tangent to the fibers of π. Then, there is a coordinate system (x,y) for Z which is adapted to the foliation defined by the fibers of π, and R-linear maps

$$a: \underset{\sim}{H} \to F(X)$$

$$b: \underset{\sim}{H} \to R$$

such that:

$$A = (a(A) + b(A)y) \frac{\partial}{\partial y} \qquad (5.5)$$

for all $A \in \underset{\sim}{H}$.

a and b must satisfy the following conditions:

$$b([\underset{\sim}{H},\underset{\sim}{H}]) = 0 \tag{5.6}$$

$$a([A,B]) = a(A)b(B) - a(B)b(A) \tag{5.7}$$

for $A,B \varepsilon \underset{\sim}{A}$.

We can now use 5.6 and 5.7 to put the algebraic structure of $\underset{\sim}{H}$ in a more useful form, similar to that given by Lie and Engel. Note that relation 5.6 means that b is a Lie algebra homomorphism from $\underset{\sim}{H}$ to the abelian Lie algebra R. Set:

$$\underset{\sim}{K} = \text{kernel } b \tag{5.8}$$

$\underset{\sim}{K}$ is then an ideal of $\underset{\sim}{H}$. It is not equal to all of $\underset{\sim}{K}$, because then $\underset{\sim}{H}$ would be abelian, and would be covered by the work of Section 2.

We can then suppose that the adapted coordinate system (x,y) is chosen so that:

$$y \frac{\partial}{\partial y} \varepsilon \underset{\sim}{H} \quad ,$$

i.e.,

$$b\left(y \frac{\partial}{\partial y}\right) = 1, \qquad a\left(y \frac{\partial}{\partial y}\right) = 0$$

$\underset{\sim}{H}$ is then a direct sum (as a vector space) of $\underset{\sim}{K}$ and the one dimensional subspace spanned by $\partial/\partial y$. Hence, $\underset{\sim}{H}$ has a basis of the following form:

$$\boxed{b_1(x)\frac{\partial}{\partial y}, \ldots, b_n(x)\frac{\partial}{\partial y}, y\frac{\partial}{\partial y}} \tag{5.9}$$

Conversely, it is readily seen that 5.9 defines, for arbitrary choice of functions $b_1,\ldots,b_n \in F(X)$, a solvable, (n+1)-dimensional non-abelian Lie algebra of vector fields on Z which are tangent to the fibers of π. Thus, we have determined "canonical forms" for the Lie equivalence classes.

Exercise. Complete this analysis by carrying out the following tasks:

a) Determine when two Lie algebras of form 5.9 are Lie equivalent under gauge transformations

$$x \to x$$
$$y \to f(x)y + g(x)$$

b) Solve for the orbits of the vector fields 5.9, and write down a Lie group acting on R^2 whose Lie algebra is given by formula 5.9.

Remark. There are certain general algebraic features to this argument that are of great interest to general Lie algebra theory. Here is a brief description of one such problem.

Let K be a field, and let

$$\underset{\sim}{G}$$

be a finite dimensional Lie algebra over K. Let F be a commutative associative algebra over K.

Consider the K-vector space

$$\underset{\sim}{G} \otimes F \quad . \tag{5.10}$$

(The tensor product in 5.10 is understood to be taken relative to the field K.) Make $\underset{\sim}{G} \otimes F$ into a Lie algebra, using the following formula:

$$[A_1 \otimes f_1, A_2 \otimes f_2] = [A_1, A_2] \otimes f_1 f_2 \tag{5.11}$$

$$\text{for } A_1, A_2 \in \underset{\sim}{G}; \quad f_1, f_2 \in F \quad .$$

$\underset{\sim}{G} \otimes F$, with the Lie algebra structure defined by 5.11, is called the <u>gauge Lie algebra based on</u> $\underset{\sim}{G}$. (The name "gauge Lie algebra" comes from physics. It is also called a <u>current Lie algebra</u>. See LAQM, Vol. VI, and my paper "Infinite dimensional Lie algebras and current algebras".)

The general problem that obviously underlies the situation treated in this section--and much of the rest of Lie's work--may be stated as follows:

<u>Determine the finite dimensional subalgebras</u> $\underset{\sim}{H}$ <u>of</u> $\underset{\sim}{G} \otimes F$.

The key condition is finite dimensional. It enables one to apply the Cayley-Hamilton theorem.

Let h be an element of $\underset{\sim}{H}$. It satisfies an algebraic equation of the form:

$$(\mathrm{Ad}\ h)^m + \lambda_{n-1}(\mathrm{Ad}\ h)^{m-1} + \cdots + \lambda_0 \quad ,$$

where:

$$m = \dim \underset{\sim}{H} \quad .$$

The λ are polynomial maps

$$\underset{\sim}{H} \to R \quad .$$

(They are invariant under the adjoint representation on $\underset{\sim}{H}$. They play a key role in Cartan's structure theory of Lie algebras.) What must be done is to examine the condition relations 5.11 imposes.

We now return to the specific situation considered by Lie to study the case where $\underset{\sim}{H}$ is semisimple.

6. $\underset{\sim}{H}$ SEMISIMPLE, $\dim (\underset{\sim}{G}/\underset{\sim}{H}) = 0$

Assuming $\underset{\sim}{H}$ semisimple, it can be written as the direct sum

$$\underset{\sim}{H} = \underset{\sim}{H}_1 \oplus \cdots \oplus \underset{\sim}{H}_n$$

of simple ideals. Each of these ideals must obviously be isomorphic to the Lie algebras of

$$SL(2,R) \quad .$$

Let A_1 be an element of $\underset{\sim}{H}_1$ which is non-zero in an open subset of Z, such that:

$$(\mathrm{Ad}\ A_1)^3 = 0 \tag{6.1}$$

(It is readily seen, given the algebraic structure of SL(2,R), that such an element exists.) We can then choose a coordinate system to

$$(x,y)$$

such that:

$$A_1 = \frac{\partial}{\partial y} \quad . \tag{6.2}$$

Now by the meaning of "direct sum" the elements B in $\underset{\sim}{H}_2,\ldots,\underset{\sim}{H}_n$ commute with A_1. They are also tangent to the fibers of π, hence are of the form

$$B = b(x,y)\frac{\partial}{\partial y} \quad .$$

Hence,

$$\frac{\partial b}{\partial y} = 0 \quad .$$

This implies that the elements in $\underset{\sim}{H}_2,\ldots,\underset{\sim}{H}_n$ commute with each other, which of course contradicts the assumption that they are semisimple. This proves:

Theorem 6.1. Let $\underset{\sim}{H}$ be a semisimple Lie algebra of vector fields on Z which is tangent to the fibers of π. Then, $\underset{\sim}{H}$ is simple, and isomorphic to the Lie algebra of SL(2,R).

We can now find the canonical form for $\underset{\sim}{H}$. Suppose that 6.1 and 6.2 are satisfied. $\underset{\sim}{H}$ has a basis composed of elements A_1, A_2, A_3 which satisfy the following algebraic relations:

$$\begin{aligned} [A_1,A_2] &= A_1 \\ [A_1,A_3] &= 2A_2 \\ [A_2,A_3] &= A_3 \end{aligned} \tag{6.3}$$

In addition to 6.2, suppose that:

A_2, A_3 take the following form in this coordinate system:

$$\begin{aligned} A_2 &= (a_2(x) + b_2(x)y + c_2(x)y^2) \frac{\partial}{\partial y} \\ A_3 &= (a_3(x) + b_3(x)y + c_3(x)y^2) \frac{\partial}{\partial y} \end{aligned} \tag{6.4}$$

(A_2, A_3 must take this general form because of 6.1 which forces the third partial derivative with respect to y of their coefficients to be zero.)

We can now use the $SL(2,R)$ structure relations 6.3 to determine the unknown functions of x in 6.4.

$$\begin{aligned} [A_1,A_2] &= (b_2 + 2c_2y) \frac{\partial}{\partial y} \\ &= A_1 = \frac{\partial}{\partial y} \quad , \end{aligned}$$

which forces:

$$b_2 = 1, \quad c_2 = 0 \tag{6.5}$$

$$\begin{aligned} [A_1, A_3] &= (b_3 + 2c_3 y) \frac{\partial}{\partial y} \\ &= 2A_2 = \text{using } 6.4 - 6.5, \\ &\qquad (2a_2 + 2y) \frac{\partial}{\partial y} \ . \end{aligned}$$

Hence,

$$c_3 = 1, \quad b_3 = 2a_2 \ . \tag{6.6}$$

Now, substitute 6.5 and 6.8 into 6.4:

$$\begin{aligned} A_2 &= (a_2 + y) \frac{\partial}{\partial y} \\ A_3 &= (a_3 + 2a_2 y + y^2) \frac{\partial}{\partial y} \end{aligned} \tag{6.7}$$

$$\begin{aligned} [A_2, A_3] &= ((a_2+y)(2a_2+2y) - (a_3+2a_2y+y^2)) \frac{\partial}{\partial y} \\ &= (2a_2^2 - a_3 + 2a_a y + y^2) \frac{\partial}{\partial y} \\ &= A_3 = (a_3 + 2a_2 y + y^2) \frac{\partial}{\partial y} \ , \end{aligned}$$

hence:

$$a_2^2 = a_3 \tag{6.8}$$

A_2 and A_3 then take the following form:

$$\begin{aligned} A_2 &= (a_2 + y) \frac{\partial}{\partial y} \\ A_3 &= (a_2^2 + 2a_a y + y^2) \frac{\partial}{\partial y} \\ &= (a_2 + y)^2 \frac{\partial}{\partial y} \end{aligned}$$

We can now make a final gauge transformation

$$
\begin{aligned}
x_1 &= x \\
y_1 &= a_2 + y \quad .
\end{aligned}
\tag{6.8}
$$

Hence,

$$
\frac{\partial}{\partial y} = \frac{\partial y_1}{\partial y} \frac{\partial}{\partial y_1} = \frac{\partial}{\partial y_1} \tag{6.9}
$$

In the new coordinates the Lie algebra $\underset{\sim}{H}$ takes its <u>canonical form</u>:

$$
\boxed{
\begin{aligned}
A_1 &= \frac{\partial}{\partial y_1} \\
A_2 &= y_1 \frac{\partial}{\partial y_1} \\
A_3 &= y_1^2 \frac{\partial}{\partial y_1}
\end{aligned}
}
\tag{6.10}
$$

Let us sum up:

<u>Theorem 6.2</u>. Let $\underset{\sim}{H}$ be a semisimple Lie algebra of vector fields which is tangent to the fibers of π. Then, there is a <u>unique</u> canonical form 6.10 for $\underset{\sim}{H}$.

<u>Remark</u>. This "rigidity" seems to be typical of semisimple Lie algebras. See the previous chapter for general comments about the relation to the theory of deformation.

7. $\underset{\sim}{H}$ NON-ABELIAN SOLVABLE, dim $(\underset{\sim}{G}/\underset{\sim}{H}) = 1$

According to previous work, there is a coordinate system

$$(x,y)$$

for Z, adapted to the foliation determined by the fibers of π, such that $\underset{\sim}{H}$ has a basis of the form:

$$\boxed{a_1(x) \frac{\partial}{\partial y}, \ldots, a_n(x) \frac{\partial}{\partial y}, y \frac{\partial}{\partial y}} \tag{7.1}$$

Let B be an element of $\underset{\sim}{G}$ which is not in $\underset{\sim}{H}$. $\underset{\sim}{G}$ is the vector space direct sum of $\underset{\sim}{H}$ and the one-dimensional vector space spanned by B. We can choose the coordinate x so that:

$$B = \frac{\partial}{\partial x} + b(x,y) \frac{\partial}{\partial y} \tag{7.2}$$

Now,

$$\left[y \frac{\partial}{\partial y}, B\right] = \left(y \frac{\partial b}{\partial y} - b\right) \frac{\partial}{\partial y} \tag{7.3}$$

The vector field on the right hand side of 7.3 must be a linear combination with real coefficients of the vector fields listed in 7.1. In particular, it must be linear in y This forces b to be linear in y_1, i.e.,

$$b = b_0(x) + b_1(x)y \quad . \tag{7.4}$$

Thus,

$$\left[B,\ y\frac{\partial}{\partial y}\right] = b_0\frac{\partial}{\partial y} \tag{7.5}$$

b_0 is a linear combination with real coefficients of the function $a_1,\ldots,a_n$	(7.6)

Now,

$$B = (b_o + b_1 y)\frac{\partial}{\partial y} + \frac{\partial}{\partial x} \tag{7.7}$$

We can now make a gauge transformation to reduce 7.6 to a canonical form. Of course, such gauge transformations should only be chosen which <u>leave invariant the canonical form</u> 7.1 <u>of</u> $\underset{\sim}{H}$. It is easy to see that such transformations are of the following form:

$$\begin{aligned} x_1 &= x \\ y_1 &= f(x)y \end{aligned} \tag{7.7}$$

Thus,

$$\frac{\partial}{\partial y} = \frac{\partial y_1}{\partial y}\frac{\partial}{\partial y_1} = f\frac{\partial}{\partial y_1}$$

$$\frac{\partial}{\partial x} = \frac{\partial y_1}{\partial x}\frac{\partial}{\partial y_1} + \frac{\partial x_1}{\partial x}\frac{\partial}{\partial x_1}$$

$$= \frac{df}{dx}y\frac{\partial}{\partial y_1} + \frac{\partial}{\partial x_1}$$

Hence,

$$B = (b_0+b_1y)f \frac{\partial}{\partial y_1} + \frac{df}{dx} y \frac{\partial}{\partial y_1} + \frac{\partial}{\partial x_1}$$

We can choose f so that B contains no term of the form:

$$(\ )y_1 \frac{\partial}{\partial y_1} \quad .$$

This proves:

Theorem 7.1. A coordinate system (x,y) adapted to the foliation can be chosen so that $\underset{\sim}{H}$ has the canonical form 7.1,

$$B = b_0(x) \frac{\partial}{\partial y} + \frac{\partial}{\partial x} \quad , \tag{7.8}$$

and condition 7.6 is satisfied.

Now, condition 7.6 means that:

$$b_0(x) \frac{\partial}{\partial y} \in \underset{\sim}{H} \quad .$$

Hence, we can redefine B --modulo $\underset{\sim}{H}$--so that:

$$B = \frac{\partial}{\partial x} \quad . \tag{7.9}$$

We have yet to fully take into account that

$$[B,\underset{\sim}{H}] \subset \underset{\sim}{H} \quad .$$

To do this, choose indices and the summation convention as follows:

$$1 \leq i,j \leq n \quad .$$

Then,

$$\left[B, a_i \frac{\partial}{\partial y}\right] = \frac{da_i}{dx} \frac{\partial}{\partial y}$$

Hence, there are real constants λ_i^j such that

$$\left[B, a_i \frac{\partial}{\partial y}\right] = \lambda_i^j a_j \frac{\partial}{\partial y}$$

The $a_j(x)$ are determined by the following differential equations

$$\boxed{\frac{da_i}{dx} = \lambda_i^j a_j} \tag{7.10}$$

These relations determine the canonical form for $\underset{\sim}{G}$. (It is left to the reader to settle the additional problem of deciding when these canonical forms are Lie equivalent.) Note that the canonical form depends on a <u>finite number of parameters</u>, namely:

$$\boxed{\lambda_i^j, \quad a_i(0)}$$

This is in contrast to $\underset{\sim}{H}$, whose canonical form 7.6 depends on "arbitrary functions."

<u>Exercise</u>. Write down the formulas for the groups G acting on R^2 whose Lie algebra $\underset{\sim}{G}$ consists of the vector fields whose canonical form we have just found.

8. $\underset{\sim}{H}$ SEMISIMPLE, $\underset{\sim}{G}/\underset{\sim}{H}$ ARBITRARY

Consider $\text{Ad}\,\underset{\sim}{H}$ acting in $\underset{\sim}{G}$. It is a linear, finite dimensional representation of a semisimple Lie algebra. Such representations are <u>completely reducible</u> (proved by Weyl, J. H. C. Whitehead). Hence, there is a subspace

$$\underset{\sim}{K} \subset \underset{\sim}{G}$$

such that:

$$\underset{\sim}{G} = \text{vector space direct sum of } \underset{\sim}{H}, \underset{\sim}{K} \tag{8.1}$$

$$\text{Ad}\,\underset{\sim}{H}(\underset{\sim}{K}) \subset \underset{\sim}{K} \tag{8.2}$$

But, $\underset{\sim}{H}$ is a <u>Lie ideal</u> of $\underset{\sim}{G}$. Hence,

$$[\underset{\sim}{H}, \underset{\sim}{G}] \subset \underset{\sim}{H} \quad . \tag{8.3}$$

8.1-8.3 are only compatible if

$$[\underset{\sim}{H}, \underset{\sim}{K}] = 0 \quad . \tag{8.4}$$

Return to the case where $\underset{\sim}{G}$ is a Lie algebra of vector fields on $Z = R^2$ which leaves invariant a foliation. $\underset{\sim}{H}$ is tangent to the foliation and is semisimple. By Section 6, there is a coordinate system

$$(x,y)$$

adapted to the foliation such that

$$\boxed{\frac{\partial}{\partial y}\ ,\ y\,\frac{\partial}{\partial y}\ ,\ y^2\,\frac{\partial}{\partial y}} \tag{8.5}$$

is a basis for $\underset{\sim}{H}$.

Let

$$A \in \underset{\sim}{K} \quad .$$

It is of the form

$$A = a(x)\,\frac{\partial}{\partial x} + b(x,y)\,\frac{\partial}{\partial y} \quad .$$

This is only possible if

$$b(x,y) \equiv 0 \quad .$$

We see that:

> $\underset{\sim}{K}$ is a Lie algebra of vector fields involving x alone.

Thus, the possibilities for canonical form are:

$$\boxed{\underset{\sim}{H},\ \frac{\partial}{\partial x}}$$

$$\boxed{\underset{\sim}{H}\ ,\ \frac{\partial}{\partial x}\ ,\ x\,\frac{\partial}{\partial x}}$$

$$\underset{\sim}{H}, \frac{\partial}{\partial x}, x\frac{\partial}{\partial x}, x^2\frac{\partial}{\partial x}$$

We can sum up as follows:

Theorem 8.1. Suppose that $\underset{\sim}{H}$ is semisimple. Then, there is an ideal $\underset{\sim}{K} \subset \underset{\sim}{G}$ such that $\underset{\sim}{G}$ is a direct sum of $\underset{\sim}{K}$ and $\underset{\sim}{H}$. Z can be written locally as a product

$$Z = X \times Y$$

so that $\underset{\sim}{G}$ is the direct sum of the Lie algebra $\underset{\sim}{K}$ of vector fields on X and the Lie algebra $\underset{\sim}{H}$ of vector fields on Y.

9. $\underset{\sim}{H}$ ABELIAN, $\dim(\underset{\sim}{G}/\underset{\sim}{H}) = 2$

Set

$$\overline{\underset{\sim}{G}} = \underset{\sim}{G}/\underset{\sim}{H}$$

$\overline{\underset{\sim}{G}}$ is a 2-dimensional, non-abelian solvable Lie algebra of vector fields on X.

Let B_1, B_2 be two elements in $\underset{\sim}{G}$ such that $\overline{B}_1, \overline{B}_2$ form a basis of $\overline{\underset{\sim}{G}}$. We can suppose the coordinate x for X chosen so that:

$$\overline{B}_1 = \frac{\partial}{\partial x}, \qquad \overline{B}_2 = x\frac{\partial}{\partial x} .$$

From previous work, we know that coordinates (x,y) can be chosen for X such that a basis for $\underset{\sim}{H}$ has the following canonical form:

$$\boxed{a_1(x)\frac{\partial}{\partial y}, \ldots, a_n(x)\frac{\partial}{\partial y}} \tag{9.1}$$

Further, B_1 can be chosen as follows:

$$\boxed{B_1 = \frac{\partial}{\partial x}} \tag{9.2}$$

We must now see what conditions this canonical form imposes on B_2.

Suppose that:

$$B_2 = x\frac{\partial}{\partial x} + b(x,y)\frac{\partial}{\partial y} \quad .$$

Then,

$$[B_1, B_2] = \frac{\partial}{\partial x} + \frac{\partial b}{\partial x}\frac{\partial}{\partial y} \tag{9.3}$$

In order that $(\underset{\sim}{H}, B_1, B_2)$ forms a Lie algebra, it is necessary that the right hand side of 9.3 be a linear combination of the elements 9.1, 9.2. This forces:

$\frac{\partial b}{\partial x}$ is a function of x alone,

or:

$$b(x,y) = b_0(x) + b_1(y) \tag{9.4}$$

Then,

$$\left[a_1 \frac{\partial}{\partial y}, B_2\right] = \left[a_1 \frac{\partial}{\partial y}, x \frac{\partial}{\partial x} + b_0 \frac{\partial}{\partial y} + b_1 \frac{\partial}{\partial y}\right]$$

$$= -x \frac{da_1}{dx} \frac{\partial}{\partial y} + a_1 \frac{db_1}{dy} \frac{\partial}{\partial y} \tag{9.5}$$

Supposing $a_1 \neq 0$, we see that the condition that the left hand side of 9.5 lie in $\underset{\sim}{G}$ forces db_1/dy to be a <u>constant</u>; say c.

Choose indices as follows, and the summation convention on these indices:

$$1 \leq i,j \leq n \quad .$$

We can then rewrite 9.5 as follows:

$$\left[a_i \frac{\partial}{\partial y}, B_2\right] = \left(-x \frac{da_i}{dx} + a_i c\right) \frac{\partial}{\partial y} \tag{9.6}$$

We also have:

$$B_2 = x \frac{\partial}{\partial x} + (b_0(x) + cy + c') \frac{\partial}{\partial y} \tag{9.7}$$

Also,

$$\left[B_1, a_i \frac{\partial}{\partial y}\right] = \frac{da_i}{dx} \frac{\partial}{\partial y}$$

$$= \lambda_i^j a_j \frac{\partial}{\partial y}$$

hence

$$\frac{da_i}{dx} = \lambda_i^j a_j \tag{9.8}$$

(c, c', λ_i^j) are real constants.)

Substitute 9.8 into 9.6:

$$\left[a_i \frac{\partial}{\partial y}, B_2\right] = (-x\lambda_i^j a_j + a_i c) \frac{\partial}{\partial y} \tag{9.9}$$

The right hand side of 9.9 must also be linear combinations of the $a_i(\partial/\partial y)$, i.e., there must be constants γ_i^j such that

$$a_i c - x a_j \lambda_i^j = \gamma_i^j a_j \tag{9.10}$$

Use vector-matrix notation:

$$\underline{a} = \begin{pmatrix} a_1 \\ \vdots \\ a_n \end{pmatrix}$$

$$\underline{\lambda} = (\lambda_1^j)$$

and so forth.

9.8 can be written as:

$$\underline{a} = e^{\underline{\lambda} x}\, \underline{a}(0) \tag{9.11}$$

9.10 means that:

$$\underline{\gamma a} + \underline{\lambda}(x\underline{a}) = \lambda \underline{a}$$

Substitute 9.11:

$$\underline{\gamma} e^{\underline{\lambda} x} a(0) + \underline{\lambda}(x e^{\underline{\lambda} x} \underline{a}(0)) = c e^{\underline{\lambda} x} \underline{a}(0) \qquad (9.12)$$

We shall now deduce from 9.12 the conditions that the n×n matrices $\underline{\gamma}$ and $\underline{\lambda}$ must satisfy. First, set $x = 0$ in 9.12:

$$\underline{\gamma}\underline{a}(0) = c\underline{a}(0) \quad , \qquad (9.13)$$

i.e.,

> $\underline{a}(0)$ <u>is an eigenvector of</u> $\underline{\gamma}$, <u>with eigenvalue</u> c.

Now, differentiate 9.12, and set $x = 0$:

$$\underline{\gamma\lambda}\underline{a}(0) + \underline{\lambda}\underline{a}(0) = c\underline{\lambda}\underline{a}(0)$$

Hence,

$$\underline{\gamma}(\underline{\lambda}\underline{a}(0)) = (c-1)(\underline{\lambda}\underline{a}(0))$$

In particular,

> $\underline{\lambda}\underline{a}(0)$ <u>is an eigenvector of</u> $\underline{\gamma}$ <u>with eigenvalue</u> (c-1).

The general relation of this type can be obtained by expanding 9.12 in a power series in x. Equating the j-th term (j=2,3,...) gives the following relation:

$$\frac{1}{j!} \underline{\gamma\lambda}^j \underline{a}(0) + \frac{1}{(j-1)!} \underline{\lambda}^j \underline{a}(0) = \frac{1}{j!} \underline{\lambda}^j \underline{a}(0)$$

or

$$j\underline{\gamma}\underline{\lambda}^j\underline{a}(0) + \underline{\lambda}^j\underline{a}(0) = cj\underline{\lambda}^j\underline{a}(0)$$

$$\underline{\gamma}(\underline{\lambda}^j\underline{a}(0)) = \left(c - \frac{1}{j}\right)(\underline{\lambda}^j\underline{a}(0)) \tag{9.14}$$

Relation 9.14 says that:

$\underline{\lambda}^j\underline{a}(0)$ <u>is an eigenvector of</u> $\underline{\gamma}$ <u>with eigenvalue</u> $\left(c - \frac{1}{j}\right)$.

Now, $\underline{\gamma}$ has <u>only a finite number of eigenvalues</u>. Hence, there is an integer m such that:

$$\underline{\lambda}^{m+1}(\underline{a}(0)) = 0 \tag{9.15}$$

$$\underline{\lambda}^{m}(\underline{a}(0)) \neq 0 \tag{9.16}$$

Thus,

$$\underline{a}(x) = e^{\underline{\lambda}x}\,\underline{a}(0)$$

$$= \underline{a}(0) + x\underline{\lambda}(\underline{a}(0)) + \cdots + \frac{1}{m!}x^m\underline{\lambda}^m(\underline{a}(0)) \tag{9.17}$$

Now, recall that

$$\underline{a}(x) = \begin{pmatrix} a_1(x) \\ \vdots \\ a_n(x) \end{pmatrix}$$

where

$$a_1(x)\frac{\partial}{\partial y}, \ldots, a_n(x)\frac{\partial}{\partial y}$$

is a basis for H. In particular,

$$a_1(x), \ldots, a_n(x)$$

are linearly independent over the real numbers.

From linear algebra, we know that:

$$\underline{a}(0), \ \underline{a}(0), \ldots, \underline{\lambda}^m \underline{a}(0)$$

are linearly independent as vectors of R^n. Hence,

$$m+1 \leq n \quad .$$

Suppose

$$m+1 < n \quad . \tag{9.18}$$

In case 9.18 is satisfied, it is clear from 9.17 that we could construct a non-zero constant vector $\underline{b} \in R^n$ such that

$$\underline{b} \cdot \underline{a}(x) = 0 \quad ,$$

i.e., the functions $a_1(x), \ldots, a_n(x)$ would not be linearly independent. 9.18 cannot be true, and we must have:

$$m + 1 = n \tag{9.19}$$

We conclude that:

Theorem 9.1. $\underline{a}(0)$ is a cyclic vector for $\underline{\lambda}$, i.e., the iterates of $\underline{a}(0)$ under $\underline{\lambda}$ span R^n. $\underline{\lambda}$ is a nilpotent matrix, i.e.,

$$\underline{\lambda}^{n+1} = 0 \tag{9.20}$$

<u>Remark</u>. If the vector $\underline{a}(0), \underline{\lambda a}(0), \ldots, \underline{\lambda}^{n-1}\underline{a}(0)$ are chosen as a basis for R^n, then the matrix of $\underline{\lambda}$ with respect to this basis is what is classically called a <u>companion matrix</u>. $\underline{\lambda}$ takes the <u>diagonal form</u> with respect to this basis, and is determined by $\underline{\lambda}$, $\underline{a}(0)$ and the real number c.

Let us now work out the rest of the Lie algebra structure for $\underset{\sim}{G}$. Return to relation 9.3.

$$[B_1, B_2] = \frac{\partial}{\partial x} + \frac{db_0}{dx}\frac{\partial}{\partial y} \tag{9.21}$$

Here, there must be constants α^i such that:

$$\frac{db_0}{dx} = \alpha^i a_i(x) \quad , \tag{9.22}$$

or, in vector notation,

$$\frac{db_0}{dx} = \langle\underline{\alpha}, \underline{a}(x)\rangle \tag{9.23}$$

where

$$\underline{\alpha} = (\alpha^1, \ldots, \alpha^n) \quad ,$$

and $\langle\ ,\ \rangle$ is the duality relation between row and column vectors of R.

Using 9.17, we have:

$$\frac{db_0}{dx} = \langle\alpha, e^{\underline{\lambda} x}\, \underline{a}(0)\rangle \quad .$$

Hence,

$$b_0(x) = \left\langle \alpha, \left(\frac{e^{\underline{\lambda} x} - 1}{\underline{x}}\right) \underline{a}(0) \right\rangle + \text{constant} \quad (9.24)$$

Theorem 9.2. The Lie canonical form for $\underset{\sim}{G}$ is parameterized by an $n \times n$ nilpotent matrix $\underline{\lambda}$, a column vector $\underline{a}(0) \in R^n$ which is cyclic for $\underline{\lambda}$, a row vector $\underline{\alpha}$, and a real constant c. The bases for $\underset{\sim}{G}$ then take the following form:

$$a_1(x) \frac{\partial}{\partial y}, \ldots, a_n(x) \frac{\partial}{\partial y}, \frac{\partial}{\partial x}, x \frac{\partial}{\partial x} + \left(\left\langle \alpha, \frac{e^{\underline{\lambda} x} - 1}{\underline{\lambda}} \underline{a}(0) \right\rangle + cy\right) \frac{\partial}{\partial y} \quad (9.25)$$

where:

$$\underline{a}(x) = e^{\underline{\lambda} x} \underline{a}(0) = \begin{pmatrix} a_1(x) \\ \vdots \\ a_n(x) \end{pmatrix} \quad (9.26)$$

Proof. These conditions were derived as necessary conditions for the existence of a Lie algebra $\underset{\sim}{G}$ with the properties described in the title of this section. It remains to show the converse, i.e., that there actually exists a Lie algebra with these properties. To this end, define vector fields by the formula 9.25. The argument given above is reversible to prove that the linear span (with constant coefficients) of the vector fields 9.25 is closed under Jacobi bracket, hence defines a Lie algebra. (The point to this remark is that, using the formula described above to define the "abstract" Lie algebra structure, it was not a priori obvious that the *Jacobi identity* was satisfied. However, this is assured because the abstract objects are realized as *vector fields*, and the Jacobi bracket is true for vector fields.)

Remark. Given the nilpotent n×n matrix $\underline{\lambda}$, the set of vectors which are cyclic with respect to $\underline{\lambda}$ forms an *open subset* of R^n. Thus, the structure of the Lie algebras parameterized in this way may be expected to change as $\underline{a}(0)$ tends to the *boundary* of this open subset. It would be interesting to calculate what happens. Clearly, there remains much to be done to elucidate the deformation-theoretic framework for these concepts.

As was proved in Volume VIII, a necessary and sufficient condition that an n×n matrix $\underline{\lambda}$ have a cyclic vector is that *the characteristic polynomial of* $\underline{\lambda}$ *be equal to its minimal polynomial*. For the case of nilpotent matrices, this means that:

$$\underline{\lambda}^n = 0, \quad \text{but} \quad \underline{\lambda}^{n-1} \neq 0 \ .$$

10. $\underset{\sim}{H}$ NON-ABELIAN SOLVABLE, $\underset{\sim}{G}/\underset{\sim}{H}$ NON-ABELIAN SOLVABLE

In Section 5, we have determined a canonical form for $\underset{\sim}{H}$, namely:

$$\boxed{a_1(x)\frac{\partial}{\partial y}, \ldots, a_n(x)\frac{\partial}{\partial y}, y\frac{\partial}{\partial y}} \qquad (10.1)$$

We can then normalize the choice of the x-coordinates so that the following vector fields

$$\begin{aligned} B_1 &= \frac{\partial}{\partial x} + b_1(x,y)\frac{\partial}{\partial y} \\ B_2 &= x\frac{\partial}{\partial x} + b_2(x,y)\frac{\partial}{\partial y} \end{aligned} \qquad (10.2)$$

together with the vector fields 10.1, span $\underset{\sim}{G}$. Then,

$$\left[y\frac{\partial}{\partial y}, B_1\right] = \left(y\frac{\partial b_1}{\partial y} - b_1\right)\frac{\partial}{\partial y} \qquad (10.3)$$

$$\left[y\frac{\partial}{\partial y}, B_2\right] = \left(y\frac{\partial b_2}{\partial y} - b_2\right)\frac{\partial}{\partial y} \qquad (10.4)$$

The right hand side of the vector fields 10.3 and 10.4 must be linear combinations of the vector fields 10.1. This implies that $b_1(x,y)$ must be, as functions of x,y, at most <u>linear functions</u> in y.

Now the canonical form 10.1 is unchanged by the following gauge transformation:

$$x_1 = x$$
$$y_1 = f(x)y \qquad (10.5)$$

Then,

$$\frac{\partial}{\partial x} = \frac{\partial y_1}{\partial x}\frac{\partial}{\partial y_1} + \frac{\partial x_1}{\partial x}\frac{\partial}{\partial x_1}$$

$$= \frac{df}{dx}\, y \frac{\partial}{\partial y_1} + \frac{\partial}{\partial x_1}$$

$$\frac{\partial}{\partial y} = \frac{\partial y_1}{\partial y}\frac{\partial}{\partial y_1} + \frac{\partial x_1}{\partial y}\frac{\partial}{\partial x_1} = f\frac{\partial}{\partial y_1}$$

Hence, after at most making a gauge transformation of the form 10.5, we may suppose that:

$$B_1 = \frac{\partial}{\partial x} + b_1(x)\frac{\partial}{\partial y} \qquad (10.6)$$

$$B_2 = x\frac{\partial}{\partial x} + (b_2(x)+c_2(x)y)\frac{\partial}{\partial y}$$

The coefficient of $y\frac{\partial}{\partial y}$ in the expansion of $[B_1,B_2]$ is:

$$\frac{dc_2}{dx}$$

This must be a constant, i.e.,

$$c_2(x) = cx + c' \quad ,$$

with constant, c'. The constant c' can be absorbed into b_2. Hence, B_2 is of the following form:

$$B_2 = x \frac{\partial}{\partial x} + (b_2(x)+cy) \frac{\partial}{\partial y} .$$

We can then substitute $-cy \frac{\partial}{\partial y}$ (which is in $\underset{\sim}{G}$, of course) to write B_2 in the following form:

$$B_2 = x \frac{\partial}{\partial x} + b_2(x) \frac{\partial}{\partial y} \tag{10.7}$$

Now,

$$\left[B_1, y \frac{\partial}{\partial y}\right] = b_1 \frac{\partial}{\partial y}$$

$$\left[B_2, y \frac{\partial}{\partial y}\right] = b_2 \frac{\partial}{\partial y}$$

Hence, b_1 and b_2 must be linear combinations of the $a_1(x),\ldots,a_n(x)$.

In particular,

$$b_1 \frac{\partial}{\partial y} , \quad b_2 \frac{\partial}{\partial y}$$

<u>belong</u> to $\underline{H}$, hence may be substituted from B_1,B_2. After the subtraction, we have:

$$\begin{aligned} B_1 &= \frac{\partial}{\partial x} \\ B_2 &= x \frac{\partial}{\partial x} \end{aligned} \tag{10.8}$$

In particular, B_1 and B_2 span a subalgebra. $\underset{\sim}{G}$ is the semidirect sum of $\underset{\sim}{H}$ and this subalgebra. We may sum up as follows:

<u>Theorem 10.1</u>. The Lie algebra extension

$$\underset{\sim}{G} \to \underset{\sim}{G}/\underset{\sim}{H}$$

splits, under the hypotheses of this section, i.e., the Lie algebra $\underset{\sim}{G}$ may be written as a <u>vector space direct sum</u> of the ideal $\underset{\sim}{H}$ and a subalgebra isomorphic to $\underset{\sim}{G}/\underset{\sim}{H}$.

This splitting property considerably simplifies the canonical form. Introduce the indices

$$1 \leq i,j \leq n \quad ,$$

and the vector-matrix notations used in previous sections. Then, there is an n×n constant matrix

$$\underline{\lambda} = (\lambda_i^j)$$

such that:

$$\left[B_1, a_i \frac{\partial}{\partial y}\right] = \lambda_i^j a_j \frac{\partial}{\partial y} \quad ,$$

or

$$\frac{da_i}{dx} = \lambda_i^j a_j$$

or

$$\underline{a}(x) = e^{\underline{\lambda} x}\, \underline{a}(0) \quad . \tag{10.9}$$

There is also an n×n constant matrix

$$\underline{\gamma} = (\gamma_i^j)$$

such that:

$$\left[B_2, a_i \frac{\partial}{\partial y}\right] = \gamma_i^j a_j \frac{\partial}{\partial y} .$$

Then,

$$x \frac{da_i}{dx} = \gamma_i^j a_j ,$$

or

$$x\underline{\lambda a} = \underline{\gamma a}$$

or

$$x\underline{\lambda} e^{\underline{\lambda} x} \underline{a}(0) = \underline{\gamma} e^{\underline{\lambda} x} \underline{a}(0) .$$

Using the power series expansion for $e^{\underline{\lambda} x}$, we have:

$$\underline{\gamma a}(0) = 0 \tag{10.10}$$

$$\frac{1}{j!} \underline{\gamma \lambda}^j \underline{a}(0) = \frac{1}{(j-1)!} \underline{\lambda}^j \underline{a}(0) \tag{10.11}$$

for $j = 1,2,\ldots$

Hence:

$(\underline{\lambda}^j \underline{a}(0))$ is an eigenvector for $\underline{\gamma}$, with eigenvalue j. (10.12)

As in Section 9, the finite dimensionality of $\underset{\sim}{H}$ means that $\underline{\gamma}$ can have only a finite number of eigenvalues, hence that:

$$\underline{\lambda}^j \underline{a}(0) = 0 \quad \text{for } j \text{ sufficiently large} \tag{10.13}$$

Again, repeating the argument given in Section 9, we see that the linear independence of the functions

$$a_1(x),\ldots,a_n(x)$$

forces $\underline{a}(0)$ to be a cyclic vector for $\underline{\lambda}$. In particular,

$$\underline{\lambda}^n = 0 , \tag{10.4}$$

i.e., $\underline{\lambda}$ is nilpotent.

Also,

$\underline{\gamma}$ is determined by $\underline{\lambda}$ and $\underline{a}(0)$ satisfying these conditions.

Let us now sum up as follows:

Theorem 10.2. $\underset{\sim}{G}$ has a canonical form parameterized by an n×n nilpotent matrix $\underline{\lambda}$ and a cyclic vector $\underline{a}(0)$ for $\underline{\lambda}$.

$$\boxed{e^{\underline{\lambda} x}\, \underline{a}(0) \frac{\partial}{\partial y},\ y \frac{\partial}{\partial y},\ \frac{\partial}{\partial x},\ x \frac{\partial}{\partial x}} \tag{10.15}$$

11. $\underset{\sim}{H}$ ABELIAN, $\dim(\underset{\sim}{G}/\underset{\sim}{H}) = 3$

$\underset{\sim}{G}/\underset{\sim}{H}$ is now semisimple. By the Levi-Malcev theorem, there is a semisimple subalgebra $\underset{\sim}{S}$ of $\underset{\sim}{G}$ isomorphic to $\underset{\sim}{G}/\underset{\sim}{H}$ such that:

$\underset{\sim}{G}$ is a semidirect sum of $\underset{\sim}{H}$ and $\underset{\sim}{S}$

$\underset{\sim}{S}$ is isomorphic to $\underset{\sim}{G}/\underset{\sim}{H}$

Of course, $\underset{\sim}{S}$ is isomorphic to the Lie algebra of $SL(2,R)$, i.e., to the Lie algebra of the vector fields

$$\frac{\partial}{\partial x},\ x\frac{\partial}{\partial x},\ x^2\frac{\partial}{\partial x} \quad .$$

Let $\underset{\sim}{S}'$ be the solvable subalgebra of $\underset{\sim}{S}$ which is isomorphic to the subalgebra generated by $\partial/\partial x$, $x(\partial/\partial x)$. Set:

$$\underset{\sim}{G}' = \underset{\sim}{H} + \underset{\sim}{S}' \quad .$$

$\underset{\sim}{G}'$ then satisfies the hypotheses of Section 9. Look at the canonical form 9.22 for $\underset{\sim}{G}'$. From the analyses given in Section 9, we see that the last two vector fields in the list 9.22 may be chosen arbitrarily in $\underset{\sim}{G}'$, so long as they are linearly independent and <u>do not belong to</u> $\underset{\sim}{H}$. In particular, they may be chosen to <u>generate the subalgebra</u> $\underset{\sim}{S}'$. This is possible only if

$$\alpha = 0 \quad ,$$

which implies the following canonical form for $\underset{\sim}{G}'$.

$$\boxed{e^{\underline{\lambda}x}\ \underline{a}(0)\ \frac{\partial}{\partial y},\ \frac{\partial}{\partial x},\ x\frac{\partial}{\partial x} + cy\frac{\partial}{\partial y}} \tag{11.1}$$

Again, $\underline{\lambda}$ is a nilpotent $n\times n$ matrix, and $\underline{a}(0)$ is a cyclic vector for $\underline{\lambda}$. c is a real constant.

We must now be able to add to this canonical form an element $B_3 \varepsilon \underset{\sim}{S}$, such that:

$$\left[\frac{\partial}{\partial x}, B_3\right] = 2\left(x\frac{\partial}{\partial x} + cy\frac{\partial}{\partial y}\right) \tag{11.2}$$

$$\left[x\frac{\partial}{\partial x} + cy\frac{\partial}{\partial y}, B_3\right] = B_3 \tag{11.3}$$

$$[\underset{\sim}{H}, B_3] \subset \underset{\sim}{H} \ . \tag{11.4}$$

Suppose that B is of the following general form:

$$B_3 = a(x)\frac{\partial}{\partial x} + b(x,y)\frac{\partial}{\partial y} \tag{11.5}$$

Using 11.2, we have:

$$\frac{da}{dx} = 2x \tag{11.6}$$

$$\frac{\partial b}{\partial x} = 2cy \tag{11.7}$$

These relations imply that:

$$a = x^2 + c_1 \tag{11.8}$$

$$b = 2cxy + b_1(y) \ , \tag{11.9}$$

hence:

$$B_3 = (x^2+c_1)\frac{\partial}{\partial x} + (2cxy+b_1(y))\frac{\partial}{\partial y} \qquad (11.10)$$

Combine 11.3 and 11.10:

$$\left[x\frac{\partial}{\partial x} + cy\frac{\partial}{\partial y}, B_3\right] = (2x^2-(x^2+c_1))\frac{\partial}{\partial x} + 2cy\frac{\partial}{\partial y} + 2cxy\frac{\partial}{\partial y}$$

$$+ cy\frac{db_1}{dy}\frac{\partial}{\partial y} - (2cxy+b_1(y))c\frac{\partial}{\partial y}$$

$$= (x^2-c_1)\frac{\partial}{\partial x} + (2cy+2xy(c-2c^2) + cy\frac{db_1}{dy}$$

$$- cb_1(y))\frac{\partial}{\partial y}$$

$$= (x^2+c_1)\frac{\partial}{\partial x} + (2exy+b_1(y))\frac{\partial}{\partial y}$$

This identity forces the following relations:

$$c_1 = 0 \qquad (11.11)$$

$$c = 0 \qquad (11.12)$$

$$b_1 = 0 \qquad (11.13)$$

Hence:

$$B_3 = x^2\frac{\partial}{\partial x}$$

The canonical form for $\underset{\sim}{G}$ is then:

$$\boxed{e^{\underline{\lambda} x}\ \underline{a}(0)\ \frac{\partial}{\partial y},\ \frac{\partial}{\partial x},\ x\,\frac{\partial}{\partial x},\ x^2\,\frac{\partial}{\partial x}} \tag{11.14}$$

There are further restrictions on $\underline{\lambda}$ now in addition to the requirement found earlier that it be <u>nilpotent</u>. In fact, we have:

$$\left[\frac{\partial}{\partial x},\ e^{\underline{\lambda} x}\underline{a}(0)\right] = \underline{\lambda} e^{\lambda x}\underline{a}(0)$$

$$\left[x\,\frac{\partial}{\partial x},\ e^{\lambda x}\underline{a}(0)\right] = \underline{\lambda}_2 e^{\lambda x}\underline{a}(0)$$

$$\left[x^2\,\frac{\partial}{\partial x},\ e^{\lambda x}\underline{a}(0)\right] = \underline{\lambda}_3 e^{\lambda x}\underline{a}(0)$$

The assignment

$$\begin{aligned} \frac{\partial}{\partial x} &\to \underline{\lambda} \\ x\,\frac{\partial}{\partial x} &\to \underline{\lambda}_2 \\ x^2\,\frac{\partial}{\partial x} &\to \underline{\lambda}_3 \end{aligned} \tag{11.15}$$

now define matrix representations of the Lie algebra $\underset{\sim}{S}$, which is of course just the Lie algebra of SL(2,R).

We know (theorem of Weyl and J. H. C. Whitehead) that such a matrix representation is <u>completely reducible</u>. This implies that $\underset{\sim}{H}$ is the direct sum

$$\underset{\sim}{H} = \underset{\sim}{H}_1 \oplus \cdots \oplus \underset{\sim}{H}_m$$

of linear subspaces, in each of which Ad $\underset{\sim}{S}$ acts irreducibly. The dimensions of the real irreducible representations are:

$$1, 2, 3, \ldots, 2\ell+1 \quad ,$$

where ℓ is an integer or a half integer. The matrices $\underline{\lambda}, \underline{\lambda}_2, \underline{\lambda}_3$ in these irreducible representations are given by formulas that are well-known in the physics literature. (See LMP, Vol. II, Miller [1]).

The canonical forms are then determined by the integers:

$$m, \dim \underset{\sim}{H}_1, \ldots, \dim \underset{\sim}{H}_m$$

In particular, notice that the canonical form now only depends on discrete parameters. Again this is a reflection of the relative "rigidity" with respect to deformations of semisimple Lie algebras. (If one allowed $\underset{\sim}{H}$ to be infinite dimensional, the possibility of continuous parameters would reappear.)

12. $\underset{\sim}{H}$ SOLVABLE, NON-ABELIAN, $\dim (\underset{\sim}{G}/\underset{\sim}{H}) = 3$

$\underset{\sim}{G}/\underset{\sim}{H}$ is again semisimple, and there is a semisimple subalgebra $\underset{\sim}{S} \subset \underset{\sim}{G}$ such that:

$\underset{\sim}{G}$ is a semidirect sum of $\underset{\sim}{S}$ and $\underset{\sim}{H}$

Set:

$$\underset{\sim}{H}_1 = [\underset{\sim}{H},\underset{\sim}{H}] = \text{first derived algebra of } \underset{\sim}{H}.$$

We know from previous "canonical form" work in this chapter that:

$$\dim (\underset{\sim}{H}/\underset{\sim}{H}_1) = 1 .$$

Set:

$$\underset{\sim}{G}_1 = \underset{\sim}{H}_1 + \underset{\sim}{S} .$$

$\underset{\sim}{H}_1$ is abelian, and $\underset{\sim}{G}_1$ is a Lie algebra <u>to which the work of Section 11 applies</u>. $\underset{\sim}{G}_1$ has the following canonical form:

$$\boxed{a_1(x)\,\frac{\partial}{\partial y}\,,\ldots,\, a_n(x)\,\frac{\partial}{\partial y},\ \frac{\partial}{\partial x},\ x\,\frac{\partial}{\partial x},\ x^2\,\frac{\partial}{\partial x}}$$

By the complete reducibility property of semisimple Lie algebra representations, $\underset{\sim}{H}$ can be written as a vector space direct sum

$$\underset{\sim}{H} = \underset{\sim}{H}' + \underset{\sim}{H}_1$$

such that

$$[\underset{\sim}{S},\underset{\sim}{H}] \subset \underset{\sim}{H}'$$

Now, $\underset{\sim}{H}'$ is one-dimensional. A semisimple Lie algebra has no non-zero one-dimensional representations. This fact forces:

$$[\underset{\sim}{S},\underset{\sim}{H}'] = 0 \ . \tag{12.1}$$

Suppose the vector field A is a generator of $\underset{\sim}{H}'$. It must be tangent to the fiber of π, i.e., if the adapted coordinate system has the form

$$A = a(x,y) \frac{\partial}{\partial y} \ . \tag{12.2}$$

$\underset{\sim}{S}$ is spanned by

$$\frac{\partial}{\partial x} , \quad x \frac{\partial}{\partial x} , \quad x^2 \frac{\partial}{\partial x} \ .$$

Hence, 12.1 forces:

$$a \text{ is a function } a(y) \text{ of } y \text{ above} \tag{12.3}$$

Now,

$$\left[a_1 \frac{\partial}{\partial y}, A\right] = a_1 \frac{da}{dy} \frac{\partial}{\partial y} \tag{12.4}$$

The right hand side of 12.4 must be a vector field in $\underset{\sim}{H}_1$, whose coefficients then are <u>functions of</u> x <u>above</u>. Hence,

$$a(y) = cy + c' \tag{12.5}$$

where c,c' are constants.

<u>Case 1</u>. $c \neq 0$

Changing variables $y \to cy + c'$, we have the following canonical form:

$$\boxed{a_1(x)\frac{\partial}{\partial y}, \ldots, a_n(x)\frac{\partial}{\partial y}, y\frac{\partial}{\partial y}, \frac{\partial}{\partial x}, x\frac{\partial}{\partial x}, x^2\frac{\partial}{\partial x}} \tag{12.6}$$

Case 2. $c = 0$.

The canonical form is:

$$\boxed{a_1(x)\frac{\partial}{\partial y}, \ldots, a_n(x)\frac{\partial}{\partial y}, \frac{\partial}{\partial y}, \frac{\partial}{\partial x}, x\frac{\partial}{\partial x}, x^2\frac{\partial}{\partial x}} \tag{12.7}$$

In each case, the $a_1(x),\ldots,a_n(x)$ are determined as before by means of a nilpotent $n \times n$ matrix $\underline{\lambda}$, which forms part of a matrix representation of $SL(2,R)$.

13. SUMMARY OF CANONICAL FORMS AND FINAL REMARKS

Here are the canonical forms for $\underset{\sim}{G}$ collected together:

$\underset{\sim}{H}$ abelian, $\underset{\sim}{G} = \underset{\sim}{H}$.

$$\boxed{a_1(x)\frac{\partial}{\partial y}, \ldots, a_n(x)\frac{\partial}{\partial y}} \tag{13.1}$$

$a_1,\ldots,a_n$ are linearly independent, but otherwise arbitrary functions of x.

$\underset{\sim}{H}$ abelian, $\dim \underset{\sim}{G}/\underset{\sim}{H} = 1$

$$\boxed{e^{\underline{\lambda} x}\ \underline{a}(0)\ \frac{\partial}{\partial y},\ \frac{\partial}{\partial x}} \tag{13.2}$$

where $\underline{\lambda}$ is an arbitrary n×n real matrix, $\underline{a}(0)$ an arbitrary n-column vector of R^n which is cyclic for $\underline{\lambda}$, i.e., such that $\underline{a}(0)$, $\underline{\lambda a}(0),\ldots,\ \underline{\lambda}^{n-1}\underline{a}(0)$ are linearly independent.

$\underset{\sim}{H}$ solvable, non-abelian, $\underset{\sim}{G} = \underset{\sim}{H}$

$$\boxed{a_1(x)\ \frac{\partial}{\partial y}\ ,\ldots,\ a_n(x)\ \frac{\partial}{\partial y},\ y\ \frac{\partial}{\partial y}} \tag{13.3}$$

The $a_1,\ldots,a_n$ are linearly independent, but otherwise arbitrary functions of x.

$\underset{\sim}{H}$ semisimple, $\underset{\sim}{G} = \underset{\sim}{H}$

$$\boxed{\frac{\partial}{\partial y},\ y\ \frac{\partial}{\partial y},\ y^2\ \frac{\partial}{\partial y}} \tag{13.4}$$

$\underset{\sim}{H}$ <u>non-abelian</u>, <u>solvable</u>, $\dim(\underset{\sim}{G}/\underset{\sim}{H}) = 1$

$$e^{\underline{\lambda} x} \underline{a}(0) \frac{\partial}{\partial y}, \; y \frac{\partial}{\partial y}, \; x \frac{\partial}{\partial x} \tag{13.5}$$

$\underline{\lambda}$ an n×n matrix, $\underline{a}(0)$ a cyclic column vector for $\underline{\lambda}$, the characteristic polynomial for $\underline{\lambda}$ equals its minimal polynomial.

$\underset{\sim}{H}$ <u>semisimple</u>, $\underset{\sim}{G}/\underset{\sim}{H}$ <u>arbitrary</u>

$$\frac{\partial}{\partial y}, \; y \frac{\partial}{\partial y}, \; y^2 \frac{\partial}{\partial y}, \; \frac{\partial}{\partial x} \tag{13.6}$$

or

$$\frac{\partial}{\partial y}, \; y \frac{\partial}{\partial y}, \; y^2 \frac{\partial}{\partial y}, \; \frac{\partial}{\partial x}, \; x \frac{\partial}{\partial x} \tag{13.7}$$

or

$$\frac{\partial}{\partial y}, \; y \frac{\partial}{\partial y}, \; y^2 \frac{\partial}{\partial y}, \; \frac{\partial}{\partial x}, \; x \frac{\partial}{\partial x}, \; x^2 \frac{\partial}{\partial x} \tag{13.8}$$

$\underset{\sim}{H}$ <u>abelian</u>, dim $\underset{\sim}{G}/\underset{\sim}{H} = 2$

$$\boxed{\begin{aligned} & e^{\underline{\lambda} x}\, \underline{a}(0)\, \frac{\partial}{\partial y},\ \frac{\partial}{\partial x},\ x\, \frac{\partial}{\partial x} \\ & + \left(\left\langle \alpha,\ \frac{e^{\underline{\lambda} x}-1}{\underline{\lambda}}\, \underline{a}(0) \right\rangle + cy \right) \frac{\partial}{\partial y} \end{aligned}} \tag{13.9}$$

where:

$\underline{\lambda}$ <u>is an n×n nilpotent matrix</u>, $\underline{\lambda}^{n-1} \neq 0$.

$\underline{a}(0)$ <u>is a cyclic vector for</u> $\underline{\lambda}$.

α <u>is a column vector of</u> R^n.

c <u>is a real constant</u>.

$\underset{\sim}{H}$ <u>non-abelian solvable</u>, dim $(\underset{\sim}{G}/\underset{\sim}{H}) = 2$

$$\boxed{e^{\underline{\lambda} x}\underline{a}(0)\, \frac{\partial}{\partial y},\ y\, \frac{\partial}{\partial y},\ \frac{\partial}{\partial x},\ x\, \frac{\partial}{\partial x}} \tag{13.10}$$

$\underline{\lambda}$ is a n×n nilpotent matrix, $\underline{a}(0)$ a cyclic column vector, $\underline{\lambda}^{n-1} \neq 0$.

$\underset{\sim}{H}$ <u>abelian</u>, dim $(\underset{\sim}{G}/\underset{\sim}{H}) = 3$

$$\boxed{e^{\underline{\lambda}x}\underline{a}(0)\ \frac{\partial}{\partial y},\ \frac{\partial}{\partial x},\ x\,\frac{\partial}{\partial x},\ x^2\,\frac{\partial}{\partial x}} \tag{13.11}$$

$\underline{\lambda}$ is a n×n nilpotent matrix, $\underline{\lambda}^{n-1} \neq 0$, $\underline{a}(0)$ is cyclic for $\underline{\lambda}$. Further, there are two additional matrices $\underline{\lambda}_2, \underline{\lambda}_3$ such that

$$\underline{\lambda}_1, \underline{\lambda}_2, \underline{\lambda}_3$$

satisfy:

$$[\underline{\lambda}, \underline{\lambda}_2] = \underline{\lambda}$$

$$[\underline{\lambda}, \underline{\lambda}_3] = 2\underline{\lambda}_2$$

$$[\underline{\lambda}_2, \underline{\lambda}_3] = \underline{\lambda}_3$$

$\underset{\sim}{H}$ <u>non-abelian</u>, <u>solvable</u> dim $(\underset{\sim}{G}/\underset{\sim}{H}) = 3$

$$\boxed{e^{\underline{\lambda}x}\underline{a}(0)\ \frac{\partial}{\partial y},\ \frac{\partial}{\partial y},\ \frac{\partial}{\partial x},\ x\,\frac{\partial}{\partial x},\ x^2\,\frac{\partial}{\partial x}} \tag{13.12}$$

or

$$\boxed{e^{\underline{\lambda}x}\underline{a}(0)\ \frac{\partial}{\partial y},\ y\,\frac{\partial}{\partial y},\ \frac{\partial}{\partial x},\ x\,\frac{\partial}{\partial x},\ x^2\,\frac{\partial}{\partial x}} \tag{13.13}$$

In either case, $\underline{\lambda}$ and $\underline{a}(0)$ satisfy the same conditions as for 13.11.

Final Remarks

There is a more elegant proof available of the fact that $\underline{\lambda}$ is nilpotent if $\underset{\sim}{G}/\underset{\sim}{H}$ is non-abelian. In fact, consider the abstract 2-dimensional, non-abelian, solvable Lie algebra spanned by elements

$$A, A_1$$

such that

$$[A, A_1] = A \quad .$$

Let

$$A \to \underline{\lambda}$$

$$A_1 \to \underline{\lambda}_1$$

be representations of this Lie algebra by (n×n) matrices. Then,

$$\underline{\lambda} = [\underline{\lambda}, \underline{\lambda}_1] \equiv \underline{\lambda\lambda}_1 - \underline{\lambda}_1\underline{\lambda}$$

Hence,

$$\text{trace}\ (\underline{\lambda}) = 0 \quad ,$$

since the trace of a commutator is zero. Now,

$$\underline{\lambda}^2 = \underline{\lambda}[\underline{\lambda}, \lambda_1] = [\underline{\lambda}, \underline{\lambda}\lambda_1]$$

Hence,

$$\text{trace } (\underline{\lambda}^2) = 0 \ .$$

Continuing in this way, we see that:

$$\text{trace } (\underline{\lambda}^2) = 0$$

for all integers j. This forces $\underline{\lambda}$ to be a nilpotent matrix. (See VB, vol. II, Chapter 3 for elaboration of this argument.)

Several tasks remain before the classification of the Lie equivalence classes can be considered as really complete. First, one must decide when two of the canonical forms listed above are really Lie equivalent. In each case, there is a group acting on the canonical form of a certain type, and what is desired is to construct a fundamental domain for this group, i.e., a subset which meets each orbit precisely once. I believe that in each of the cases listed above it is not too difficult to do this, but I have not carried it out completely.

Another interesting and (perhaps) important problem is to see canonical forms of various types fit in and deform into each other. Here is one simple problem of this sort.

Suppose $\underset{\sim}{G}, \underset{\sim}{G}'$ are Lie algebras of vector fields with

$$\underset{\sim}{G} \subset \underset{\sim}{G}' \ .$$

Can one choose the canonical forms listed above so that canonical coordinate systems agree?

Is there some "master" Lie algebra of vector fields so that all others are obtained by taking its subalgebras or deformations?

There are clearly many problems of this sort, which could keep an army of mathematicians busy for many years, studying Lie algebras of vector fields (perhaps only in the "formal power series sense"), their subalgebras, and deformations.

Gelfand and his coworkers have developed methods for calculating Lie algebra cohomology of Lie algebras of vector fields whose coefficients are "formal" (i.e., possibly non-convergent) power series in the underlying variables $x_1, x_2, \ldots$. Notice that a good deal of Lie's work may be interpreted in this framework. Here is one such problem.

Let $X = R^n$, and let

$$V_\omega(X)$$

be the Lie algebra of vector fields whose coefficients are formal power series in the variables of x. Develop the notion of "Lie equivalence" in terms of changes of variables that are given by formal power series. It would then be interesting to see which Lie subalgebras of $V_\omega(X)$ are

Lie equivalent to Lie subalgebras of $V(X)$, i.e., to vector fields with convergent power series expansions for the coefficients.

It is clear from the work in this paper that this is so if $n = 1$ and the Lie algebra is finite dimensional. I have not done so, but I would think that the analogous problem is readily accessible for $n = 2$ by the methods developed in this paper. $n = 3$ or 4 might be challenging.

In general, it seems feasible to study the higher dimensional and global analogs of the problems treated here in the case of one and two dimensional manifolds. The generalization of the basic classification problem itself is probably too complicated to be interesting, but it would be very interesting to study, say, the ways finite dimensional Lie algebras of vector fields can act on manifolds to preserve a foliation, for example. There are many problems in physics, control and system theory, etc., which await treatment by Lie's methods.

As I stated in the beginning to Vol. IX, it is to my view one of the great mysteries of the history of science why only the most trivial parts of Lie's thought survived into the twentieth century. Clearly, Lie thought in terms of a very grandiose generalization of Galois theory, to cover the whole area of differential equations. The only one to pick up Lie's ideas in a strong way was another genius, E. Cartan, who developed his own mysticism and was equally incomprehensible to his contemporaries. My opinion is that there are many scientific problems of the highest importance which await the Lie touch. Certainly, one would think that elementary particle physicists trying to understand the role that groups like SU(3) play in their discipline would be proclaiming (and studying) Lie as their Prophet, but it seems that it will take a new Einstein for that to happen!

Chapter G

RELATIONS WITH SYSTEMS-CONTROL THEORY

1. INTRODUCTION

One of my main reasons for translating and interpreting Lie's work in this way is that I believe there are important and fruitful interelations between Lie's work on groups and differential equations and the modern disciplines of systems and control theory. In this chapter I will briefly indicate some of these interconnections. For more detail, see the other volumes in the Interdisciplinary series, particularly III, VIII, and IX. The further volumes concerned with Lie's work will also deal with these applications in more detail.

2. INPUT-OUTPUT SYSTEMS MODELLED BY DIFFERENTIAL EQUATIONS AND LIE ALGEBRAS OF VECTOR FIELDS

As explained in Volume VIII one aim of "systems theory" is to study the mathematical properties of devices which connect a set of "inputs" into a set of "outputs". Such a system is typically denoted--at least in the engineering literature--by a block diagram of the form:

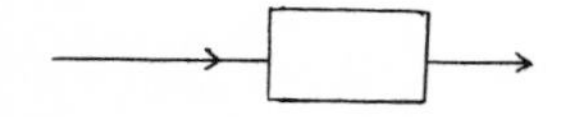

Many of the systems of greatest interest in engineering involve inputs and outputs which are scalar or vector valued functions of a time variable t, with the relation between input and output described by differential equations. From the mathematical point of view this viewpoint is also very convenient, since it enables one to begin work with a minimum of distraction and generalized nonsense. (Notice that the best texts and treatises on systems-control theory, e.g., Anderson-Moore [1], Brockett [1], Rosenbrock [1], usually take this point of view.)

Here is one basic framework. X, U, Y are real, finite dimensional vector spaces, called the <u>state input</u>, and <u>output</u> spaces. An <u>input-output system</u> is defined by a system of ordinary differential equations of the following form:

$$\begin{aligned} \frac{dx}{dt} &= f(x,u,t) \\ y &= g(x,u,t) \end{aligned} \qquad (2.1)$$

x denotes a point of X; u denotes a point of U; y denotes a point of Y. t denotes a point of a one-dimensional time interval, T. f and g are maps with the following domains and ranges:

$$f: X \times U \times T \to X$$

$$g: X \times U \times T \to Y \quad .$$

We shall introduce a set of vector fields on $X \times T$, parameterized by elements of U. Symbolically, to each $u \in U$ we associate the vector field:

$$A_u = f(x,u,t) \frac{\partial}{\partial x} + \frac{\partial}{\partial t} \tag{2.2}$$

To understand what 2.2 means in terms of manifold theory, let

$$(x^i)$$

be a set of <u>linear</u> coordinates for the vector space X. (Choose indices, and the summation convention, as follows:

$$i \leq i,j \leq n = \dim X .)$$

Then, the first of Equations 2.1 takes the following form:

$$\frac{dx^i}{dt} = f^i(x,u,t) \tag{2.3}$$

The vector field A_u has the form:

$$A_u = f^i(x,u,t) \frac{\partial}{\partial x^i} + \frac{\partial}{\partial t} \tag{2.4}$$

There are <u>three</u> useful and significant meanings that can be given to the symbol "u" in 2.4.

<u>First meaning</u>: u does not depend on x or t, but is an arbitrary element of the input (or "control") space U. Thus, an <u>orbit curve</u> of A_u, in this case, is a curve $t \to x(t)$ which is a solution of 2.1 with <u>constant input</u> or <u>constant control</u>.

Second meaning: u is a function of t. It is called the control. There is then a curve

$$t \to (x(t), y(t))$$

in $X \times Y$, a solution of 2.1, determined by u, and an initial vector $x_0 \in X$. The curve $t \to x(t)$ in X is called the state curve. The curve $t \to y(t)$ is called the output curve. The "system" as a whole may be thought of as determining a map:

(control curves) × (initial state vector) → (output curves)

Third meaning: Let u be a map

$$X \times T \to U \quad .$$

We can then solve the differential Equations 2.1, which take the form:

$$\frac{dx}{dt} = f(x, u(x(t), t), t)$$
$$y = g(x, u(x(t), t), t) \tag{2.5}$$

We call such a map a control law or a control strategy. From either point a view, a "system" is a map

(control strategies) × (initial state vector) → (input curves)

(Such a function is also sometimes called a feedback control law.)

3. SYSTEMS ASSOCIATED WITH LIE ALGEBRAS

Let X, U, Y, T be as in Section 2. Let $\underset{\sim}{G}$ be a Lie algebra of vector fields on X. Consider a system, with U as <u>inputs</u>, X as <u>states</u>, Y as <u>outputs</u>, of the form:

$$\begin{aligned} \frac{dx}{dt} &= f(x,u,t) \\ y &= h(x,u,t) \end{aligned} \tag{3.1}$$

For each $(u,t) \in U \times T$, let:

$$A_{(u,t)} = f(x,u,t) \frac{\partial}{\partial x} \in V(X) \tag{3.2}$$

<u>Definition</u>. Let $\underset{\sim}{G}$ be a Lie algebra of vector fields on the state space X. $\underset{\sim}{G}$ is said to be <u>associated</u> to the system 2.1 if

$$A_{(u,t)} \in \underset{\sim}{G}$$

$$\text{for each } (u,t) \in U \times T .$$

To see the meaning of this condition, suppose that G is a Lie group acting on X, whose infinitesimal action is $\underset{\sim}{G}$. In particular, we suppose that $\underset{\sim}{G}$ is <u>finite dimensional</u> as a real vector space. Let $B_1,\ldots,B_m$ be vector fields on X that form a basis for $\underset{\sim}{G}$. Then, there are relations of the form

$$A_{(u,t)} = a_1(u,t)B_1+\cdots+a_m(u,t)B_m$$

Let $t \to u(t)$ be a <u>control curve</u>, i.e., a curve in the input space U. Consider the curve

$$t \to A_{u(t),t}$$

in $\underset{\sim}{G}$. It generates a <u>flow</u> on X. The curve $t \to u(t)$, together with the initial state vector x_0, generates a curve $t \to x(t)$ which is a solution of 3.1. Let $t \to g(t)$ be the curve in G such that the infinitesimal generator of the flow

$$x \to g(t)$$

is $t \to A_{u(t),t}$. We see that:

$$x(t) = g(t)(x_0) \quad ,$$

i.e.,

$x(t)$ <u>is the orbit of</u> $g(t)$.

We can then sum up this discussion as follows:

<u>Theorem 3.1</u>. Suppose the Lie algebra $\underset{\sim}{G}$ is associated to the system 3.1, and $\underset{\sim}{G}$ arises from the action of a Lie group G on X. Given a control curve $u(t)$ and an initial state vector x_0, there is a curve $t \to g(t)$ in G such that the solution of the Equations 3.1 may be written in the following form:

$$\begin{aligned} x(t) &= g(t)(x_0) \\ y(t) &= h(g(t)(x_0), u(t), t) \end{aligned} \tag{3.3}$$

In other words, the Lie group G can be used to write down an explicit formula for the solution of the system. As we shall see later on, the systems associated with Lie groups in this way form a very natural class of systems, generalizing the linear systems about which so much is known, and which are so important for applications. One may hope (and expect) that this larger class of systems will also be useful and important.

Example. G = group of linear fractional transformations

Suppose

$$X = R ,$$

with the system 3.1 of the following form:

$$\frac{dx}{dt} = \alpha(u,t)x^2 + \beta(u,t)x + \gamma(u,t) \tag{3.4}$$

Then, we know that the solution of 3.4 is of the form of a linear fractional transformation, i.e., the relations between input-state-output take the following form:

$$x(t) = \frac{a(t)x_0 + b(t)}{c(t)x_0 + d(t)}$$

$$y(t) = h(x(t),u(t),t) \tag{3.5}$$

(Of course, the coefficients a,b,c,d in 3.5 depend also on $u(t)$; in fact, in a rather complicated way. Unless the control is constant, the explicit formulas cannot be written down, since they depend on a solution of a Riccati differential equation.)

Remark. A possibly useful terminology for these systems is Lie systems. The only trouble with it is that there is a possibility of confusion with a similar use of the term in the classical literature. (See Volume IX.)

4. STATE EQUIVALENCE AND HOMOMORPHISMS OF SYSTEMS: APPLICATION OF LIE'S CLASSIFICATION THEOREMS

As we have seen, the basic work in this paper by Lie involves the enumeration of (local) equivalence classes of actions of local Lie groups acting on manifolds. Unfortunately, Lie never pushed his methods beyond the case of 2 dimensions (there are partial results on 3-dimensions in "Transformationsgruppen") and the subject has not really been worked on since with more modern tools. However, the basic idea of "equivalence" inherent in Lie's work provides an interesting systems-theoretic concept.

Let

$$\frac{dx}{dt} = f(x,u,t) \qquad y = h(x,u,t) \tag{4.1}$$

$$\frac{dx'}{dt} = f'(x',u',t) \qquad y' = h'(x',u',t) \tag{4.2}$$

be two systems. A map

$$\phi: x \to x' = \phi(x)$$

from $X \to X'$ is said to be a state homomorphism of one system to the other if the following condition is satisfied:

For each solution

$$t \to (x(t),u(t),y(t))$$

of 4.1, the curve

$$t \to (\phi(x(t)),u(t),y(t)$$

is a solution of 4.2.

If, in addition, ϕ is a diffeomorphism, it is said to define a state equivalence.

Remarks. More general sorts of equivalence are possible and important. For example, feedback equivalence is a diffeomorphism map

$$x \to x'(x)$$
$$u \to u'(x,u)$$

which takes solutions of one system into solutions of the other.

Similarly, one can define local state equivalence of two systems, as a diffeomorphism of an open subset of X with an open subset of X' which takes solutions into solutions. This is an appropriate notion to tie in with Lie's work.

Consider the system 4.1. Let $\underset{\sim}{G}$ be a Lie algebra of vector fields on X-space. For each $(u,t) \in U \times T$, set:

$$A_{(u,t)} = f(x,u,t) \frac{\partial}{\partial x} \tag{4.3}$$

Suppose that the system 4.1 is associated with $\underset{\sim}{G}$, in the sense that, for each pair (u,t), the vector field 4.3 belongs to $\underset{\sim}{G}$.

Following Lie's ideas, we can now seek (local) canonical forms for $\underset{\sim}{G}$. This can either be thought of as a new coordinate system for X in which the formulas for $\underset{\sim}{G}$ take an especially simple form, or representatives of Lie subalgebras of $V(X)$ which are equivalent in the sense of Lie. Such a new coordinate system can be thought of as defining a local equivalence of the system with another.

Here is a typical result:

Theorem 4.1. Suppose $\dim X = 1$, and that $\underset{\sim}{G}$ is a finite dimensional Lie algebra of vector fields. Then, the system 4.1 is locally equivalent to a system of the following form:

$$\begin{aligned} \frac{dx'}{dt} &= a(u,t)(x')^2 + b(u,t)x' + c(u,t) \\ y &= h(x',u,t) \end{aligned} \tag{4.4}$$

Here is an alternate way of stating the result.

Theorem 4.2. Keep the hypothesis of Theorem 4.1. Then, there is a function $x \to \phi(x)$ such that the input-output relations of the system 4.1 take the following form:

$$\begin{aligned} x(t) &= \phi^{-1}\left(\frac{\alpha_u(t)\phi(x_0) + \beta_u(t)}{\gamma_u(t)\phi(x_0) + \delta_u(t)}\right) \\ y(t) &= h(x(t),u(t),t) \end{aligned} \tag{4.5}$$

Remark. The coefficients $\alpha_u,\dots,\delta_u$ in 4.5 depend on the input control curve $t \to u(t)$ in a relatively complicated way. However, there is an evident interest for applications involving the computer that a representation of this form <u>exists</u>. The success of applications of the theory of <u>linear</u> systems often depends on the possibility of such explicit solution, with "unknown" coefficients which can be estimated in terms of the observed data. Thus, the systems associated with finite dimensional Lie algebras form a natural class generalizing the linear systems.

The results proved in this paper concerning the canonical forms for Lie algebras of vector fields on 2-dimensional manifolds then can be interpreted as proving local state-equivalence results for systems associated with Lie algebras of vector fields with 2-dimensional state spaces.

5. FEEDBACK EQUIVALENCE FOR SYSTEMS ASSOCIATED WITH FINITE DIMENSIONAL LIE ALGEBRAS

In the last section we have briefly discussed state equivalence of systems, and its relation to Lie theory in general, and the results of this paper in particular. Now we consider the other natural system concept, <u>feedback equivalence</u>.

Consider two systems:

$$\frac{dx}{dt} = f(x,u,t) \qquad y = h(x,u,t) \tag{5.1}$$

$$\frac{dx'}{dt} = f'(x',u',t)$$

$$y' = h'(x',u',t) \qquad (5.2)$$

System 5.1 is said to be *feedback equivalent to system 5.2* if there is a change of variable of the form:

$$x = x'$$

$$u = u' + \alpha(x,t) \qquad (5.3)$$

$$y = y'$$

which takes 5.1 into 5.2.

One can illustrate this concept by a block diagram. Suppose, for simplicity, that

$$x = y$$

$$x' = y'$$

System 5.1 can be taken of the form:

u → □ — x (5.4)

Attach a *feedback loop* to this system:

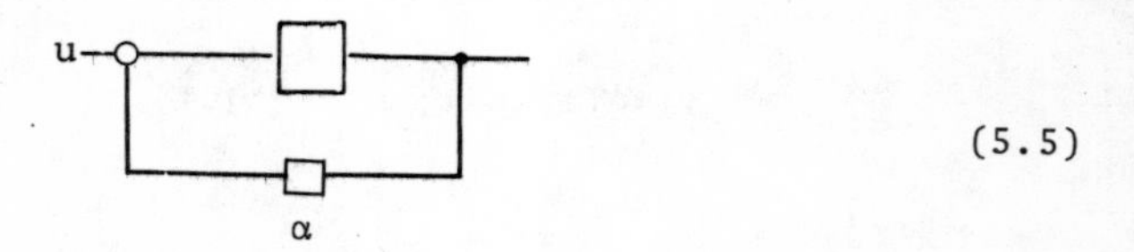

(5.5)

Here is the meaning of this: An input $u(t)$ comes in, goes through the box, with initial state vector set at x_0, and comes out as

$$x(t) \quad ,$$

uniquely defined by the following differential equation and initial condition

$$\frac{dx}{dt} = f(x,u(t),t); \qquad x(0) = x_0$$

Then, the signal $x(t)$ is passed through the <u>filter</u> α, resulting in the signal

$$\alpha(x(t),t) \quad .$$

This signal is then <u>fed back</u> and subtracted from the initial input signal u, resulting in a new input signal

$$u'(t) = u(t) - \alpha(x(t),t) \quad .$$

This input is then fed into the system, resulting in an output

$$t \to x'(t)$$

satisfying the following differential equation:

$$\frac{dx'}{dt} = f(x'(t),u(t)-\alpha(x'(t),t),t)$$
$$x'(0) = x_0 \tag{5.6}$$

<u>Define</u> $f'(x',u',t)$ by the following formula:

$$f'(x',u',t) = f(x',u'-\alpha(x',t),t) \quad (5.7)$$

We shall now show that the system resulting from modification of the system 5.4 by the feedback loop as indicated in 5.5 is precisely the system 5.2, with f' _defined_ by 5.7.

As we have calculated above in deriving Equation 5.6, we see that the response of the system 5.5 to the input $t \to u'(t)$ is the curve $t \to x'(t)$ satisfying:

$$\frac{dx'}{dt} = f(x'(t),u'(t)-\alpha(x'(t),t),t) \quad (5.8)$$

Insert 5.7 into the right hand side of 5.8:

$$\frac{dx'}{dt} = f'(x'(t),u'(t),t) \quad .$$

This shows that the diagram 5.5 is the correct one to represent the system 5.2.

In this way, we have motivated the terminology "feedback equivalent" in saying that two systems modelled by Equations 5.1, 5.2 are related via a change of variable of form 5.3. Let us now turn to the study of the Lie-theoretic aspects of this correspondence. Suppose that $\underset{\sim}{G}$ is a finite dimensional Lie algebra of vector fields on the state manifold X which is associated to the system 5.1. Recall that this means that, for each $(u,t) \in U \times T$, the vector field on X:

$$A_{(u,t)} = f(x,u,t)\frac{\partial}{\partial x} \qquad (5.9)$$

belongs to $\underset{\sim}{G}$.

We then set:

$$A'_{(u,t)} = f(x,u+\alpha(x,t),t)\frac{\partial}{\partial x} \qquad (5.10)$$

A basic question is then:

What are the conditions the "feedback filter" α must satisfy in order that the vector fields 5.10 belong to a finite dimensional Lie algebra of vector fields $\underset{\sim}{G}'$ on X? What possibilities are allowed for $\underset{\sim}{G}'$? These ideas might suggest a new area of study of Lie algebras of vector fields, the feedback equivalence of two such Lie algebras. The most important property of "feedback" from the point of view of applications is the possibility of stabilization of an unstable system by means of feedback. For example, this is how we drive a car or fly an airplane.

Example. Feedback equivalence of bilinear Lie systems with one-dimensional state spaces:

Suppose that the system is of the form:

$$\frac{dx}{dt} = (a_1u+a_0)x^2 + (b_1u+b_0)x + (c_1u+c_0) \qquad (5.11)$$

Let us change variables as follows:

$$x' = x$$
$$u' = u+\alpha(x) \quad .$$

After this change, 5.11 becomes the following system:

$$\frac{dx}{dt} = (a_1u+a_0)x^2 + (b_1u+b_0)x + (c_1u+c_0)$$
$$+ (a_1d(x)x^2+b_1d(x)+c_1\alpha)$$

We must look for the conditions that this system be associated with a finite dimensional Lie algebra. Given Lie's results, in this paper, on finite dimensional Lie algebras acting on a 1-dimensional manifold, it is evident that the condition is that:

$$a_1\alpha(x)x^2 + b_1\alpha(x) + c_1\alpha(x)$$

is a polynomial of degree at most <u>two</u>

Of course, this restriction on the feedback is evident in this simple case. The interest in the remark mainly lies in the possibility for its generalization to higher dimensional state spaces. An important property of linear systems is the possibility of changing an unstable system by feedback into a stable one. (See Wonham [1]). This sort of problem then has a natural generalization to Lie systems.

6. REMARKS ABOUT STABILITY OF SYSTEMS ASSOCIATED WITH LIE ALGEBRAS

"Stability" is obviously a topic of prime importance for the application of differential equations. For ordinary differential equations in their traditional "dynamic system" form, there is an extensive classical and modern literature. The key work was done by Poincaré and Liapounov, in the 19-th century, and only relatively minor details have been added since.

The introduction in recent times of input-output systems as interesting objects to study has suggested new formulations and methods for consideration of "stability". For example, from the practical point of view, what is often important is not "stability" in the way considered by the classical authors, but the possibility of "stabilization" by means of "feedback". For example, this is how we drive a car. Biological systems probably work in this way. (The eminent mathematician René Thom has attempted to provide an underlying mathematical metaphysics for biological systems--besides its fault of pomposity, it probably is too tied to classical mathematical stability ideas, and does not take into account--in its underlying intuition--these more recent ideas interrelating "stability" and "feedback".)

Unfortunately, very little is known about "input-output stability" for systems that go much beyond the linear time-invariant case. I hope that Lie theory may provide some useful ideas. This is the motivation for the brief remarks of this section.

Start with a "dynamical system", in its traditional form associated with physics:

$$\frac{dx}{dt} = f(x,t) \tag{6.1}$$

$$x \in X \equiv R^n; \quad t \in T \equiv R \quad .$$

The initial data $x_0 \equiv x(0)$ defines a map

$$X \to (\text{solutions of } 6.1) \subset (\text{curves in } X) \tag{6.2}$$

The system is thought of as "stable" if the map 6.2 is "continuous", with respect to some sort of natural topology on X and the space of curves in X. (Physically, "stability" means that "small changes in initial conditions" propogate into "small changes in the system at large times".)

Suppose the differential equation 6.1 generates a flow

$$t \to \phi_t$$

on X, i.e., the solutions of 6.1 are the <u>orbits</u> of the flow. "Stability" is often associated with the existence of a limit $\phi_\infty: X \to X$,

$$\lim_{t\to\infty} \phi_t = \phi_\infty \tag{6.3}$$

Now, in what precise sense this limit 6.3 is to be considered is not at all obvious. ϕ_∞ need not be a diffeomorphism of X. What is important (from the point of view of Lie theory) is that the naive physical ideas of "stability" and the naive geometric ideas of Lie's theory have some interconnection.

To see what might be involved here, consider the only situation which is really well understood, a <u>linear</u>, <u>time invariant</u> system:

$$\frac{dx}{dt} = Ax \quad , \tag{6.4}$$

$$x \in R^n, \qquad A \in L(R^n, R^n) \quad .$$

Thus,

$$x(t) = e^{At}(x_0) \quad , \tag{6.5}$$

and the flow it generates is given by:

$$\phi_t = e^{At} \tag{6.6}$$

If x_0' is another initial vector,

$$x'(t) = e^{At}(x_0') \quad ,$$

then

$$|x(t) - x'(t)| = |e^{At}(x_0 - x_0')| \leq e^{|A|t} \; x_0 - x_0'|$$

(| | denotes an appropriate norm on vectors and matrices.) We see that if the norm $|A|$ on matrices can be chosen so that:

$$|A| \leq 0 \quad , \tag{6.7}$$

then the system is <u>stable</u> in the usual sense. If

$$|A| < 0 \quad , \tag{6.8}$$

it is <u>asymptotically stable</u>, in the sense that:

$$|x(t) - c'(t)| \to 0 \tag{6.9}$$

as $t \to \infty$

For the usual sort of norms that can be chosen for matrices, 6.7 will be satisfied if:

All eigenvalues of A have non-positive real parts (6.10)

6.8 will be satisfied if:

All eigenvalues of A have negative real parts (6.11)

(The <u>Routh-Hurwitz criterion</u> (see Volume II of IM) describes polynomial inequality conditions on the matrix A which are equivalent to 6.11.)

These conditions have a Lie-theoretic interpretation. 6.11 means that:

$$\lim_{t\to\infty} \phi_t = 0 \tag{6.12}$$

Here, the limit is <u>pointwise</u>, and "0" on the right hand side means the map $0:R^n \to R^n$ which takes each element into the zero element of R^n.

In case condition 6.10 is satisfied,

$$\lim_{t\to\infty} (\phi_t \equiv e^{At})$$

may not exist <u>pointwise</u>. However, the limit can be considered to exist in a generalized, and/or ergodic, sense.

Return to the general system 6.1. Here is another Lie-theoretic situation which is associated with "stability". Suppose G is a Lie group acting on X, and that the flow ϕ_t generated by the Equation 6.1 has the property that:

$$\phi_t \in G \quad \text{for each} \quad t.$$

<u>Suppose</u> that K is a <u>compact</u> subgroup of G, and that:

$$\phi_t \in K \quad \text{for all} \quad t. \tag{6.13}$$

Then, we have stability in a weaker sense that:

$$|x(t)-x'(t)| \quad \text{remains bounded as} \quad t \to \infty$$

For, if

$$x_0 = x(0), \qquad x'(0) = x' \quad ,$$

then for all t

$$x(t), x'(t) \in K(x_0) \cup K(x_0') \quad ,$$

which is a <u>compact</u> subset of X.

Although this condition is completely trivial mathematically, it is often useful in practice, particularly if supplemented with Cartan's theorem that maximal compact connected subgroups of semi-simple connected Lie groups are conjugate.

Here is a typical simple application. Suppose:

$$X = R^{2m} ,$$

G = group of linear symplectic automorphisms of R^{2m}

The maximal compact subgroup of G is then

$$U(m) ,$$

the group of $m \times m$ unitary matrices considered as a group of $2m \times 2m$ real matrices. Suppose the system 6.1 is autonomous (i.e., t does not appear explicitly on the right hand side of 6.1) and the flow ϕ_t it generates belongs to G. (It is then a one-parameter subgroup of G. Physically, it represents the motion of a conservative, time-independent, linear mechanical system of a finite number of degrees of freedom.) Then if

$$\phi_t = e^{At}$$

$$A \in L(R^{2m}, R^{2m}) ,$$

we see (using Cartan's theorem) that $\{\phi_t\}$ lies in a compact subset of G if and only if the infinitesimal generator A satisfies the following property:

All eigenvalues of A are pure imaginary and A has simple elementary divisors.

Turn now to input-output systems. For simplicity, we suppose that:

output = state vectors.

Such a system then takes the following form:

$$\frac{dx}{dt} = f(x,u,t) \tag{6.14}$$

It defines a map:

$$X \times (\text{curves in } U) \to \text{curves in } X. \tag{6.15}$$

As a natural generalization of the classical notion of stability for classical systems 6.1 (with no controls) one may think of "stability" of 6.12 as expressing some continuity and/or boundedness property of the map 6.13.

For example, the simplest such notion is suggested by the obvious engineering condition that:

<u>bounded inputs give bounded outputs</u> (6.16)

This is called <u>input-output stability</u>. An excellent treatment of the relations between the classical and modern stability ideas is to be found in the book "Stability Theory of Dynamical Systems" by J. L. Willems.

Let us begin to think about the Lie-theoretic meaning of these ideas by consideration of a <u>linear time invariant system</u>, of the form:

$$\frac{dx}{dt} = Ax + Bu$$

$$A \in L(R^n, R^n), \qquad B \in L(R^m, R^n) \tag{6.17}$$

Then, the map 6.13 is determined by the following formula:

$$x(t) = \int_0^t e^{A(t-s)} Bu(s)\, ds + e^{At}x(0) \tag{6.18}$$

$s \to u(s)$ is the <u>input curve</u>, $x(0)$ is the <u>initial state vector</u>. $t \to x(t)$, given by formula 6.18, is the output curve, associated to the map 6.13. To investigate stability properties, let $\{u'(s)\}$ and $\{x'(0)\}$ be another input curve, $x'(t)$ the corresponding solution. Then

$$|x(t)-x'(t)|$$

$$\leq \int_0^t e^{|A|(t-s)}|B||u(s)-u'(s)|\, ds + e^{|A|t}|x(0)-x'(0)|$$

It is readily seen that a sufficient condition for $|x(t)-x'(t)|$ to remain <u>bounded</u> as $t \to \infty$ is that

$$|A| < 0 \, , \tag{6.19}$$

i.e., that A is a <u>stability matrix</u>.

Let us examine this from the Lie point of view. We can write the solution 6.16 of the system 6.15 in the form:

$$x(t) = g(t)(x(0) ,$$

where $t \to g(t)$ is a curve in the group of <u>affine automorphisms</u> of the vector space X. We can read off from 6.16 the formula for $g(t)$:

$$g(t)(x) = e^{At}(x) + z(t) , \tag{6.20}$$

with:

$$z(t) = \int_0^t e^{A(t-s)} Bu(s)\, ds \tag{6.21}$$

For example, suppose:

$$u(s) \equiv \text{constant} = u_0 .$$

Then,

$$z(t) = A^{-1}(1-e^{tA})Bu_0 \tag{6.22}$$

We see that:

If A is a stability matrix, i.e., if all its eigenvalues have negative real parts, then

$$\lim_{t\to\infty} z(t) = A^{-1}B(u_0)$$

In particular, the response of the system to a constant input is <u>bounded</u>.

In case the input is more general, time-varying, the argument can be refined to show that:

$$\lim_{t\to\infty} z(t) \quad \text{exists,}$$

so long as A is a stability matrix. In particular, the system 6.18 is <u>input-output stable</u>. In terms of Lie theory, this translates into the following condition:

$$\lim_{t\to\infty} g(t)(x) = z(\infty) \tag{6.23}$$

$$\text{for all} \quad x \in X$$

In particular, notice that the limiting map

$$g(\infty) = \lim_{t\to\infty} g(t)$$

is no longer a <u>diffeomorphism</u> of X, but rather a constant map. It is interesting, however, that "stability" means, in this case, that the curve in the Lie group determining the time-evaluation of the system approaches a limit which is a map of the state space onto itself. I believe (and conjecture) that this phenomenon generalizes to more complicated, nonlinear systems!

7. STABILITY OF A CONSTANT-COEFFICIENT, ONE-STATE, VARIABLE BILINEAR SYSTEM

The next-simplest examples of systems after the linear systems are the <u>bilinear systems</u>. Since their theory is considerably more complicated than for the linear case (and, in fact, requires Lie group theory in an essential way) I will restrict attention to the following very simple example.

$$\frac{dx}{dt} = u_1(t)x + u_2(t) \tag{7.1}$$

Here, "x" is a <u>one-dimensional</u> variable; $u_1(t)$, $u_2(t)$ are the <u>control functions</u>. The solution of 7.1 can, of course, be written down explicitly, but in a rather complicated formula. Let us first consider the simplest case, where the inputs u_1, u_2 are <u>constant</u>. The solution is then a special case of formulas 6.16 and 6.12:

$$\begin{aligned} x(t) &= e^{u_1 t}\, x(0) + u_1^{-1}\left(1-e^{tu_1}\right)u_2 \qquad (7.2) \\ &= g(t)(x(0)) \quad , \end{aligned}$$

where:

$$g(t)(x) = e^{u_1 t}\, x + u_1^{-1}\left(1-e^{tu_1}\right)u_2 \tag{7.3}$$

Hence, constant inputs go into <u>bounded</u> outputs if and only if:

$$u_1 \leq 0 \quad . \tag{7.4}$$

In this case,

$$\lim_{t\to\infty} g(t) = g(\infty) \ ,$$

where:

$$g(\infty)(x) = u_1^{-1} u_2 \tag{7.5}$$

if $u_1 < 0$

$$g(\infty)(x) = x + u_1^{-1}u_2 \qquad (7.6)$$
$$\text{if } u_1 = 0 .$$

Again, we see that "stability" is associated with a natural Lie group theoretic condition.

$g(t)$ is a one parameter subgroup of the group of affine automorphisms of the real line R. Notice that condition 7.4 defines a sub-semigroup of this group. Again, I believe that this is typical of a more general situation!

Let us now consider the case of time-varying controls. Here is the formula for the solution to 7.1:

$$x(t) = e^{-\int_0^t u_1(s)\, ds} x(0)$$

$$+ \left(\int_0^t e^{-\int_0^v u_1(s)\, ds} u_2(v)\, dv \right) e^{\int_0^t u_1(t)\, dt}$$

Again, we see that:

$$u_1(t) \leq 0$$

is essentially the condition for input-output stability.

8. STABILITY FOR SYSTEMS WHICH CAN BE SOLVED ITERATIVELY IN TERMS OF ONE-STATE DIMENSION SYSTEMS

Anyone who reads Lie's work will find it evident that one of his major aims was to systematize the study of the solutions of differential equations by quasi-algebraic methods.

A typical question was: How may a "complicated" system be solved as an iteration of "simple" ones? Of course, what one means by "simpler" is relative. In Lie's day, the "simplest" differential equations were those which could be solved "by quadratures", i.e., by an iteration of the operations of integral calculus. In this section I present a brief discussion of certain systems-theoretic material which is relevant to Lie's program. To give it a modern flavor, I show how it can be used, in a very crude way, to study "input-output stability".

Consider a system of the following form:

$$\frac{dx_1}{dt} = f_1(x_1,u,t)$$
$$\frac{dx_2}{dt} = f_2(x_1,x_2,u,t) \tag{8.1}$$

x_1 is an element of a vector space X_1, x_2 an element of X_2. u is an element of a vector space U. The state space of the system 8.1 is:

$$X = X_1 \times X_2 \quad .$$

The input space is U.

Also, consider the following system:

$$\frac{dx_1}{dt} = f_1(x_1,u,t) \tag{8.2}$$

Its state space is X_1, control space is U.

There is obviously a <u>state-homomorphism</u> of system 8.1 onto system 8.2. Namely, map

$$X \equiv X_1 \times X_2 \to X_1$$

via the Cartesian product projection. It should be clear from 8.1 that this has the required systems-theoretic "homomorphism" property, i.e., that the projection of a solution

$$t \to (x_1(t), x_1(t), u(t))$$

of 8.1 is a solution of system 8.2.

We can construct from 8.1 another input-output system, whose state space is X_2, and whose input space is

$$X_1 \times U \equiv U_2 \quad .$$

Namely, if

$$u_2 = (x_1, u) \quad ,$$

consider the system:

$$\frac{dx_2}{dt} = h(x_2, u_2, t) \quad , \tag{8.3}$$

where

$$h(x_2, u_2, t) = f_2(x_1, x_2, u_1, t)$$

<u>Definition</u>. The system 8.1 is said to result from the <u>iteration of systems</u> 8.2 <u>and</u> 8.3.

This definition is merely a formalization of how one would, in practice, solve the system 8.1. First choose an input

$$t \to u(t) \quad .$$

Find $t \to x_1(t)$ as a solution of

$$\frac{dx_1}{dt} = f_1(x_1, u(t), t) \quad ,$$

then use this curve, together with u, as <u>input</u> into the system:

$$\frac{dx_2}{dt} = f_2(x_1(t), x_2, u(t), t) \qquad (8.4)$$

Here is a typical application of these ideas:

<u>Theorem 8.1</u>. Suppose the systems 8.2 and 8.3 are input-output stable. Then, the system 8.1--which results from the iteration of 8.1 and 8.2--is also input-output stable.

<u>Proof</u>. Suppose $t \to u(t)$ is bounded input to system 8.1. Each output $t \to x_1(t)$ to 8.2, with given input u, is, by hypothesis, bounded. Note that this property is inherited under iteration.

One can, in a similar way, iterate more systems. Systems built up in this way take the following form:

$$
\begin{aligned}
\frac{dx_1}{dt} &= f_1(x_1,u_1,t) \\
\frac{dx_2}{dt} &= f_2(x_1,x_2,u,t) \\
\frac{dx_3}{dt} &= f_3(x_1,x_2,x_3,u,t) \\
&\vdots
\end{aligned}
\tag{8.5}
$$

Feed $t \to (x_1(t),u(t)) \equiv u_2(t)$ as input into the system 8.3. Its output $t \to x_2(t)$ is, by hypothesis, bounded. Hence, the total output

$$t \to (x_1(t),x_2(t))$$

of system 8.1 is bounded.

Remark. Another variant of the stability property is to say that:

> The output $t \to x(t)$ has a limit as $t \to \infty$ if the input $t \to u(t)$ has a limit as $t \to \infty$.

How can one tell whether a system given a priori is equivalent to one in form 8.5? Lie's methods give a partial answer to this type of question. In the next section we consider a case that is important for applications, the bilinear case.

9. LIE ALGEBRAS ASSOCIATED WITH BILINEAR SYSTEMS

Consider an input-output system of the form:

$$\frac{dx}{dt} = f(x,u,t) \quad . \tag{9.1}$$

x is a point of a vector space X. We suppose X is a finite dimensional complex vector space.

Definition. We say that the system 9.1 is a bilinear (stationary) system if it is of the following form:

$$\frac{dx}{dt} = Ax + Bu + C(u,x) \tag{9.2}$$

Here,

$$A \in L(X) \equiv \text{space of linear maps } X \to X$$

$$B \in L(U,X) \equiv \text{space of linear maps } U \to X$$

$$C \in L(U \otimes X, X) = \text{space of bilinear maps } U \times X \to X \ .$$

We suppose (for simplicity), that all such linear maps are linear with respect to multiplication by complex scalars. $L(X)$ is then a Lie algebra, with the complex numbers as field of scalars. The Lie bracket operation [,] is then the commutator of linear maps.

Definition. Let $\underset{\sim}{G}$ be a Lie subalgebra of $L(X)$. The system 9.2 is said to be associated to $\underset{\sim}{G}$ if the following condition is satisfied:

For *each* $u \in U$, the linear maps A and $x \to C(u,x)$ belong to $\underset{\sim}{G}$. (9.3)

Remark. Suppose G is the connected Lie group of linear maps on X whose Lie algebra is $\underset{\sim}{G}$. Let

$$\mathrm{AFF}(G)$$

be the group of diffeomorphisms of X consisting of the elements of G, composed with the abelian group of translations. Algebraically, $\mathrm{AFF}(G)$ is the *semidirect product* of G and the invariant abelian subgroup consisting of the additive group of the vector space X. It is called the *affine group generated by* G.

Consider now the system 9.2. It is a *Lie group system associated with* $\mathrm{AFF}(G)$, in the sense that there is a curve

$$t \to g(t)$$

in $\mathrm{AFF}(G)$ such that each solution of 9.2 can be written in the following form:

$$x(t) = g(t)(x_0) \tag{9.3}$$

Now, we turn to the study of the algebraic structure of the system 9.2.

Definition. Consider a bilinear system of form 9.2, and one of the following form:

$$\frac{dx'}{dt} = A'X' + B'u + C'(u,x') \qquad (9.4)$$

(x' is an element of a vector space X'.) A linear map

$$\phi\colon X' \to X$$

is said to be a <u>state homomorphism</u> of the system 9.4 into the system 9.2 if the following condition is satisfied:

> For each solution $t \to (x'(t),u(t))$ of 9.4, the curve
>
> $$t \to (\phi(x'(t)) \equiv x(t),u(t))$$
>
> is a solution of 9.2.

Let us work out the conditions this relation imposes on ϕ.

$$\frac{d}{dt}\,\phi(x'(t)) = \phi\left(\frac{dx'}{dt}\right)$$

$$= \phi(A'x'+X'u+X'(u,x')$$

$$= A\phi x' + Bu + C(u,\phi x')$$

Hence, the conditions are:

$$A\phi = \phi A \qquad (9.4)$$

$$\phi B' = B \qquad (9.5)$$

$$\phi C'(u) = C(u)\ , \qquad (9.6)$$

where $C(u)$, $C'(u)$ are the maps $X \to X$, $X' \to X'$ defined as:

$$C(u)(x) = C(u,x) \tag{9.7}$$

$$C'(u),(x') = C'(u,x') \tag{9.8}$$

Suppose now that the system 9.2 is associated with the Lie algebra $\underset{\sim}{G}$ of linear maps $X \to X$.

Let X' be a linear subspace of X which is invariant under $\underset{\sim}{G}$. If, in addition,

$$B(U) \subset X' \quad , \tag{9.9}$$

i.e., that the system 9.2 "restricts" to X, to define a subsystem with X' as state space. In other words, if

$$A' = A \text{ restricted to } X',$$

$$B' = B$$

$$C'(u) = C(u) \text{ restricted to } X' ,$$

then the inclusion map

$$X' \to X$$

defines a homomorphism of the system 9.4 into the system 9.2.

We can now also define a quotient system, by setting:

$$X'' = X/X'$$

$(A'',B'',X''(u))$ the maps

$(A,B,C(u))$ acting in the quotient

The system 9.2 is then obviously a <u>composite</u>, as described in Section 8, of the primed and double-primed systems.

If $B \equiv 0$, we see that the reducibility of the system 9.2 into subsystems is determined by the reducibility of $\underset{\sim}{G}$ acting on X. In particular, if $\underset{\sim}{G}$ is a <u>semi-simple</u> Lie algebra, then the system is a <u>direct</u> sum of subsystems. If $\underset{\sim}{G}$ is <u>solvable</u>, then the system is a composite of systems with one-dimensional state space.

The algebraic study of bilinear systems with non-zero B has not really been carried out in adequate generality, even in the case where $\underset{\sim}{G}$ is semi-simple. What is probably needed is a generalization of the Kronecker "pencil" theory. (See Volume IX of IM. For further information about the algebraic and geometric properties of bilinear systems, see the articles on this subject in Brockett and Wayne [1].

BIBLIOGRAPHY

[1] B. Anderson and J. Moore, Linear optimal control, Prentice-Hall, 1971.

[1] R. Bishop and S. Goldberg, Tensor analysis on manifolds, MacMillan, New York, 1968.

[1] R. Brockett, Lie theory and control systems defined on spheres, SIAM J. App. Math. 25 (1973), 213-225.

[1] R. Brockett and D. Mayne, Geometric and Algebraic Systems Theory, D. Reidel, 1974.

[1] C. Chevalley, Theory of Lie Groups, Princeton Univ. Press, 1946.

[1] E. Dynkin, Maximal subgroups of the classical groups, Am. Math. Soc. Transl., 6, 245-378 (1957).

[1] V. Guillemin and S. Sternberg, An algebraic model of transitive differential geometry, Bull. Am. Math. Soc., 70 (1964), 16-47.

[2] V. Guillemin and S. Sternberg, Remarks on a paper of Hermann, Trans. Amer. Math. Soc., 130 (1968), 110-116.

[1] R. Hermann, The differential geometry of foliations, II, J. Math. and Mech., 11 (1962), 303-316.

[2] R. Hermann, On the accessibility problem of control theory, Proc. of the Symposium on Differential Equations, J. Lasalle and S. Lefshetz, eds., Academic Press, 1961.

[3] R. Hermann, Cartan connections and the equivalence problem for geometric structures, Contributions to differential equations, 3 (1964), 199-248.

[4] R. Hermann, Analytic continuation of group representations, Pts I-VI, Comm. Math. Phys., vols. 2,3,5,6, (1966-67).

[5] R. Hermann, Formal linearization of a semisimple Lie algebra of vector fields about a singular point, Trans. Amer. Math. Soc., 130 (1968), 105-109.

[6] R. Hermann, Formal linearization of Lie algebras of vector fields near an invariant submanifold, J. of Diff. Geom., 1973.

[7] R. Hermann, Lie groups for physicists, W.A. Benjamin, 1971 (Abbreviation: LGP).

[8] R. Hermann, Differential geometry and the calculus of variations, Academic Press, New York, 1969 (Abbreviation: DGCV).

[9] R. Hermann, Geometry, physics and systems, Marcel Dekker, New York, 1973 (Abbreviation: GPS).

[10] R. Hermann, Interdisciplinary mathematics, Vols. 1-10, Math Sci Press, Brookline, Mass., 1973-75 (Abbreviation: IM).

[1] G. Hochshild, The structure of Lie groups, Holden-Day, S. Francisco, 1965.

[1] N. Jacobson, Lie algebras, Interscience, New York, 1962.

[1] I. Kaplansky, An introduction to differential algebra, Hermann, Paris, 1957.

[1] E. Kolchin, Differential algebra and algebraic groups, Academic Press, New York, 1973.

[1] L. Loomis and S. Sternberg, Advanced Calculus, Addison-Wesley, Reading, Mass., 1968.

[1] S. MacLane, Categories for the working mathematician, Springer-Verlag, 1971.

[1] C. Misner, K. Thorne, and J. Wheeler, Gravitation, Freeman, San Francisco, 1973.

[1] D. Montgomery and L. Zippin, Topological transformation groups, Interscience, New York, 1955.

[1] B. O'Neill, Elementary differential geometry, Academic Press, New York, 1966.

[1] R. Palais, A global formulation of the Lie theory of transformation groups, Mem. Amer. Math. Soc., No. 22 (1957).

[1] A. Sagle and R. Walde, Introduction to Lie groups and Lie algebras, Academic Press, New York, 1973.

[1] H. Samelson, Notes on Lie algebras, Van Nostrand-Reinhold, New York, 1969.

[1] J. P. Serre, Lie algebras and Lie groups, W.A. Benjamin, New York, 1965.

[1] M. Spivak, A comprehensive introduction to differential geometry, Publish or Perish, Boston, 1970.

[1] S. Sternberg, Lectures on differential geometry, Prentice-Hall, 1965.

[1] H. Sussman, A generalization of the closed subgroup theorem to quotients of arbitrary manifolds, J. Diff. Geom., 10 (1975), 151-165.

[2] H. Sussman, Orbits of families of vector fields and integrability of systems with singularities, Bull. Amer. Math. Soc., 79 (1973), 197-199.

[1] N. Wallach, Harmonic analysis on homogeneous spaces, M. Dekker, New York, 1973.

[1] F. Warner, Foundations of differential geometry and Lie groups, Scott, Foresman Co., Glenview, Ill., 1971.

[1] J.L. Willems, Stability theory of dynamical systems, J. Wiley, New York, 1970.

[1] M. Wonham, Linear multivariable control, Lecture notes in Economics and Math. Systems, Vol. 101, Springer-Verlag, 1974.